UNDERSTANDING
VOLTAMMETRY

2nd Edition

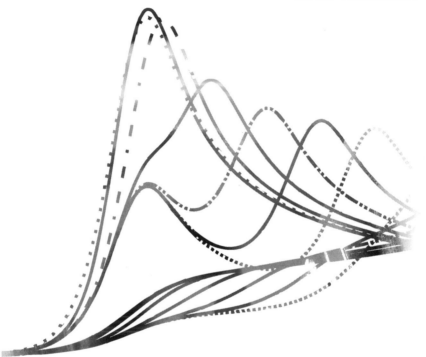

RICHARD G COMPTON
University of Oxford, UK

CRAIG E BANKS
Manchester Metropolitan University, UK

Imperial College Press

ICP

Published by

Imperial College Press
57 Shelton Street
Covent Garden
London WC2H 9HE

Distributed by

World Scientific Publishing Co. Pte. Ltd.
5 Toh Tuck Link, Singapore 596224
USA office: 27 Warren Street, Suite 401-402, Hackensack, NJ 07601
UK office: 57 Shelton Street, Covent Garden, London WC2H 9HE

British Library Cataloguing-in-Publication Data
A catalogue record for this book is available from the British Library.

First published 2011
Reprinted 2014

UNDERSTANDING VOLTAMMETRY (2nd Edition)

ISBN 978-1-84816-585-4
ISBN 978-1-84816-586-1 (pbk)

Typeset by Stallion Press
Email: enquiries@stallionpress.com

Printed in Singapore by World Scientific Printers.

"First we'll make enough sausages and then we won't have any dissidents."

Yuri Andropov (1914–1984)

General Secretary of the Communist Party of the Soviet Union,
November 1982–February 1984, and formerly head of the KGB

Preface

This book is not a research monograph, nor is it a reference book. Rather, it is a book designed for those who wish to understand and very likely undertake voltammetric experiments. The power of electrochemical measurements in respect of thermodynamics, kinetics and analysis is widely recognised and their importance ever growing as scientists seek to explore the links between the molecular, the nano-, the micro- and the macro scales. However, electrochemistry can be unpredictable to the novice even if they have a strong physical and chemical background, especially if they wish to pursue quantitative measurements. Accordingly, some possible significant experiments are never undertaken, whilst the literature is sadly replete with flawed attempts at rigorous voltammetry.

The aim of our book is to provide the reader with a largely self-contained account of the design, explanation and interpretation of experiments centred around various forms of voltammetry (cyclic, pulse, microelectrode, hydrodynamic, etc.). We assume a knowledge of Physical Chemistry, but relatively little exposure to electrochemistry in general, or voltammetry in particular. We seek to generate understanding plus insight into the design of real experiments. We hope you grow to share our fascination of the subject!

RGC, CEB, October 2006

The second edition of our book contains two new chapters and a few additional sections as well as corrections to the first edition. We thank all those who have commented so positively on the approach taken in the book and especially those who have encouraged us to refine and enlarge the content.

RGC, CEB, June 2010

Contents

1

Equilibrium Electrochemistry and the Nernst Equation

This chapter presents fundamental thermodynamic insights into electrochemical processes.

1.1 Chemical Equilibrium

Thermodynamics predicts the direction (but not the rate) of chemical change. Consider the chemical reaction,

$$aA + bB + \cdots \rightleftarrows xX + yY + \cdots, \tag{1.1}$$

where the reactants $A, B, \ldots$ and products $X, Y, \ldots$ may be solid, liquid, or gaseous. Thermodynamics tells us that the Gibbs energy of the system, G_{sys}, is minimised when it has attained equilibrium, as shown in Fig. 1.1.

Mathematically, at equilibrium, under conditions of constant temperature and pressure, this minimisation is given by

$$dG_{sys} = 0. \tag{1.2}$$

Consider the Gibbs energy change associated with dn moles of reaction (1.1) proceeding from left to right

$$dG = \{\text{Gain in Gibbs energy of products}\}$$
$$+ \{\text{Loss of Gibbs energy of reactants}\}$$

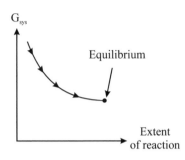

Fig. 1.1

$$= \{x\mu_X dn + y\mu_Y dn \cdots\} - \{a\mu_A dn + b\mu_B dn + \cdots\}$$
$$= \{x\mu_X + y\mu_Y \cdots - a\mu_A - b\mu_B\}dn, \tag{1.3}$$

where

$$\mu_j = \left(\frac{\partial G}{\partial n_j}\right)_{T,n_i \neq n_j} \tag{1.4}$$

is the chemical potential of species j ($j = A, B, \ldots X, Y, \ldots$), T is the absolute temperature (K) and n_i is the number of moles of i ($i = A, B, \ldots X, Y, \ldots$). The chemical potential of j is therefore the Gibbs energy per mole of j. It follows at equilibrium that

$$a\mu_A + b\mu_B + \cdots = x\mu_X + y\mu_Y + \cdots, \tag{1.5}$$

so that under conditions of constant temperature and pressure, the sum of the chemical potential of the reactants (weighted by their stoichiometric coefficients $a, b, \ldots x, y, \ldots$) equals that of the products. If this were not the case, then the Gibbs energy of the system would not be a minimum, since the Gibbs energy could be further lowered by either more reactants turning into products, or vice versa.

For an ideal gas,

$$\mu_j = \mu_j^o + RT \ln\left(\frac{P_j}{P^o}\right), \tag{1.6}$$

where μ_j^o is the standard chemical potential of j, R is the universal gas constant ($8.313\ \mathrm{J\ K^{-1}\ mol^{-1}}$), P_j is the pressure of gas j and P^o is a standard pressure conventionally taken to be $10^5\ \mathrm{Nm^{-2}}$ approximating to 1 atmosphere (atm), although strictly speaking 1 atm $= 1.01325 \times 10^5\ \mathrm{Nm^{-2}}$. It follows that μ_j^o is the Gibbs energy of one mole of j when it has a pressure of $1.01325 \times 10^5\ \mathrm{Nm^{-2}}$. It follows from Eqs. (1.5) and (1.6) that at equilibrium

$$x\mu_X^o + y\mu_Y^o + \cdots - a\mu_A^o - b\mu_B^o = -xRT \ln \frac{P_X}{P^o} - yRT \ln \frac{P_Y}{P^o} + \cdots$$
$$+ aRT \ln \frac{P_A}{P^o} + bRT \ln \frac{P_B}{P^o}, \quad (1.7)$$

so that

$$\Delta G^o = -RT \ln K_p, \quad (1.8)$$

where $\Delta G^o = x\mu_X^o + y\mu_Y^o + \cdots - a\mu_A^o - b\mu_B^o \cdots$ is the standard Gibbs energy change accompanying the reaction and

$$K_p = \frac{\left(\frac{P_X}{P^o}\right)^x \left(\frac{P_Y}{P^o}\right)^y \cdots}{\left(\frac{P_A}{P^o}\right)^a \left(\frac{P_B}{P^o}\right)^b \cdots} \quad (1.9)$$

is a constant at a particular temperature, because the standard chemical potential μ^o depends only on this parameter (unless the gases are not ideal, in which case K_p may become pressure dependant). Thus, for the gas phase reaction

$$2O_2(g) + N_2(g) \rightleftharpoons 2NO_2(g) \quad (1.10)$$

equilibrium is denoted by the equilibrium constant

$$K_p = \frac{\left(\frac{P_{NO_2}}{P^o}\right)^2}{\left(\frac{P_{O_2}}{P^o}\right)^2 \left(\frac{P_{N_2}}{P^o}\right)}. \quad (1.11)$$

Note that if some of the reactants and/or products in reaction (1.10) are in solution, then the pertinent ideal expression for their chemical potentials are

$$\mu_j = \mu_j^o + RT \ln \frac{[j]}{[\ ]^o}, \quad (1.12)$$

where $[\ ]^o$ is a standard concentration taken to be one molar (one mole per cubic decimetre). Applied to Eq. (1.1) this leads to a general equilibrium constant

$$K_c = \frac{\left(\frac{[X]}{[\ ]^o}\right)^x \left(\frac{[Y]}{[\ ]^o}\right)^y}{\left(\frac{[A]}{[\ ]^o}\right)^a \left(\frac{[B]}{[\ ]^o}\right)^b}. \quad (1.13)$$

It follows that for the equilibrium

$$HA(aq) \rightleftharpoons H^+(aq) + A^-(aq), \quad (1.14)$$

where *HA* is, say, a carboxylic acid and A^- a carboxylate anion, the equilibrium constant, K_c is given in terms of concentrations by

$$K_c = \frac{\left(\frac{[H^+]}{[\ ]^o}\right)\left(\frac{[A^-]}{[\ ]^o}\right)}{\left(\frac{[HA]}{[\ ]^o}\right)}. \tag{1.15}$$

In common usage, Eqs. (1.11) and (1.1) take the more familiar forms of

$$K_p = \frac{P_X^x P_Y^y \cdots}{P_A^a P_B^b \cdots} \quad \text{and} \quad K_c = \frac{[X]^x[Y]^y \cdots}{[A]^a[B]^b \cdots},$$

where it is implicitly understood that pressure is measured in units of $10^5\ \mathrm{Nm}^{-2}$ (or strictly $1.01325 \times 10^5\ \mathrm{Nm}^{-2}$) and concentrations in M (mol dm^{-3}) units.

In the case that the reactants in Eq. (1.1) are pure solids or pure liquids,

$$\mu_j \simeq \mu_j^o. \tag{1.16}$$

That is to say, the chemical potential approximates (well) to a standard chemical potential. Note that unlike gases or solutions, Gibbs energy *per mole* depends only on the temperature and pressure; changing the amount of material changes the total Gibbs energy, but not the Gibbs energy *per mole*.

It follows from Eq. (1.16) that, since the chemical potentials of pure liquids and solids are independent of the amount of material present, there are no corresponding terms in the expression for equilibrium constants in which these species participate. So for the general case

$$aA(g) + bB(aq) + cC(s) + dD(l) \rightleftharpoons wW(g) + xX(aq) + yY(s) + zZ(l), \tag{1.17}$$

the equilibrium constant will be

$$K = \frac{P_W^w\ [X]^x}{P_A^a\ [B]^b}, \tag{1.18}$$

where it is understood that the pressures are measured in units of $10^5\ \mathrm{Nm}^{-2}$ and the concentrations in units of moles dm^{-3}. The pure solids C and Y, and pure liquids D and Z do not appear. Illustrative real examples follow:

First, for

$$AgCl(s) \rightleftharpoons Ag^+(aq) + Cl^-(aq),$$
$$K = [Ag^+][Cl^-]. \tag{1.19}$$

Second, for

$$CaCO_3(s) \rightleftharpoons CaO(s) + CO_2(g), \tag{1.20}$$

$$K = P_{CO_2}.$$

Last,

$$Fe^{2+}(aq) + 1/2Cl_2(g) \rightleftharpoons Fe^{3+}(aq) + Cl^-(aq) \tag{1.21}$$

leads to

$$K = \frac{[Fe^{3+}][Cl^-]}{[Fe^{2+}]P_{Cl_2}^{1/2}}.$$

1.2 Electrochemical Equilibrium: Introduction

In the previous section, we considered various forms of chemical equilibrium involving gaseous, liquid, solution phase and solid species. We now turn to electrochemical equilibrium and, as a paradigm case, focus on the following process

$$Fe(CN)_6^{3-}(aq) + e^- \rightleftharpoons Fe(CN)_6^{4-}(aq). \tag{1.22}$$

Such an equilibrium can be established by preparing a solution containing both potassium hexacyanoferrate(II), $K_4Fe(CN)_6$, and potassium hexacyanoferrate(III), $K_3Fe(CN)_6$ dissolved in water and then inserting a wire (an 'electrode') made of platinum or another inert metal into the solution (Fig. 1.2).

The equilibrium in Eq. (1.22) is established at the surface of the electrode and involves the two dissolved anions and the electrons in the metal electrode. The establishment of equilibrium implies that the rate at which $Fe(CN)_6^{4-}$ gives

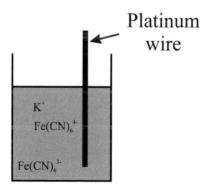

Fig. 1.2 A platinum wire immersed into an aqueous solution containing both ferrocyanide and ferricyanide.

up electrons to the metal wire or 'electrode' is exactly balanced by the rate at which electrons are released by the wire to the $Fe(CN)_6^{3-}$ anions, which are said to be 'reduced'. Correspondingly, the $Fe(CN)_6^{4-}$ ions losing electrons are said to be 'oxidised'. That a dynamic equilibrium of this type is established implies that, once established, no further change occurs. Moreover, the net number of electrons that are transferred in one direction or another is infinitesimally small, such that the concentrations of $Fe(CN)_6^{4-}$ and $Fe(CN)_6^{3-}$ are not measurably changed from their values before the electrode is introduced into the solution.

Equation (1.22) has a significant difference from the chemical equilibrium considered in section 1.1. In particular, the reaction involves the transfer of charged particles, electrons, between the metal and solution phases. As a result, when equilibrium is attained, there is likely to be a net electrical charge on each of these two phases. If reaction (1.22) lies to the left when equilibrium is reached, in favour of $Fe(CN)_6^{3-}$ and electrons, then the electrode will bear a net negative charge and the solution a net positive charge of equal magnitude. Conversely, if the equilibrium favours $Fe(CN)_6^{4-}$ and lies to the right, then the electrode will be positive and the solution negatively charged.

Irrespective of the position of the equilibrium of reaction (1.22), it can be recognised that there will likely exist a charge separation between the electrode and the solution phases. Accordingly there will be a potential difference (difference in electrical potential) between the metal and the solution. In other words, an *electrode potential* has been established at the metal wire relatively to the solution phase. The (electro)chemical reaction given in (1.22) is the basis of this electrode potential and it is helpful to refer to the chemical processes which establish electrode potentials as *potential determining equilibria*.

Other examples of electrochemical processes capable of establishing a potential on an electrode in solution include the following:

(a) The hydrogen electrode, shown in Fig. 1.3, comprises a platinum black electrode dipping into a solution of hydrochloric acid.

The electrode may be formed by taking a bright platinum 'flag' electrode and electro-depositing a fine deposit of 'platinum black' from a solution containing a soluble platinum compound, typically K_2PtCl_6. Hydrogen gas is bubbled over the surface of the electrode and the following potential determining equilibrium is established:

$$H^+(aq) + e^-(m) \rightleftarrows 1/2H_2(g), \tag{1.23}$$

where (m) reminds us that the source of electrons resides in the *metal* electrode.

(b) The silver/silver chloride electrode comprises a silver wire coated with a porous layer of silver chloride. The latter is almost insoluble in water and can be formed

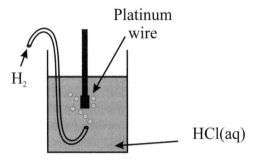

Fig. 1.3 A hydrogen electrode.

on the surface of the wire by electro-oxidation of the wire in a medium containing chloride ions such as an aqueous solution of KCl. The coated wire is then dipped into a fresh KCl solution, as shown in Fig. 1.4.

The following potential determining equilibrium is rapidly established:

$$AgCl(s) + e^-(m) \rightleftarrows Ag(s) + Cl^-(aq). \tag{1.24}$$

The equilibrium is established at the silver/silver chloride boundary. It is crucial that the layer of silver chloride is porous so that the aqueous solution bathing the electrodes penetrates the layer allowing equilibrium to be established at the three-phase boundary comprising the silver metal, the solid silver chloride and the aqueous solution.

(c) The calomel electrode is depicted in Fig. 1.5. It comprises a column of liquid mercury, contacting insoluble di-mercury (I) chloride (known traditionally as 'calomel'). Both contact an aqueous solution containing chloride ions, usually in the form of KCl.

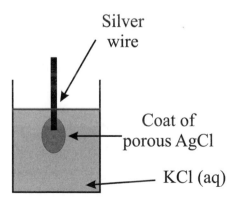

Fig. 1.4 A silver/silver chloride electrode.

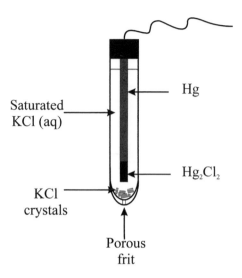

Fig. 1.5 A saturated calomel reference electrode (SCE).

The potential determining equilibrium is established at the three-phase boundary:

$$1/2 Hg_2 Cl_2(s) + e^-(m) \rightleftarrows Hg(l) + Cl^-(aq). \qquad (1.25)$$

(d) Finally, we consider an example of a potential determining equilibrium, which is not based on an aqueous solution but rather the aprotic solvent acetonitrile. The equilibrium involves an acetonitrile solution containing the molecule ferrocene, $Cp_2 Fe$ and a ferrocenium ($Cp_2 Fe^+$) salt such as ferrocenium hexafluorophosphate, $Cp_2 Fe^+ PF_6^-$:

$$Cp_2 Fe^+ + e^-(m) \rightleftarrows Cp_2 Fe. \qquad (1.26)$$

Figure 1.6 illustrates the structure of ferrocene, $Cp_2 Fe$.

In all of the examples considered above, a dynamic equilibrium is rapidly established between the chemical species involved in the potential determining equilibrium, a charge separation established between the solution and the metal

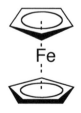

Fig. 1.6 The structure of ferrocene.

wire, and an electrode potential is set up on the latter. We now consider one further example which is that of a platinum wire dipping into a solution containing nitrate (NO_3^-) and nitrite (NO_2^-) anions. At first sight, it is tempting to assume that the following potential determining equilibrium will be set up:

$$1/2NO_3^-(aq) + H^+(aq) + e^-(m) \rightleftarrows 1/2NO_2^-(aq) + 1/2H_2O. \qquad (1.27)$$

However, the *rates* of electron transfer in both the forward (reducing) and back (oxidising) directions are so slow that no equilibrium is set up. Accordingly, there is no charge separation at the solution–metal wire interface and no electrode potential is established.

It is clear from the above discussion that fast rates of electron transfer between solution phase species and the electrode are *essential* for an electrode potential to be developed. In the absence of so-called 'fast electrode kinetics', no fixed potential is developed and any attempt at measurement of the potential discovers a floating variable value reflecting the failure to establish an electrode potential. The case of the hydrogen electrode discussed above nicely illustrates the importance of fast electrode kinetics. We have already noted that the electrodes are fabricated from platinised platinum rather than bright platinum metal. This difference is key to ensuring fast electrode kinetics. In particular the purpose of depositing a layer of fine platinum black is to provide catalytic sites, which ensures that the potential determining equilibrium

$$H^+(aq) + e^-(m) \rightleftarrows 1/2H_2(g) \qquad (1.28)$$

is *rapidly* established. In the absence of this catalysis, on a bright platinum electrode, the electrode kinetics are sluggish and cannot be guaranteed to establish the desired electrode potential. Figure 1.7 illustrates the effect of the platinum black in reducing the activation energy of the reaction and hence speeding up reaction (1.28). The catalyst binds the intermediate $H^\bullet$ atoms leading to the transition state for the reaction being lowered in energy.

1.3 Electrochemical Equilibrium: Electron Transfer at the Solution–Electrode Interface

The ideas behind the development of electrode potentials at the solution–electrode interface can be usefully re-examined considering the energy levels associated with the species involved in the potential determining equilibrium. We return to our paradigm case,

$$Fe(CN)_6^{3-}(aq) + e^-(m) \rightleftarrows Fe(CN)_6^{4-}(aq). \qquad (1.29)$$

The pertinent energy levels are shown in Fig. 1.8.

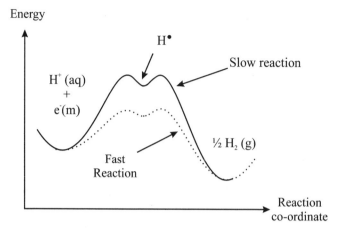

Fig. 1.7 The effect of platinum black on the H^+/H_2 equilibrium. The barrier to reaction is lowered with Pt black present (dotted line).

The electronic structure of a metal involves electronic conduction 'bands' in which the electrons are free to move throughout the solid, binding the (metal) cations together. The energy levels in these bonds form an effective continuum of levels which are filled up to an energy maximum known as the Fermi level. In contrast, the electronic energy levels associated with the solution phase $Fe(CN)_6^{3-}$ and $Fe(CN)_6^{4-}(aq)$ ions are discrete and relate to an unfilled molecular orbital in $Fe(CN)_6^{3-}$, which gains an electron to form $Fe(CN)_6^{4-}$. Note that, although not shown in Fig. 1.8, adding an electron to $Fe(CN)_6^{3-}$ alters the solvation of the ion so that the electron energy has a different value in the two complexes even though the same molecular orbital is involved. Figure 1.8 shows that before electron transfer takes place between the electrode and the solution, the Fermi level is higher than the vacant orbital in the $Fe(CN)_6^{3-}$ ion. It is accordingly energetically favourable for electrons to leave the Fermi level and join the $Fe(CN)_6^{3-}$ species converting them to $Fe(CN)_6^{4-}$ ions. This energy difference is the driving force of the electron transfer discussed in the previous section. As this electron transfer proceeds, positive charge must build up on the electrode (metal) and corresponding negative charge in the solution phase. Accordingly, since the energy scale in Fig. 1.8 measures that of the electron, then the electronic energy in the metal must be lowered and so the Fermi level becomes progressively lower in the diagram, as shown in Fig. 1.8. Correspondingly, the generation of negative charge in the solution must raise the (electronic) energy levels of the solution phase species. Ultimately, a situation is reached when the Fermi level lies in between the energy levels of the two ions, so that the rate at which electrons leave the electrode and reduce $Fe(CN)_6^{3-}$ ions is

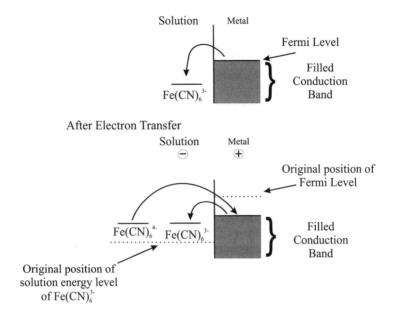

Fig. 1.8 The energy of electrons in the ions in solution and in the metal wire.

exactly matched by the rate at which electrons join the metal from the $Fe(CN)_6^{4-}$ ions which become oxidised. As we have noted before, this situation corresponds to dynamic equilibrium and once it is attained, no further net charge is possible. However, at the point of equilibrium, there is a charge separation between the electrode and the solution, and this is the origin of the electrode potential established on the metal.

1.4 Electrochemical Equilibrium: The Nernst Equation

We saw in Section 1.1 that the position of chemical equilibrium was controlled by the chemical potentials of the reactants and products. In the case of an electrochemical equilibrium such as reaction 1.29, the position of equilibrium represents a balance between chemical energies (quantified via the chemical potentials) *and* electrical energies. The reason for this is that electrochemical equilibrium involves the transfer of a charged particle, the electron, between two phases, the solution and the electrode, which may have two different electrical potentials. Accordingly, the electrical energy of the electrons differs from one phase to another.

In order to account for both chemical and electrical energies, we introduce the electrochemical potential, $\overline{\mu}_j$, of a species, j,

$$\overline{\mu}_j = \mu_j + Z_j F \phi, \tag{1.30}$$

where Z_j is the charge on molecule j, F is the Faraday constant corresponding to the charge on one mole of electrons (96487 Coulombs) and ϕ is the potential of the particular phase — electrode or solution — in which species j is found. The electrochemical potential of j is thus comprised of two terms. The first is the chemical potential, μ_j. The second term, $Z_j F \phi$, describes the electrical energy of species j. The latter is of the form of charge (Z_j) multiplied by potential ϕ and the constant F puts the electrical energy on a 'per mole' basis in the same way that the chemical potential is Gibbs energy *per mole*.

Equation (1.30) allows us to analyse electrochemical equilibrium recognising that when this is attained under conditions of constant temperature and pressure there will be a balance (equality) between the electrochemical potentials of the reactants and those of the products. Returning to the example we have considered throughout this chapter,

$$Fe(CN)_6^{3-}(aq) + e^-(m) \rightleftarrows Fe(CN)_6^{4-}(aq)$$

we note that this implies that at equilibrium

$$\overline{\mu}_{Fe(III)} + \overline{\mu}_{e^-} = \overline{\mu}_{Fe(II)}$$

where $Fe(III)$ denotes $Fe(CN)_6^{3-}$ and $Fe(II)$ indicates $Fe(CN)_6^{4-}$. Applying Eq. (1.30) we obtain

$$(\mu_{Fe(III)} + 3F\phi_S) + (\mu_{e^-} - F\phi_M) = (\mu_{Fe(II)} + 2F\phi_S)$$

where ϕ_M and ϕ_S refer to the electrical potential of the metal electrode and of the solution respectively. Rearranging

$$F(\phi_M - \phi_S) = \mu_{Fe(III)} + \mu_{e^-} - \mu_{Fe(II)}.$$

But

$$\mu_{Fe(III)} = \mu_{Fe(III)}^0 + RT \ln \left(\frac{[Fe(CN)_6^{3-}]}{[\ ]^0} \right)$$

$$\mu_{Fe(II)} = \mu_{Fe(III)}^0 + RT \ln \left(\frac{[Fe(CN)_6^{4-}]}{[\ ]^0} \right)$$

and hence

$$\phi_M - \phi_S = \frac{\Delta\mu^0}{F} + \frac{RT}{F} \ln \left(\frac{[Fe(CN)_6^{3-}]}{[Fe(CN)_6^{4-}]} \right) \tag{1.31}$$

where

$$\Delta\mu^0 = \mu^0_{Fe(III)} + \mu_{e^-} - \mu^0_{Fe(III)}$$

which is a constant at a given temperature and pressure. Equation (1.31) is the famous Nernst equation written in a form appropriate to a single electrode–solution interface. It is helpful to examine Eq. (1.31) in the light of the electrochemical equilibrium (1.29). First the ions $Fe(CN)_6^{4-}$ and $Fe(CN)_6^{3-}$ feature in the potential determining equilibrium given in Eqs. (1.29) and (1.31). Unsurprisingly therefore they determine the magnitude and sign of the electrode potential established on the platinum wire shown in Fig. 1.2. Second, to explain this dependence, consider what happens if a further amount of $Fe(CN)_6^{3-}$ is added to the solution shown in Fig. 1.2, whilst maintaining the *same* concentration of $Fe(CN)_6^{4-}(aq)$, so perturbing the equilibrium

$$Fe(CN)_6^{3-}(aq) + e^-(m) \rightleftarrows Fe(CN)_6^{4-}(aq).$$

The effect of the addition may be thought of as a consequence of Le Chatelier's Principle. Henri Louis Le Chatelier (1850–1936), shown left, was an industrial chemist and is famous for his work on the principle of equilibrium. Le Chatelier was educated at the École Polytechnique followed by the École des Mines, and elected to the Académie des science.[a]

LE CHÂTELIER (Henry)
PROFESSEUR DE CHIMIE INDUSTRIELLE
1850

Bibliothèque de l'École
des mines de Paris

Le Chatelier originally stated the principle of equilibrium as:

'*Any system in stable chemical equilibrium, subjected to the influence of an external cause which tends to change either its temperature or its condensation (pressure, concentration, number of molecules in unit volume), either as a whole or in some of its parts, can only undergo such internal modifications as would, if produced alone, bring about a change of temperature or of condensation of opposite sign to that resulting from the external cause*'.[1]

Le Chatelier later changed this rather awkward statement to:

'*Every change of one of the factors of an equilibrium occasions a rearrangement of the system in such a direction that the factor in question experiences a change in a sense opposite to the original change.*'[2] Le Chatelier's Principle can

---

[a] A full biography can be found at www.annales.org/archives/x/lc.html

be summarised as 'if a change (of temperature, pressure, concentration, ...) is imposed on a system previously at chemical equilibrium, then the system will respond in a way so as to oppose or counteract the imposed perturbation'.

Applying the principle to the electrochemical equilibrium of interest, Eq. (1.29), this equilibrium will become 'pushed' to the right and electrons will be taken from the electrode. Consequently, the metal electrode will become more positive relative to the solution and the potential difference $\phi_M - \phi_S$ will similarly be more positive. Conversely, addition of $Fe(CN)_6^{4-}$ will shift the equilibrium to the left and the electrode will gain electrons and so become relatively more negatively charged in comparison with the solution, thus making $\phi_M - \phi_S$ more negative (less positive). Both these shifts are qualitatively exactly as predicted by Eq. (1.31).

As a further illustration of the application of the electrochemical potential concept to the description of electrochemical equilibria, we consider the examples examined in the previous section.

(a) For the hydrogen electrode based on the equilibrium

$$H^+(aq) + e^-(m) \rightleftharpoons 1/2H_2(g)$$

at equilibrium

$$\overline{\mu}_{H^+} + \overline{\mu}_{e^-} = 1/2\overline{\mu}_{H_2}$$

so that

$$(\mu_{H^+} + F\phi_S) + (\mu_{e^-} - F\phi_M) = 1/2\mu_{H_2}.$$

Then, expanding the chemical potential terms using

$$\mu_{H^+} = \mu_{H^+}^0 + RT \ln \left(\frac{[H^+]}{[\ ]^0} \right)$$

and

$$\mu_{H_2} = \mu_{H_2}^0 + RT \ln \left(\frac{P_{H_2}}{p^0} \right)$$

we obtain the Nernst equation:

$$\phi_M - \phi_S = \frac{\Delta\mu^0}{F} + \frac{RT}{F} \ln \left(\frac{[H^+]}{P_{H_2}^{1/2}} \right), \tag{1.32}$$

where we have assumed that $[H^+]$ will be measured in units of M and P_{H_2} in units of 10^5 Nm^{-2}.

Note that the constant

$$\Delta\mu^0 = \mu_{H^+}^0 + \mu_{e^-}^0 - 1/2\mu_{H_2}^0.$$

Further note that Eq. (1.32) predicts that $\phi_M - \phi_S$ will become more positive as the H^+ concentration increases and/or the H_2 pressure decrease: both of these and/or predictions are consistent with the application of Le Chatelier's Principle to the potential determining equilibrium

$$H^+(aq) + e^-(m) \rightleftharpoons 1/2H_2(g).$$

(b) The silver/silver chloride electrode is based on the equilibrium

$$AgCl(s) + e^-(m) \rightleftharpoons Ag(s) + Cl^-(aq).$$

Equating the electrochemical potential of the reactants and products,

$$\bar{\mu}_{AgCl} + \bar{\mu}_{e^-} = \bar{\mu}_{Ag} + \bar{\mu}_{Cl^-}$$

so that

$$(\mu_{AgCl}) + (\mu_{e^-} - F\phi_M) = (\mu_{Ag}) + (\mu_{Cl^-} - F\phi_S),$$

noting that

$$\mu_{Cl^-} = \mu_{Cl^-}^o + RT \ln \left(\frac{[Cl^-]}{[\ ]^o} \right)$$

but for the pure solids AgCl and Ag,

$$\mu_{AgCl} = \mu_{AgCl}^o$$

and

$$\mu_{Ag} = \mu_{Ag}^o,$$

we obtain the Nernst equation:

$$\phi_M - \phi_S = \frac{\Delta\mu^o}{F} + \frac{RT}{F} \ln \left(\frac{1}{[Cl^-]} \right)$$

$$= \frac{\Delta\mu^o}{F} - \frac{RT}{F} \ln[Cl^-],$$

where we are presuming that $[Cl^-]$ is measured in unit of moles dm^{-3}. The constant

$$\Delta\mu^o = \mu_{Ag}^o + \mu_{Cl^-}^o - \mu_{e^-} - \mu_{AgCl}^o.$$

(c) The calomel electrode is based on the equilibrium

$$1/2Hg_2Cl_2(s) + e^-(m) \rightleftharpoons Hg(l) + Cl^-(aq).$$

The application of the electrochemical potential concept together with the chemical potentials of the pure solids and liquids,

$$\mu_{Hg_2Cl_2} = \mu^o_{Hg_2Cl_2}$$

and

$$\mu_{Hg} = \mu^o_{Hg}$$

leads to

$$\phi_M - \phi_S = \frac{\Delta\mu^o}{F} - \frac{RT}{F}\ln\,[Cl^-],$$

as the Nernst equation for this electrode, where

$$\Delta\mu^o = \mu^o_{Hg} + \mu^o_{Cl^-} - \mu_{e^-} - 1/2\mu^o_{Hg_2Cl_2}.$$

(d) For the ferrocene/ferrocenium couple in acetonitrile

$$Cp_2Fe^+ + e^-(m) \rightleftharpoons Cp_2Fe,$$

analogy with the $Fe(CN)_6^{4-}/Fe(CN)_6^{3-}$ equilibrium discussed fully above shows

$$\phi_M - \phi_S = \frac{\Delta\mu^o}{F} + \frac{RT}{F}\ln\left(\frac{[Cp_2Fe^+]}{[Cp_2Fe]}\right),$$

where $[Cp_2Fe^+]$ is measured in M and

$$\Delta\mu^o = \mu^o_{Cp_2Fe} - \mu^o_{e^-} - \mu^o_{Cp_2Fe^+}.$$

Having established the Nernst equation for several specific examples, we consider the general case and focus on the following electrochemical equilibrium

$$\upsilon_A A + \upsilon_B B + \cdots e^-(m) \rightleftharpoons \upsilon_X X + \upsilon_Y Y + \cdots.$$

The terms υ_j ($j = A, B, \ldots, X, Y, \ldots$) are the so-called stoichiometric coefficients. Since the reaction is assumed to be equilibrium

$$\upsilon_A\overline{\mu}_A + \upsilon_B\overline{\mu}_B + \cdots \overline{\mu}_{e^-} = \upsilon_X\overline{\mu}_X + \upsilon_Y\overline{\mu}_Y + \cdots$$

or,

$$\upsilon_A(\mu_A + Z_A F\phi_S) + \upsilon_B(\mu_B + Z_B F\phi_S) + \cdots (\mu_{e^-} - F\phi_M)$$
$$= \upsilon_X(\mu_X + Z_X F\phi_S) + \upsilon_Y(\mu_Y + Z_Y F\phi_S) + \cdots,$$

where Z_j is the charge on the species j. Conservation of electrical charge requires that

$$v_A Z_A + v_B Z_B + \cdots - 1 = v_X Z_X + v_Y Z_Y + \cdots .$$

Hence,

$$F(\phi_M - \phi_S) = v_A \mu_A + v_B \mu_B + \cdots + \mu_{e^-} - v_X \mu_X - v_Y \mu_Y - \cdots .$$

We now write

$$\mu_j = \mu_j^o + RT \ln a_j,$$

where, if j is the solution phase

$$a_j = \frac{[j]}{[\ ]^o},$$

but

$$a_j = \frac{P_j}{P^o}$$

if j is gaseous, whilst if j is a pure solid or liquid

$$a_j = 1,$$

and we obtain

$$\phi_M - \phi_S = \frac{\Delta\mu^o}{F} + \frac{RT}{F} \ln \left(\frac{a_A^{v_A} a_B^{v_B} \cdots}{a_X^{v_X} a_Y^{v_Y} \cdots} \right), \tag{1.33}$$

which is a general statement of the Nernst equation with

$$\Delta\mu^o = v_A \mu_A^o + v_B \mu_B^o + \cdots + \mu_{e^-} - v_X \mu_X^o - v_Y \mu_Y^o - \cdots .$$

1.5 Walther Hermann Nernst

Walther Hermann Nernst was born in Briesen, West Prussia (now Wabrzezno, Poland) on the 25th June 1864. Nernst studied physics and mathematics at the universities of Zurich, Berlin and Graz (Ludwig Boltzmann and Albert von Ettingshausen). Whilst at Graz, he worked with von Ettingshausen and published work in 1886 which formed part of the experimental foundation of the modern electronic theory of metals the Nernst–Ettingshausen effect).

NERNST, Walther Hermann
Nobel Laureate CHEMISTRY 1920
© Nobelstiftelsen

Copyright ©
The Nobel Foundation 1920

He undertook his PhD at Wurzburg with Friedrich Kohlrausch and graduated in 1887 with a thesis on electromotive forces produced by magnetism in heated metal plates. Nernst then joined Ostwald at Leipzig University, where Van't Hoff and Arrhenius were already established. In 1888, Nernst developed work on the theory of electromotive force of voltaic cells. He devised methods for measuring dielectric constants and first showed that solvents of high dielectric constant promote the ionisation of substances. Nernst also proposed the theory of solubility products, generalised the distribution law and offered a theory of heterogeneous reactions. In 1889, he elucidated the theory of galvanic cells via assuming an 'electrolytic pressure of dissolution', which forces ions from electrode into solution and which was opposed to the osmotic pressure of the dissolved ions.

In 1894, Nernst received invitations to the Physics Chairs in Munich and in Berlin, as well as to the Physical Chemistry Chair in Göttingen. He accepted this latter invitation, and founded the *Physikalisch-Technisches Reichsanstalt* in Göttingen (now the Institute for Physical Chemistry and Electrochemistry) and became its Director in 1922, a position he retained until his retirement in 1933. His transition to chemistry actually began in Leipzig, but developed fully in his subsequent position as an associate professor of physics at Göttingen.

By 1900, Nernst realised that interfacial potential differences between different phases were not individually measurable and concluded that electrochemical potential could only be measured relative to another and proposed the hydrogen electrode as the standard. This allowed Nernst to formulate his equation for the potential of a general cell.

In 1906, Nernst developed his heat theorem, known as the Third Law of Thermodynamics; in addition to its theoretical implications, the theorem was soon applied to industrial problems, including calculations in ammonia synthesis. In 1918, his studies of photochemistry led him to his atom chain reaction theory. Nernst was awarded the Nobel prize for his heat theorem work of 1906.

Nernst, as well as his substantial contributions to the physical sciences developed an improved electric light, the Nernst Lamp, which was commercialised by George Westinghouse. The 'Nernst Lamp Company' was founded in 1901 in Pittsburg, USA and by 1904, 130,000 Nernst glowers had been sold. However, the Nernst lamp, which contained oxides such as Y_2O_3, lost competition when the more conve-

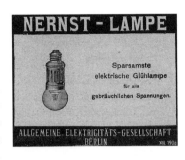

nient incandescent light bulbs containing metal (tungsten) filaments became available. Nernst also conceived an electric piano, the 'Neo-Bechstein-Flügel' in 1930 in association with the Bechstein and Siemens companies, replacing the sounding board with radio amplifiers. The piano used electromagnetic pickups to produce electronically modified and amplified sound.

After winning the highest scientific accolade possible, Nernst also received the Benjamin Franklin Medal in Chemistry (1928) and was elected a Fellow of the Royal Society (London) in 1932. Nernst retired in 1933 to breed carp and to hunt. Nernst died in 1941 and a tomb was erected at Göttingen Stradtfriedhof. Post-mortem events include a crater on the far side of the moon (coordinates 35.3°N/94.8°W, diameter 116 km) and several roads named after Nernst.[3]

1.6 Reference Electrodes and the Measurement of Electrode Potentials

Equation (1.33) is the Nernst equation for an arbitrary electrochemical equilibrium involving an electrode and the reaction components, $A, B, \ldots X, Y, \ldots$. It relates the quantity $\phi_M - \phi_S$ to the concentrations and/or pressures of these species. However, a little thought shows that although this quantity can be discussed conceptually, *it is impossible to measure an absolute value for the quantity $\phi_M - \phi_S$ relating to a single electrode–solution interface.*

Measurements of potential are usually carried out using a digital voltammeter ('DVM'), a device which measures the potential between the two test leads as shown in Fig. 1.9.

It does so by passing a tiny current ($\sim$ pico-amperes, 10^{-12}A) through the external circuit under test. Measurement of the potential drop $\phi_M - \phi_S$ at a single electrode–solution interface such as that developed at a platinum wire dipping into a solution of $Fe(CN)_6^{3-}$ and $Fe(CN)_6^{4-}$ ions is clearly impossible, since there will necessarily be two metal solution interfaces created if the measurement is attempted

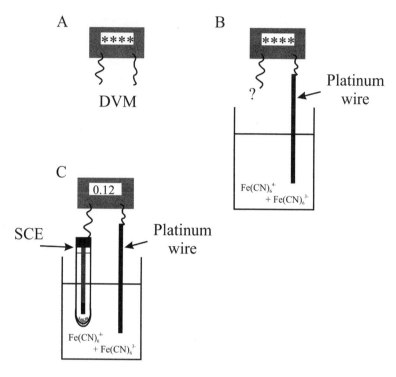

Fig. 1.9 Measurement of electrode potentials.

(Fig. 1.9). However, if a second electrode, for example a calomel electrode, is introduced into the solution (Fig. 1.9C) the measurement becomes feasible and the DVM records the difference of two quantities of the form $\phi_M - \phi_S$ pertaining to the two electrodes:

Measured potential difference $= (\phi_M - \phi_S)_{Pt\ wire} - (\phi_M - \phi_S)_{calomel}.$

On the basis of the discussion in the previous two sections we can recognise that, for a fixed temperature and pressure,

$$(\phi_M - \phi_S)_{Pt\ wire} = A + \frac{RT}{F} \ln \left(\frac{[Fe(CN)_6^{3-}]}{[Fe(CN)_6^{4-}]} \right),$$

where A is a constant.
    Also,

$$(\phi_M - \phi_S)_{calomel} = B - \frac{RT}{F} \ln \left([Cl^-] \right),$$

where B is another constant. The measured potential difference is then

$$(\phi_M - \phi_S)_{Pt\,wire} - (\phi_M - \phi_S)_{calomel} = C + \frac{RT}{F} \ln \left(\frac{[Fe(CN)_6^{3-}][Cl^-]}{[Fe(CN)_6^{4-}]} \right),$$

where C is a further constant, equal to $(A - B)$.

The observed potential difference thus depends on the concentrations $[Cl^-]$, $[Fe(CN)_6^{4-}]$ and $[Fe(CN)_6^{3-}]$. The introduction of the second electrode, the calomel electrode, has facilitated the successful measurement in contrast to the hopeless situation of Fig. 1.9B.

In the above measurement, the calomel electrode can be thought of as acting as a reference electrode. If the concentration of chloride ions inside the calomel electrode (see Fig. 1.9) is maintained constant, then $(\phi_M - \phi_S)_{calomel}$ is also constant, so that

$$(\phi_M - \phi_S)_{Pt\,wire} = D + \frac{RT}{F} \ln \left(\frac{[Fe(CN)_6^{3-}]}{[Fe(CN)_6^{4-}]} \right),$$

where D is yet a further constant, equal to $C + (\phi_M - \phi_S)_{calomel} + \frac{RT}{F} \ln[Cl^-]$.

Accordingly, the reference electrode allows us to establish changes in the electrode potential of the platinum wire, for examples induced by changes in the concentration of $Fe(CN)_6^{4-}$ and $Fe(CN)_6^{3-}$. Since we can write

$$(\phi_M - \phi_S)_{Pt\,wire} = D + \frac{2.3RT}{F} \log_{10} \left(\frac{[Fe(CN)_6^{3-}]}{[Fe(CN)_6^{4-}]} \right) \qquad (1.34)$$

and the ratio $2.3RT/F$ has the value of *ca.* 59 mV at room temperature, it follows that if the concentration of $Fe(CN)_6^{3-}$ is changed by a factor of ten whilst the concentration of $Fe(CN)_6^{4-}$ is kept constant, then the measured potential on the DVM will change by 59 mV. Note however that because of the constant term, D, in Eq. (1.34) we can only measure *changes* in the electrode potential, not absolute values.

It is instructive to consider what is happening when the DVM in Fig. 1.10 makes the measurement of the difference in potential between the platinum wire and the calomel electrode.

As already discussed, this requires a tiny current to be passed through the meter and hence, through the external circuit involving our two electrodes, the calomel and the platinum wire. The tiny current corresponds to an almost infinitesimal flow of electrons around the external circuit. Suppose that this is in the direction shown in Fig. 1.10. Then, the way the charge is passed through the two electrode–solution interfaces is by the occurrence of an almost infinitesimal amount of the following two reactions:

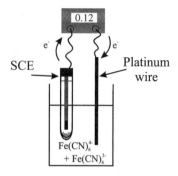

Fig. 1.10 A tiny flow of electrons (current) is required to measure the potential difference between the two electrodes.

At the Pt wire

$$Fe(CN)_6^{3-} + e^- \rightleftarrows Fe(CN)_6^{4-}$$

and at the calomel electrode

$$Hg + Cl^- - e^- \rightarrow 1/2 Hg_2 Cl_2.$$

In practice, the magnitude of the current passed is so small that the concentrations of the species in the cell are effectively unperturbed from the values pertaining before the measurement was conducted, but enough current flows to secure the measurement. Given that the reactions above allow charge to pass through the two solution interfaces the passage of the measuring current through the bulk solution is carried by the 'conduction' of the ionic species (K^+, Cl^-, $Fe(CN)_6^{4-}$, $Fe(CN)_6^{3-}$) in the solution phases, both inside the calomel electrode and inside the solution bathing the platinum wire.

The term 'conduction' implies that there is an electric field (potential drop) within the solution phased to 'drive' ion motion. However, since the current, I, being passed is so tiny, this is almost negligible. Algebraically,

$$\begin{aligned} \textit{Measured potential difference} &= \lim_{I \to 0} [(\phi_M - \phi_S)_{Pt\ wire} \\ &\quad + IR - (\phi_M - \phi_S)_{calomel}] \\ &= (\phi_M - \phi_S)_{Pt\ wire} - (\phi_M - \phi_S)_{calomel}, \end{aligned}$$

where R corresponds to the resistance of the electrolyte solution.

Finally, it is instructive to consider the liquid–liquid interface formed at the frit of the calomel electrode, where it enters the solution bathing the platinum wire electrode. On one side of this interface is a high concentration of aqueous KCl (see

Fig. 1.5), whilst on the other side is a solution containing $Fe(CN)_6^{4-}$ and $Fe(CN)_6^{3-}$ ions. Before we fully address this rather complex situation, let us consider the simpler case where two solutions of the same chemical composition but of different concentrations are put into contact. Let us consider solutions of KCl and HCl, both fully dissociated electrolytes. The relative rates of ionic movement are similar regardless of whether this is induced by an electric field or by a concentration difference. Measurements of aqueous ionic conductivity suggest the relative rates to be

$$H^+ : Cl^- : K^+ \sim 350 : 76 : 74.$$

The proton moves much more quickly in water than the other two ions. This experimental fact is often interpreted in terms of the Grotthus mechanism of proton (and hydroxide anion) conduction shown in Fig. 1.11, which allows the more rapid movement of H^+ and OH^- ions in comparison with the motion of species such as K^+ and Cl^-, which must displace solvent molecules from their path.

Consider first two solutions of HCl of different concentrations, C_1 and C_2, put in contact, initially as shown diagrammatically in Fig. 1.12. Diffusion of both H^+ and Cl^- ions from the higher to the lower concentration will occur. But we know from our discussion above that the H^+ ions will move faster than the Cl^- ions. As a result, a charge difference and hence, a potential difference will be set up across the interface between the two solutions. The lower concentration solution will become positively charged having 'gained' protons, whilst the higher concentrated solution will become negatively charged having lost more protons than chloride ions. This charge separation creates a local electric field and has the effect that the rate of chloride movement is increased, whilst that of the proton movement is slowed. Ultimately, a steady state is rapidly reached, as shown in Fig. 1.12, in

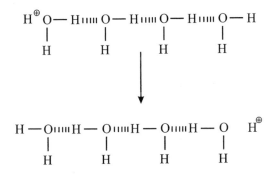

Fig. 1.11 Synchronous proton movement in aqueous media.

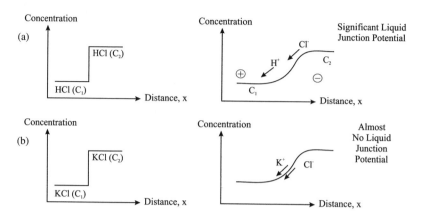

Fig. 1.12 Interfacial diffusion between two phases can lead to liquid junction potentials.

which both ions move at a steady state through the liquid–liquid interface across which the steady state charge separation leads to a potential difference, known as a *liquid junction potential*. Typically, these potentials are no more than some tens of millivolts.

Next, consider Fig. 1.12(b), in which two solutions of KCl of different concentrations are put in contact. Again, there will be diffusion of K^+ and Cl^- ions from the higher to the lower concentration. However, as we noted above, these two ions move with almost exactly the same speed and so now the diffusion leads to no charge separation and hence, no liquid junction potential is established.

It follows from the above that if a charge is carried through a liquid–liquid interface by ions of closely similar mobility, then no liquid junction potentials will be established. In contrast, if the ions have different mobilities, a significant liquid junction potential may be established. Table 1.1 reports the single ion conductivities for various cations and anions at 25°C. These can be thought to represent the relative speeds of movement of the species under the same potential gradient (electric field).

It is evident that liquid–liquid interfaces between solutions of different concentrations of HCl, Li_2SO_4 or NaOH will experience significant liquid junction potentials, whereas electrolytes such as NH_4NO_3 or KCl will be relatively liquid junction potential free.

We next return the measurement of the potential difference between the platinum wire and the calomel electrode shown in Fig. 1 and focus on the transport of current through the liquid–liquid interface formed at the frit of the calomel electrode (see Fig. 1.13).

Table 1.1. Single ion conductivities in water $(25°C)/\Omega^{-1}\,cm^2\,mol^{-1}$.

Ion	Λ_+	Ion	Λ_-
H^+	350	$Fe(CN)_6^{4-}$	442
Ba^{2+}	127	OH^-	199
Ca^{2+}	119	SO_4^{2-}	158
Mg^{2+}	106	Br^-	78
NH_4^+	74	I^-	77
K^+	74	Cl^-	76
Ag^+	62	NO_3^-	71
Na^+	50	F^-	55
Li^+	39	CH_3COO^-	41

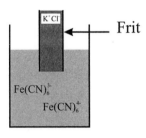

Fig. 1.13 The liquid–liquid interface formed at the frit of the calomel electrode shown in Fig. 1.10.

This interface is depicted in Fig. 1.13. The solution above the frit inside the calomel electrode is of very high concentrations ($> 1\,M$), since it is saturated with KCl. The concentrations outside of the frit are typically much smaller. Accordingly, the dominant diffusive fluxes across the liquid–liquid interfaces are from K^+ and Cl^- ions, rather than from $Fe(CN)_6^{4-}$ or $Fe(CN)_6^{3-}$ ions. For this reason, no significant liquid junction potentials will be established unless unusually large ($\sim M$) concentrations of $Fe(CN)_6^{4-}$ or $Fe(CN)_6^{3-}$ ions are being studied.

To summarise, the measurement of the potential difference between the platinum wire electrode shown in Fig. 1.10 and the calomel electrode acting as a reference electrode gives the following result:

$$\text{Measured potential difference} = A' + \frac{RT}{F} \ln\left(\frac{[Fe(CN)_6^{3-}]}{[Fe(CN)_6^{4-}]}\right),$$

where A' is a constant.

Key features of the experiment shown in Fig. 1.9(C) are as follows:

- The passage of tiny currents through the DVM and hence the external circuit means that the 'IR' term in the bulk solution is negligible.
- The presence of a saturated solution of KCl inside the calomel electrode caused by the presence of solid KCl inside the electrode, coupled with the passage of tiny currents only, leads to the pinning of the quantity $(\phi_M - \phi_S)_{calomel}$ at a constant value since the chloride ion concentration is constant. Under these conditions, the calomel electrode acts as a suitable 'reference' electrode.
- The use of KCl inside the calomel electrode ensures that negligibly small liquid junction potentials are established at the liquid–liquid interface at the tip of the frit unless extremely high concentrations of $Fe(CN)_6^{4-}$ and/or $Fe(CN)_6^{3-}$ are used.

1.7 The Hydrogen Electrode as a Reference Electrode

The discussion in the previous section identified the key features required of a 'reference' electrode and showed the merits of the calomel electrode for this purpose. Indeed, the calomel electrode is a widely employed reference electrode. However, the primary reference electrode against which data is conveniently reported is the standard hydrogen electrode. This is shown in Fig. 1.3 and for the electrode to be 'standard' the pressure of hydrogen gas, P_{H_2}, must be close to 10^5 Nm^{-2} and the concentration of protons, $[H^+]$ to be close to 1 mol dm^{-3}. In practice, because the solution is not ideal, Eq. 1.32 does not hold exactly, so that a concentration of 1.18 M is required at 25°C for the protons to behave as if they were an ideal solution of 1 M. The origins of this deviation lie in Debye–Huckel theory and its extension to concentrated solutions.[4,5] However, for our present purpose, it is sufficient to note that provided $P_{H_2} = 1.01325 \times 10^5$ Nm^{-2} and $[H^+] = 1.18$ M at 25°C, then the hydrogen electrode is 'standard', and it is this particular reference electrode against which the International Union of Pure and Applied Chemistry (IUPAC) formally require that potentials are reported.[6]

If a measurement such of the type shown in Fig. 1.10 is carried out, in which the potential of a calomel electrode containing saturated KCl is measured with respect to the standard hydrogen electrode (SHE), it is found to be 0.242 V positive of the SHE. Accordingly, measurements made relative to the saturated calomel electrode are readily correlated to the SHE by adding this value to the number measured. For example, returning to Fig. 1.10, it is found experimentally that the measured potential difference on the DVM for case when $[Fe(CN)_6^{4-}] = [Fe(CN)_6^{3-}]$ is 0.118 V (at 25°C). On the SHE scale, this becomes 0.118 V + 0.242 V = 0.36 V.

1.8 Standard Electrode Potentials and Formal Potentials

The potential of 0.36 V obtained at the end of the previous section is the standard electrode potential of the following potential determining equilibrium:

$$Fe(CN)_6^{3-}(aq) + e^-(m) \rightleftharpoons Fe(CN)_6^{4-}(aq).$$

In order to understand this quantity more generally, it is necessary to briefly address the issue of solution non-ideality. In establishing the Nernst equation of the form of Eqs. (1.31)–(1.33) for single electrode–solution interfaces, we relied on the following relationship between chemical potential and concentration, which is correct for an ideal solution:

$$\mu_j = \mu_j^o + RT \ln \left(\frac{[j]}{[\ ]^o} \right). \tag{1.35}$$

However, concentrated solutions of electrolytes in particular, are not ideal. Accordingly, it is necessary to introduce the 'activity coefficient'. In order to modify Eq. (1.34) to allow for non-ideality, we write

$$\mu_j = \mu_j^o + RT \ln \left(\frac{\gamma_j [j]}{[\ ]^o} \right),$$

where γ_j is the activity coefficient of species j. For electrolyte solutions

$$\gamma_j \rightarrow 1,$$

as the solute becomes highly dilute, *viz*

$$[j] \rightarrow 0.$$

Accordingly, under these conditions the solution becomes ideal. For more concentrated solutions, γ_j deviates from unity and the extent of the deviation measures the degrees of non-ideality reflecting ion–ion and ion–solvent interactions in the electrolyte. We saw in section 1.6 that, for the hydrogen electrode to be standard, it was necessary that

$$[H^+] = 1.18\,M,$$

implying under these conditions that

$$\gamma_{H^+} = \frac{1}{1.18} = 0.85.$$

The standard electrode potential of the $[Fe(CN)_6^{4-}]/[Fe(CN)_6^{3-}]$ couple involves an arrangement similar to that of Fig. 1.10, except that the reference electrode must be the *standard* hydrogen electrode and the concentrations of the two anions must be chosen carefully, so that

$$\mu_{Fe(III)} = \mu^o_{Fe(III)}$$

and

$$\mu_{Fe(II)} = \mu^o_{Fe(II)}.$$

That is the terms,

$$\frac{\gamma_{Fe(II)}[Fe(II)]}{[\ ]^0} = \frac{\gamma_{Fe(III)}[Fe(III)]}{[\ ]^0} = 1.$$

Under these conditions, the 'activity' of the ions is said to be unity and the 'ln' term in the expression for the chemical potentials disappears (Eq. (1.33)). Extensive tables of standard electrode potentials exist; Table 1.2 provides a fragment of the data that is available.

Each of the entries in the table relate to the standard electrode potential of the redox couple shown, measured relative to a standard hydrogen electrode under conditions where all the species involved in the potential determining equilibrium are at unit activity. These potentials are given the symbol E^0. For the general electrochemical equilibrium

$$\upsilon_A A + \upsilon_B B + \cdots e^-(m) \rightleftarrows \upsilon_X X + \upsilon_Y Y + \cdots,$$

it follows that for arbitrary concentrations of $A, B, \ldots X, Y, \ldots$, we have

$$E = E^0(A, B, \ldots / X, Y, \ldots) + \frac{RT}{F} \ln\left(\frac{\gamma_A^{\upsilon_A} \gamma_B^{\upsilon_B} \cdots [A]^{\upsilon_A}[B]^{\upsilon_B}}{\gamma_X^{\upsilon_X} \gamma_Y^{\upsilon_Y} \cdots [X]^{\upsilon_X}[Y]^{\upsilon_Y}}\right),$$

where E is again measured against the standard hydrogen electrode and $E^0(A, B, \ldots / X, Y, \ldots)$ is the standard electrode potential of the $A, B, \ldots / X, Y, \ldots$ couple. It is evident that $E = E^0(A, B, \ldots / X, Y, \ldots)$ when the 'ln' term vanishes corresponding to each ion having unit activity. Needless to say, however, that it is far from trivial to arrange this situation experientially since a knowledge of the relevant activity coefficients, γ_j, and their concentration dependence is typically lacking. Accordingly, the concept of the formal potential, E_f^0 is introduced where

$$E_f^0(A, B, \ldots / X, Y, \ldots) = E^0(A, B, \ldots / X, Y, \ldots) + \frac{RT}{F} \ln\left(\frac{\gamma_A^{\upsilon_A} \gamma_B^{\upsilon_B} \cdots}{\gamma_X^{\upsilon_X} \gamma_Y^{\upsilon_Y} \cdots}\right),$$

so that

$$E = E_f^0(A, B, \ldots / X, Y, \ldots) + \frac{RT}{F} \ln\left(\frac{[A]^{\upsilon_A}[B]^{\upsilon_B} \cdots}{[X]^{\upsilon_X}[Y]^{\upsilon_Y} \cdots}\right).$$

Table 1.2. Standard electrode potentials for aqueous solutions (25°C).

Half Reaction	E / V
$Li^+ + e^- \rightarrow Li$	−3.04
$K^+ + e^- \rightarrow K$	−2.92
$1/2Ca^{2+} + e^- \rightarrow 1/2Ca$	−2.76
$Na^+ + e^- \rightarrow Na$	−2.71
$1/2Mg^{2+} + e^- \rightarrow 1/2Mg$	−2.37
$1/3Al^{3+} + e^- \rightarrow 1/3Al$ (0.1 M NaOH)	−1.71
$1/2Mn^{2+} + e^- \rightarrow 1/2Mn$	−1.18
$H_2O + e^- \rightarrow 1/2H_2 + OH^-$	−0.83
$1/2Zn^{2+} + e^- \rightarrow 1/2Zn$	−0.76
$1/2Fe^{2+} + e^- \rightarrow 1/2Fe$	−0.44
$1/3Cr^{3+} + e^- \rightarrow 1/3Cr$	−0.41
$1/2Cd^{2+} + e^- \rightarrow 1/2\,Cd$	−0.40
$1/2Co^{2+} + e^- \rightarrow 1/2Co$	−0.28
$1/2Ni^{2+} + e^- \rightarrow 1/2Ni$	−0.23
$1/2Sn^{2+} + e^- \rightarrow 1/2Sn$	−0.14
$1/2Pb^{2+} + e^- \rightarrow 1/2Pb$	−0.13
$1/3Fe^{3+} + e^- \rightarrow 1/3Fe$	−0.04
$H^+ + e^- \rightarrow 1/2H_2$	0.00
$1/2Sn^{4+} + e^- \rightarrow 1/2Sn^{2+}$	+0.15
$Cu^{2+} + e^- \rightarrow Cu^+$	+0.16
$1/2Cu^{2+} + e^- \rightarrow 1/2Cu$	+0.34
$1/2H_2O + 1/4\,O_2 + e^- \rightarrow OH^-$	+0.40
$Cu^+ + e^- \rightarrow Cu$	+0.52
$1/2I_2 + e^- \rightarrow I^-$	+0.54
$1/2O_2 + H^+ + e^- \rightarrow 1/2H_2O_2$	+0.68
$Fe^{3+} + e^- \rightarrow Fe^{2+}$	+0.77
$1/2Hg^{2+} + e^- \rightarrow 1/2Hg$	+0.79
$Ag^+ + e^- \rightarrow Ag$	+0.80
$1/3NO_3^- + 4/3H^+ + e^- \rightarrow 1/3NO + 2/3H_2O$	+0.96
$1/2Br_2(l) + e^- \rightarrow Br^-$	+1.06
$1/4O_2 + H^+ + e^- \rightarrow 1/2H_2O$	+1.23
$1/2MnO_2 + 2H^+ + e^- \rightarrow 1/2Mn^{2+} + H_2O$	+1.21
$1/2Cl_2 + e^- \rightarrow Cl^-$	+1.36
$1/3Au^{3+} + e^- \rightarrow 1/3Au$	+1.52
$Co^{3+} + e^- \rightarrow Co^{2+}$ (3M HNO_3)	+1.84

The formal potentials will depend on temperature and pressure, as do the standard potentials, but will also have a dependence on electrolyte concentrations, not only on these of the species involved in the potential determining equilibrium but also on other electrolytes present in the solution bathing the electrode on which the potential is established, since these influence ion activities.

The formal potential loses the thermodynamic generality of the standard potential being only on applicable to very specific conditions, but enables the experimentalist to proceed with meaningful voltammetric measurements.

1.9 Formal Potentials and Experimental Voltammetry

For practical purposes, *viz* voltammetry, the Nernst equation can be written for the general electrochemical equilibrium of Section 1.4,

$$E = E_f^0(A, B, \ldots / X, Y, \ldots) + \frac{RT}{F} \ln \left(\frac{[A]^{\upsilon_A}[B]^{\upsilon_B} \ldots}{[X]^{\upsilon_X}[Y]^{\upsilon_Y} \ldots} \right)$$

or

$$\left(\frac{[A]^{\upsilon_A}[B]^{\upsilon_B} \ldots}{[X]^{\upsilon_X}[Y]^{\upsilon_Y} \ldots} \right) = e^\Theta,$$

where

$$\Theta = \frac{F}{RT}[(E - E_f^0(A, B, \ldots / X, Y, \ldots))].$$

The formal potential, E_f^0 will *approximate* to the standard potentials given in Table 1.2 but will typically show differences reflecting the precise composition of the solution under study which leads to deviations from solution ideality.

In using Table 1.2 as a guide for approximate values of formal potentials, it should be noted that if protons (or hydroxide ions) are involved in the potential determining equilibrium, then the values are only relevant near pH $= 0$ corresponding to unit activity of protons. Therefore, if the solution under study deviates from this pH value, then the estimate of the formal potential must be correspondingly adjusted. For the general reaction

$$A + mH^+ + e^- \rightleftarrows B \quad E_{A/B}^0,$$

it follows that

$$E = E^0(A/B) + \frac{RT}{F} \ln \left(\frac{\gamma_A}{\gamma_B} (\gamma_{H^+})^m \frac{[A][H^+]^m}{[B]} \right)$$

becomes

$$E = E^0(A/B) + \frac{RT}{F} \ln \frac{\gamma_A}{\gamma_B} + m\frac{RT}{F} \ln (\gamma_{H^+})[H^+] + \frac{RT}{F} \ln \frac{[A]}{[B]}$$

$$E = E_f^0(A/B) + \frac{RT}{F} \ln \frac{[A]}{[B]} - 2.303m\frac{RT}{F} pH,$$

since

$$E_f^0(A/B) = E^0(A/B) + \frac{RT}{F} \ln \frac{\gamma_A}{\gamma_B},$$

$$\log_{10} N = \frac{\ln N}{\ln 10} = \frac{\ln N}{2.303} \quad \text{and}$$

$$pH = - \log_{10} \gamma_{H^+}[H^+],$$

and pH is defined by IUPAC in terms of the single ion activity of H^+.

From an experimental point of view, in buffered media especially, the electrochemical equilibrium can be rewritten as

$$A + e^- \rightleftarrows B$$

for which the *effective* formal potential is

$$E_{f,eff}^0(A/B) = E_f^0(A/B) - 2.303m\frac{RT}{F} pH,$$

so that

$$E = E_{f,eff}^0(A/B) + \frac{RT}{F} \ln \frac{[A]}{[B]}.$$

The above equation shows that at 25°C the formal potential will change by m times 0.059 mV per pH unit. Accordingly, for a pH shift from 0 (corresponding to standard conditions) to 7 (neutral) the formal potential will change by a little more than 400 mV if one proton is taken up per electron in the electrochemical equilibrium. This is a very large change given the size of potential windows typically explored in voltammetry.

To summarise:

- The most useful form of the Nernst equation, relating to a process $A + e^- B$ for voltammetry is

$$\frac{[A]}{[B]} = e^\Theta,$$

where $\Theta = \frac{F}{RT}(E - E_f^\vartheta)$ and E_f^0 is the *formal* electrode potential of the A/B couple.

- Standard electrode potentials provide approximate values for formal potentials. Helpful extensive tables of standard potentials are available, most notably and authoritatively 'Standard potentials in aqueous solutions'.[b] The latter are reported on the standard hydrogen electrode scale so, if for example, your reference electrode is a saturated calomel electrode or a saturated silver/silver chloride electrode then you need to *subtract* a value of 0.242 V or 0.197 V to put the estimate derived from the tables on the correct scale.
- If working in aqueous media at a pH different from the 'standard' condition of pH 0, estimates of formal potentials need to be corrected by 59 mV per proton per pH unit (at 25°C).

Finally, it must be stressed the estimates of the formal potential are thermodynamic quantities and presume electrochemical equilibration and hence, fast electrode kinetics. This is often not a correct assumption. We address this issue in the next section.

1.10 Electrode Processes: Kinetics vs. Thermodynamics

Classically, electrode potentials were measured using a potentiometer such as that shown schematically in Fig. 1.14.

The principle of the method is as follows. A battery C, of constant voltage larger than any to be measured, is connected across a wire AB of high electrical resistance. The cell X under study is connected to the point A and then through a current measuring device, G, to a sliding contact, D, which can be moved along the wire AB. The position of D is adjusted so that no current flows through G. At this point,

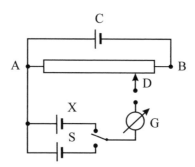

Fig. 1.14 Schematic diagram of a Poggendorf potentiometer.

---

[b] By A.J. Bard, R. Parsons and J. Jordan, Marcel Dekker, New York, 1985.

the drop of potential along AD due to the cell, C, is exactly compensated by the voltage of X, E_x. By means of a switch, the cell X is then replaced by a standard cell, S of known voltage E_s. The point of contact D is re-adjusted to a new point, D' until no current flows, The drop of voltage across AD' then matches E_s. It follows then that the unknown voltage

$$E_x = \frac{AD}{AD'} \times E_s.$$

A merit of the potentiometer approach is that the electrode kinetics of the electrochemical equilibrium of interest rapidly becomes apparent. For example, let us consider the following two electrode processes:

$$Fe(CN)_6^{3-}(aq) + e^-(m) \rightleftarrows Fe(CN)_6^{4-}(aq)$$

and

$$1/2C_2H_6(g) + CO_2(g) + e^-(m) \rightleftarrows CH_3COO^-(aq).$$

Figure 1.15 shows possible experimental arrangements of their study using a calomel electrode as a reference electrode. Also shown is the current measured on a potentiometer such as that shown in Fig. 1.14. In the case of the $Fe(CN)_6^{4-}/Fe(CN)_6^{3-}$ process, true electrochemical equilibrium is rapidly established at the platinum wire electrode.

If the sliding contact is moved either side of the balance point then significant current will flow, since the applied potential no longer balances the potential from the cell under study and either $Fe(CN)_6^{4-}$ ions are oxidised or $Fe(CN)_6^{3-}$ ions are reduced, depending on which direction the contact is shifted. The fact that significant currents flow in both the oxidising and reducing senses when there is only a slight imbalance of the potentiometer shows that the electrode kinetics are fast and hence, that a true electrochemical equilibrium has been set up on the platinum wire. In contrast in the case of acetate/CO_2/C_2H_6, *no* current flows at any point of contact with the potentiometer. This reflects the fact that the electrode kinetics in this case are very slow, so that no electrochemical equilibrium is established on the platinum wire and, even if significant positive or negative potentials are applied to the cell (by varying the contact point), no current flows.

In conclusion, electrode potentials are only established when an electrochemical equilibrium is truly established at the electrode and this requires fast electrode kinetics. The next chapter first considers a model by which electrode kinetics can be interpreted and then sets out to rationalise why some electrode processes are fast and some slow.

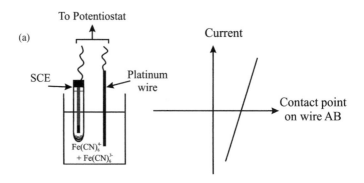

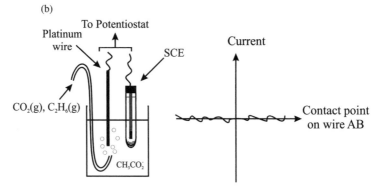

Fig. 1.15 Potentiometric measurements of systems with (a) fast electron kinetics and (b) slow electrode kinetics.

References

[1] H. L. Le Chatelier, *Comptes Rendus* **99** (1884) 786.

[2] H. L. Le Chatelier, *Annales des Mines* **13** (1888) 157.

[3] Text adapted from: *Nobel Lectures, Chemistry 1901–1921*, Elsevier Publishing Company, Amsterdam, 1966. The text and picture of Nernst are © The Nobel Foundation. See also: www.nernst.de (accessed Dec 2006).

[4] P. W. Atkins, J. De Pavla, *Atkins Physical Chemistry*, 7th edn, Oxford University Press, 2002.

[5] R. G. Compton, G. H. W. Sanders, *Electrode Potentials*, Oxford University Press, 1996.

[6] IUPAC Compendium of Chemical Terminology 2nd edn (1997); see also: http://www.iupac.org/publications/compendium/index.html.

[7] A. J. Bard, R. Parsons, J. Jordan, *Standard Potentials in Aqueous Solutions*, Marcel Dekker, New York, 1985.

2

Electrode Kinetics

This chapter is concerned with establishing both phenomenological and molecular models for electron transfer reactions and in particular, their dependence on the electrode potential. We start by relating currents and fluxes.

2.1 Currents and Reaction Fluxes

Consider the electro-reduction of hexacyanoferrate (III) to hexacyanoferrate (II) in aqueous solution:

$$Fe(CN)_6^{3-}(aq) + e^-(m) \rightleftharpoons Fe(CN)_6^{4-}(aq) \tag{2.1}$$

at an electrode of area, A cm^2, brought about through the application of a suitable negative potential to the electrode. Note that of course a second electrode, at least, will be needed somewhere in the solution to facilitate the passage of the required electrical current through the solution. The necessary experimental approach for this will be considered in the next section; for the present, we simply focus on the interface shown in Fig. 2.1.

The electrode transfer process between the electrode and $Fe(CN)_6^{3-}$ species takes place via quantum mechanical tunnelling between the two locations. Accordingly, the $Fe(CN)_6^{3-}$ undergoing electro-reduction must be located within ca 10–20 Å of the electrode surface, since the rate of tunnelling falls off strongly with separation, as it requires overlap of the quantum mechanical wavefunctions describing the electron in the donor (electrode) and acceptor $(Fe(CN)_6^{3-})$ locations. It follows that for $Fe(CN)_6^{3-}$ anion to undergo reduction at the electrode, it

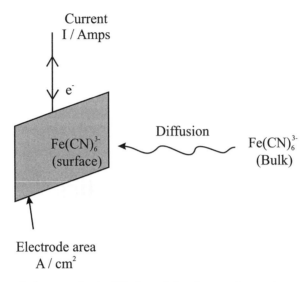

Current
I / Amps

e⁻

Fe(CN)₆³⁻
(surface)

Diffusion

Fe(CN)₆³⁻
(Bulk)

Electrode area
A / cm²

Fig. 2.1 Electrolysis proceeds via diffusion of the electroactive species to the electrode surface.

must first diffuse from bulk solution to within this critical distance for electron tunnelling before electron transfer takes place.

The electron transfers lead to the passage of an electrical current, I (amps) through the electrode. This is related to the flux of reactant undergoing electrolysis, j/mol cm^{-2} s^{-1} via the following equation

$$I = FAj, \tag{2.2}$$

where F is the Faraday constant and A is the electrode area. The flux can be thought of as a measure of the rate of the (heterogeneous) interfacial electrochemical reaction, in the same way that the quantity $d[reactant]/dt$ measures the rate of a homogeneous chemical reaction.

Rate laws can be written for interfacial reactions in much the same manner as for homogeneous chemical processes:

$$j = k(n)[reactant]_0^n, \tag{2.3}$$

where n is the order of the reactant, $k(n)$ is a nth order rate constant and the subscript '0' on the concentration term donates it is the surface concentration of reactant, namely that within the critical electron tunnelling distance, which is pertinent.

Three points of importance relate to Eq. (2.3). First the case of $n = 1$ is very commonly encountered corresponding to a first order heterogeneous reaction. Note that in this situation the units of $k(1)$ are cm s^{-1}, those of velocity. It is helpful for the reader to become familiar with the notion that first order heterogeneous rate constants have these dimensions; of course this contrasts with the case of first order homogeneous rate constants where the units are s^{-1}.

Second, the fact that the concentration term in Eq. (2.3) has been subscripted to emphasise that the relevant concentration of reactant is that adjacent to the electrode, rather than, say, that in bulk solution, implicitly suggests that in general

$$[reactant]_0 \neq [reactant]_{bulk}.$$

A little reflection shows that this is necessarily true, since the passage of current through the electrode–solution interface will lead to the depletion of reactants local to the electrode surface. Material will then diffuse from further away by virtue of the concentration gradient established. Accordingly, it follows that the concentrations of electrolysis reactants (and products) can be quite different near the electrode surface from those in bulk solution. As will become more fully evident in Chapters 3 and 4, one of the features of electrolysis is that concentrations change both spatially and temporally during the process. Keeping track of the evolution of concentrations with both distance from the electrode, and in time, is the key challenge in understanding voltammetry.

Third, the heterogeneous rate constants $k(n)$ depend on temperature and pressure, just as familiar homogeneous rate constants are sensitive to these parameters. However, the heterogeneous (electrochemical) rate constants are extremely sensitive to the electrical potential. In particular, as we will see in Section 2.3, there is often an exponential dependence of $k(n)$ on the difference in potential between the electrical (ϕ_M) and the solution (ϕ_S). Moreover, in many cases, if the difference $\phi_M - \phi_S$ is changed by just one volt then the corresponding change in $k(n)$ is a factor of the order of 10^9, a huge number!

Before developing the basis for the exponential dependence of electrochemical rate constants on $\phi_M - \phi_S$, we will first consider the experimental way in which changes in this quantity can be imposed on a single electrode–solution interface.

2.2 Studying Electrode Kinetics Requires Three Electrodes

Quantitative controlled studies of electrode kinetics are conducted using three electrodes:

- The working electrode possesses the interface of interest and under study.
- The reference electrode, for example, a saturated calomel electrode, acts in the manner described in Section 1.5.
- The counter (or auxiliary) electrode is the third electrode.

All three electrodes are controlled by a 'potentiostat' which acts in the manner shown in Fig. 2.2

First, the device imposes a fixed potential, E, between the working and reference electrodes. Since the potentiostat draws negligible current through the latter electrode,

$$E = (\phi_M - \phi_S)_{working} - (\phi_M - \phi_S)_{reference}. \qquad (2.4)$$

It follows that, since the reference electrode serves to provide a constant value of $(\phi_M - \phi_S)_{reference}$ as detailed in Section 1.7, any changes in E are reflected as changes in $(\phi_M - \phi_S)_{working}$; changing E by, say, 1 Volt, also changes $(\phi_M - \phi_S)_{working}$ by 1 Volt.

Second, the imposition of the potential drop $(\phi_M - \phi_S)_{working}$ on the working electrode–solution interface will typically cause a current to flow; this is the aim of the experiment to study the current through the working electrode–solution interface as a function of a controlled, applied potential. The counter electrode serves to pass the same current as that induced to flow through the working electrode. Accordingly, the potentiostat 'drives' the counter electrode to whatever voltage is required to pass this current. Note that the introduction of the third electrode, the counter electrode acting in this manner, is the only reason why a controlled potential can be applied to the working electrode as expressed by Eq. (2.4). If

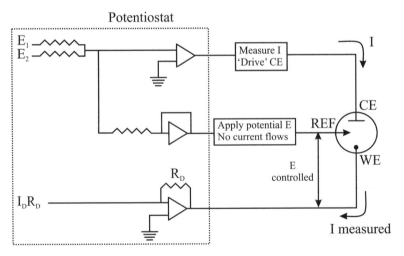

Fig. 2.2 A potentiostat is required for running electrochemical experiments. Note all the resistances are equal except R_D which is variable. CE is the counter electrode, WE is the working electrode and REF is the reference electrode.

only two electrodes were present with one attempting to act simultaneously as a counter and a reference electrode, then Eq. (2.4) would be compromised in two ways. First, appreciable current might be drawn through the reference electrode inducing chemical changes within it and so altering $(\phi_M - \phi_S)_{reference}$ in accordance with the Nernst equation (Section 1.8). Second, Eq. (2.4) would gain a third term:

$$E = (\phi_M - \phi_S)_{working} + IR - (\phi_M - \phi_S)_{reference}, \qquad (2.5)$$

where the IR term reflects the electrical resistance of the bulk solution between the working and reference electrodes. In this situation, changes in E would cause unknown changes in IR, as well as altering $(\phi_M - \phi_S)_{working}$, so that the latter would no longer be controlled.

An exception to the need for the use of three electrodes arises when the working electrode is a 'microelectrode'. Such electrodes have dimensions of the micron scale or less and so pass very low currents ($\sim 10^{-9}$ A). Accordingly, a two-electrode system of a working electrode and a reference/counter electrode may suffice since first, the tiny currents lead to negligible electrolytic change within the reference electrode and second, the iR term in Eq. (2.5) becomes small enough to neglect.

Finally, it is worth making two practical points in connection with the presence of the electrode, the counter electrode. We have stressed that the role of this electrode is to pass the same current as that flowing through the working electrode. It follows that this current will electrolytically induce chemical changes in the solution in the vicinity of the counter electrode. Experiments need to be conducted recognizing that chemical products must necessarily form as a result of the counter electrode process. If necessary, the electrochemical cell can be designed to separate the counter electrode behind a frit, perhaps in a side arm of the cell, to reduce possible contamination of the solution bathing the working electrode. The issue of counter electrode products is particularly significant in electrosynthesis, where large scale conversion of the cell contents is usually desirable. In voltammetric studies, relatively tiny quantities of material are consumed and so the problem may be more easily controlled. Equally, the passage of current through the counter electrode requires that this electrode is driven by the potentiostat to the necessary voltage for this current flow. The capacity of any potentiostat in terms of the voltage it can apply to the counter electrode is limited and finite; many commercial potentiostats can only deliver *ca.* ±15 V, whilst those designed for synthetic purposes may deliver up to *ca.* ±70 V. It is good experimental practice to check the counter electrode voltage to ensure that maximum values has not be reached; if it has, the working electrode is no longer 'potentiostatted' and the experiment will need re-designing.

2.3 Butler–Volmer Kinetics

In this section, we develop the phenomenological model of electrode kinetics which underpins most current interpretations of electrode kinetics: the Butler–Volmer model.

As in Section 2.1, we consider the reduction of hexacyanoferrate (III) and oxidation of hexacyanoferrate (II):

$$Fe(CN)_6^{3-}(aq) + e^-(m) \underset{k_a}{\overset{k_c}{\rightleftharpoons}} Fe(CN)_6^{4-}(aq),$$

where the rate constants k_c and k_a describe the reduction and oxidation respectively. The subscripts reflect the terms 'cathodic' and 'anodic' respectively: a cathodic process is one at an electrode (a cathode) supplying electrons and hence causing a reduction, whilst an anodic process is one at an electrode (an 'anode') which removes electrons and hence causes an oxidation process in the solution.

It follows from Section 2.1 that the following rate law can be written for the *net* process:

$$j = k_c[Fe(CN)_6^{4-}]_0 - k_a[Fe(CN)_6^{3-}]_0. \tag{2.6}$$

The rate constants k_c and k_a will be potential dependant; specifically, we might expect the cathodic reduction to predominate at relatively negative electrode potentials, whilst the anodic oxidation would represent the dominant term in Eq. (2.6) at relatively positive potentials. The precise interplay will emerge below. Figure 2.3 shows a reaction profile for the process of interest.

Since $Fe(CN)_6^{3-}$, $Fe(CN)_6^{4-}$ and e^- are all charged species it follows that the profile shown must be a function of both ϕ_M and ϕ_S. For example, making ϕ_M more negative whilst keeping ϕ_S constant would raise the energy of the reactants, which include $e^-(m)$, whilst keeping the product energy constant. Conversely, keeping ϕ_M constant whilst making the solution more negative would change both the reactant and product energies, with the $Fe(CN)_6^{4-}$ energy being raised more than $Fe(CN)_6^{3-}$ the energy on account of the different charges of the ions. So, by altering ϕ_M and/or ϕ_S, the reactant profile is changed so that either the reduction of $Fe(CN)_6^{3-}$ (ϕ_M moves negative, ϕ_S positive) or the oxidation of $Fe(CN)_6^{4-}$ (ϕ_M moves positive, ϕ_S negative) is the downhill, thermodynamically driven reaction as shown in Fig. 2.4.

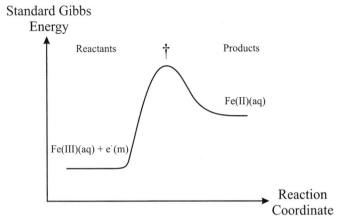

Fig. 2.3 A reaction profile for the $Fe(CN)_6^{4-}/Fe(CN)_6^{3-}$ electrode process. Note that both ϕ_M and ϕ_S are fixed.

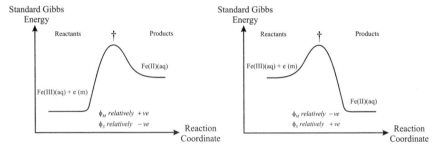

Fig. 2.4 Changing ϕ_M and ϕ_S alters the energy profile for an electrode process.

Considering Fig. 2.3 we can write

$$k_c = A_c \exp \frac{-\Delta G_c^{(\dagger)}}{RT} \tag{2.7}$$

and

$$k_a = A_a \exp \frac{-\Delta G_a^{(\dagger)}}{RT}, \tag{2.8}$$

where the Arrhenius equation relates the rate constants k_c and k_a to the Gibbs energies of activation,

$$\Delta G_c^0(\dagger) = G^0(\dagger) - G^0(R) \tag{2.9}$$

and

$$\Delta G_a^0(\dagger) = G^0(\dagger) - G^0(P), \tag{2.10}$$

and the pre-exponential factors A_c and A_a. The latter will, amongst other factors, depend on a 'frequency factor' describing the number of collisions per second with the electrode surface. In Eqs. (2.9) and (2.10) ($\dagger$) corresponds to the transition state, R to the 'reactants' in Fig. 2.4 and P to the products. $G^0(x)$ for $x = R$, ($\dagger$) or P is the standard molar Gibbs energy of species x. We know from Chapter 1 that

$$G^0(R) = a\ constant - F\phi_M - 3F\phi_S, \tag{2.11}$$

leading to

$$G^0(R) = a\ constant - 4F\phi_S - F(\phi_M - \phi_S)$$

and

$$G^0(P) = another\ constant - 4F\phi_S. \tag{2.12}$$

Inspection of Eqs. (2.11) and (2.12) shows that, in respect of potential dependence, the equations differ by the term $F(\phi_M - \phi_S)$. If we then assume that the transition state has a standard Gibbs energy 'in between' that of the reactants and produces:

$$\Delta G^0(\dagger) = a\ further\ constant - 4F\phi_S - \beta F(\phi_M - \phi_S), \tag{2.13}$$

where β, known as a transfer coefficient, is such that

$$0 < \beta < 1, \tag{2.14}$$

so that if β is close to zero, the transition state ($\dagger$) is product (P) like whilst if β is near unity the transition state is reactant (R) like, at least in terms of the potential dependence of its standard Gibbs energy.

Equations (2.11)–(2.13) allow us to evaluate the Gibbs energies of activation and hence the electrochemical rate constant via Eqs. (2.7) and (2.8). The result is that

$$k_c \propto \exp\left[\frac{-(1-\beta)F(\phi_M - \phi_S)}{RT}\right] \tag{2.15}$$

$$k_a \propto \exp\left[\frac{\beta F(\phi_M - \phi_S)}{RT}\right]. \tag{2.16}$$

The term, $1 - \beta$, is usually replaced by α, so that

$$\alpha + \beta = 1$$

and both α and β are known as transfer coefficients. It follows that Eq. (2.16) can be re-written as

$$k_c \propto \exp\left[\frac{-\alpha F(\phi_M - \phi_S)}{RT}\right]. \tag{2.17}$$

Equations (2.16) and (2.17) tell us that the electrochemical rate constants for the one oxidation of $Fe(CN)_6^{4-}$ (k_a) and for the reduction $Fe(CN)_6^{3-}$ (k_c) depend exponentially on the electrode potential: (k_a) increases as the electrode is made more positive relative to the solution $[(\phi_M - \phi_S)$ becomes more positive] whilst (k_c) increases as the electrode is made more negative relative to the solution $[(\phi_M - \phi_S)$ becomes more negative].

It was indicated above that the transfer coefficient β (and hence also α) lay in the range zero to unity. Often, but not always, however,

$$\alpha \sim \beta \sim 0.5, \tag{2.18}$$

implying that the transition state lies intermediate in its electrical behaviour between the reactants and products. If we substitute the values of Eq. (2.18) into Eqs. (2.16) and (2.17), we find that if the term $(\phi_M - \phi_S)$ is changed by 1 Volt then the rate constants (k_a) and (k_b) are changed by factor of *ca.* 10^9. It is this overwhelming sensitivity of the electrochemical rate constant on the electrode potential that dominates the kinetic behaviour of electrode processes.

2.4 Standard Electrochemical Rate Constants and Formal Potentials

In Eqs. (2.16) and (2.17), we related the electrochemical rate constants to the quantity $(\phi_M - \phi_S)$. However, we saw in Chapter 1 that such quantities relating to a single electrode–solution interface cannot be measured, since there is always a need for a reference electrode in any complete measuring circuit. It is therefore helpful to modify the two expressions to the form

$$k_c = k_c^0 \exp\left[\frac{-\alpha F\left(E - E_f^0\right)}{RT}\right] \tag{2.19}$$

and

$$k_a = k_a^0 \exp\left[\frac{\beta F\left(E - E_f^0\right)}{RT}\right], \tag{2.20}$$

where $E - E_f^0$ measures the potential applied to working electrode relative to the formal potential of the $Fe(CN)_6^{3-}/Fe(CN)_6^{4-}$ couple with *both* potentials measured relative to the same reference electrode. It is clear from the discussion in Section 1.9 that

$$E - E_f^0 = (\phi_M - \phi_S)_{working} + constant \qquad (2.21)$$

so that shifting between Eqs. (2.16), (2.17) to (2.19) and (2.20) is algebraically consistent since the constant term of Eq. (2.21) is incorporated into the terms k_c^0 and k_a^0. We next focus on these expressions in a little more detail and, specifically, return to Eq. (2.6):

$$j = k_c^0 \exp\left[\frac{-\alpha F(E - E_f^0)}{RT}\right][Fe(CN)_6^{3-}]_0 - k_a^0 \exp\left[\frac{+\beta F(E - E_f^0)}{RT}\right][Fe(CN)_6^{4-}]_0 .$$

$$(2.22)$$

If we consider the case of a dynamic equilibrium at the working electrode such that the oxidation and reduction currents exactly balance each other, then, since no net current flows,

$$j = 0. \qquad (2.23)$$

From Eqs. (2.21) and (2.22), and the fact that $\alpha + \beta = 1$, we see that

$$E = E_f^0 + \frac{RT}{F} \ln\left(\frac{[Fe(CN)_6^{4-}]}{[Fe(CN)_6^{3-}]}\right) + \frac{RT}{F} \ln\left(\frac{k_a^0}{k_c^0}\right). \qquad (2.24)$$

From the discussion in Section 1.8, it is clear that the potential when no net current flows is given by

$$E = E_f^0 + \frac{RT}{F} \ln\left(\frac{[Fe(CN)_6^{4-}]}{[Fe(CN)_6^{3-}]}\right), \qquad (2.25)$$

so that $k_a^0 = k_c^0 = k^0$.

We can therefore write

$$k_c = k^0 \exp\left[\frac{-\alpha F(E - E_f^0)}{RT}\right] \qquad (2.26)$$

$$k_a = k^0 \exp\left[\frac{\beta F(E - E_f^0)}{RT}\right]. \qquad (2.27)$$

Equations (2.26) and (2.27) are the most convenient forms of the Butler–Volmer expression for the electrochemical rate constants k_c^0 and k_a^0. The quantity k^0, with units of cm s^{-1}, is the *standard electrochemical rate constant*.

Equations (2.26) and (2.27) are general expressions for any electrode process of the form

$$A^{Z_A} + e^- \rightleftharpoons B^{Z_B},$$

where Z_A and Z_B are the charges on A and B, respectively. This follows since

$$G^0(R) = G^0(A) + G^0(e^-) = constant + Z_A F \phi_S - F(\phi_M)$$
$$G^0(R) = G^0(A) + G^0(e^-) = constant + (Z_A - 1)F\phi_S - F(\phi_M - \phi_S) \qquad (2.28)$$

and

$$G^0(P) = G^0(B) = another\ constant + Z_B F \phi_S$$
$$G^0(P) = G^0(B) = another\ constant + (Z_A - 1)F\phi_S$$

since $Z_A = Z_B + 1$. Accordingly,

$$G^0(\dagger) = a\ further\ constant + (Z_A - 1)F\phi_S - \beta F(\phi_M - \phi_S)$$
$$G^0(\dagger) = a\ further\ constant + (Z_A - 1)F\phi_S - (1 - \alpha)F(\phi_M - \phi_S).$$

Exactly the same arguments applied to the specific case of $B = Fe(CN)_6^{4-}$ and $A = Fe(CN)_6^{3-}$ then show that

$$k_c = k^0 \exp\left[\frac{-\alpha F\left(E - E_f^0\right)}{RT}\right] \qquad (2.29)$$

and

$$k_a = k^0 \exp\left[\frac{\beta F\left(E - E_f^0\right)}{RT}\right], \qquad (2.30)$$

where E_f^0 is the formal potential of the A/B redox couple.

2.5 The Need for Supporting Electrolyte

Equations (2.29) and (2.30) show the importance of the quantity $\phi_M - \phi_S$ in controlling the magnitude of the electrochemical rate constants k_a and k_c. Section 2.1 pointed out that, because electron transfer between the electrode and the solution phase species occurs by quantum mechanical tunneling, it is essential that the solution species is located within a distance of some 10–20 Å of the electrode surface. It follows that the drop in potential between the electrode and the bulk solution must occur within a similar distance if we are to use the rate constants in the form we have derived. In the case that the potential drop from ϕ_M to ϕ_S occurs over a larger distance than 10–20 Å, then the rate constant expressions (2.29) and (2.30) will need modification to allow for the fact that at the sites of electron transfer (tunneling) only a fraction of the drop ($\phi_M - \phi_S$) will be available to 'drive' the electron transfer reaction. Such 'Frumkin corrections' are described more fully elsewhere[1] and are *essential* if experiments are done in poorly conducting media. However, in most experimental situations, it is common to work with a deliberately added large quantity of electrolyte — so-called 'supporting' or 'background' electrolyte — the purpose of which is to ensure that the potential drop ($\phi_M - \phi_S$) is compressed to a distance within 10–20 Å of the working electrode surface. Typical concentrations for the supporting electrolyte are $> 10^{-1}$ M and the chosen electrolytes are selected for their electrochemical inertness at the potentials of interest in the experiments conducted. For example KCl, might be used in aqueous media, or tetra-n-butylammonium tetrafluoroborate in acetonitrile solution. With suitable additions of supporting electrolyte, the rate laws are as given in Eqs. (2.29) and (2.30).

2.6 The Tafel Law

We saw earlier in this chapter that for the simple electrode process

$$A + e^- \underset{k_a}{\overset{k_c}{\rightleftarrows}} B,$$

the net rate (flux) of reaction was given by

$$j/\text{mol cm}^{-2}\text{s}^{-1} = k_c[A]_0 - k_a[B]_0 \tag{2.31}$$

$$j = k_c^0 \exp\left[\frac{-\alpha F(E - E_f^0)}{RT}\right][A]_0 - k_a^0 \exp\left[\frac{+\beta F(E - E_f^0)}{RT}\right][B]_0. \tag{2.32}$$

The expression shows that with an arbitrary potential, E, imposed on the working electrode the net flux (and hence current) is a balance between the reduction current due to the addition of an electron to A, and the oxidation current due to the removal of an electron from B. Under extreme potentials, such as

$$E \gg E_f^0 \quad \text{or} \quad E \ll E_f^0$$

it is possible to neglect one term or another. In this case, for a reductive electrochemical process

$$A + e^- \rightarrow B$$

$$j = k^0 \, \exp\left[\frac{-\alpha F\left(E - E_f^0\right)}{RT} \right] [A]_0, \qquad (2.33)$$

whilst for an oxidation

$$B - e^- \rightarrow A$$

$$j = k^0 \exp\left[\frac{\beta F\left(E - E_f^0\right)}{RT} \right] [B]_0. \qquad (2.34)$$

It follows that, provided $[A]_0$ and $[B]_0$ are not substantially changed from their constant bulk values,

$$\ln |I_{red}| = -\frac{\alpha FE}{RT} + constant \qquad (2.35)$$

and

$$\ln |I_{ox}| = \frac{\beta FE}{RT} + constant, \qquad (2.36)$$

where I_{red} and I_{ox} are the reduction and oxidation currents respectively. Accordingly, plots of $\ln |I_{ox}|$ vs E, or $\ln |I_{red}|$ vs E, so called Tafel plots, can provide information about the magnitude of the transfer coefficients, α and β, as shown in Fig. 2.5.

Note that this procedure requires the concentration terms $[A]_0$ and $[B]_0$ in Eqs. (2.33) and (2.34) to be constant over the range of potential values studied. This presents a considerable restriction. Accordingly, we return to Tafel analysis in Chapter 3 after a consideration of reactant depletion effects and a discussion of diffusion.

2.7 Julius Tafel

© Klaus Müller

Julius Tafel (1862–1918) was an organic and physical chemist who studied the electrochemistry of organic compounds and is known for one of the most important equations of electrochemical kinetics: the Tafel equation.

Tafel was born in Switzerland and studied in Zürich, München and Erlangen. In 1882, Tafel received his PhD with Emil Fischer in Erlangen with a thesis on the isomerisation of indazole. In 1885, Tafel moved to Wurzburg following Fischer. Working with Fischer, Tafel researched into developing routes to amines via the reduction of phenylhydrazones of aldehydes and ketones. However, such work with Fischer distracted him from his own research geared toward elucidating the structure of the alkaloids strychnine

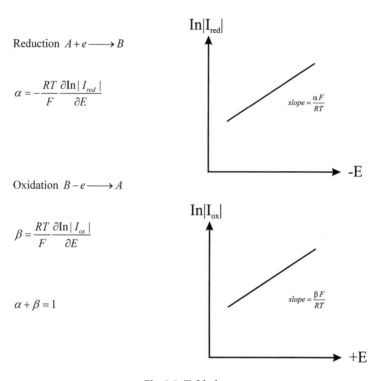

Reduction $A + e \longrightarrow B$

$$\alpha = -\frac{RT}{F} \frac{\partial \ln|I_{red}|}{\partial E}$$

Oxidation $B - e \longrightarrow A$

$$\beta = \frac{RT}{F} \frac{\partial \ln|I_{ox}|}{\partial E}$$

$$\alpha + \beta = 1$$

$\ln|I_{red}|$

$slope = \dfrac{\alpha F}{RT}$

$-E$

$\ln|I_{ox}|$

$slope = \dfrac{\beta F}{RT}$

$+E$

Fig. 2.5 Tafel plots.

and brucine, which was not completed during his lifetime. In 1892, Fischer moved to Berlin but there were no positions for his assistants, so Tafel stayed at Wurzburg and carried out his own research full-time. Tafel spent some time with Ostwald in Leipzig, the result of which was his first publication in the area of physical chemistry in 1896. Tafel explored the use of lead cathodes for the electrochemical reduction of strychnine and this was the subject of his first electrochemistry paper. The discovery of the use of lead electrodes for the electrochemical reduction of organic compounds was a seminal contribution.

Tafel proposed the catalytic mechanism for hydrogen evolution bearing his name. By combining the measurements of current with analysis of the overpotential for electrochemical reaction, Tafel empirically discovered the first formulation of the electrochemical kinetics law (see above, Section 2.6). Tafel's observations were on the studies of irreversible electrochemical reactions where thermodynamics cannot be applied, and thus his studies were the first to separate electrochemical kinetics from thermodynamics allowing irreversible reactions to be studied. Tafel also discovered that hydrocarbons with isomerised structures could be generated via electrochemical reduction of the respective acetoacetic esters, a reaction which is known as the Tafel re-arrangement. Tafel retired at the age of 48 due to his poor health but was still frequented by his pupils attending his bedside during fever spells! During his last years (1911–1918) he wrote up to 60 book reviews. Sadly, Tafel suffered from insomnia which resulted in a complete nervous breakdown and he committed suicide in 1918 at the age of 56.[2]

2.8 Multistep Electron Transfer Processes

Consider the following electrode process

$$A + e^-(m) \rightleftarrows B$$
$$B + e^-(m) \rightleftarrows C$$

and focus on the overall two-electron reduction of A to C. Chemical experience tells us that in the two-step mechanism shown, either the first step or the second step might be rate-determining. We consider each case in turn, having first established some general equations as follows. We assume that the rate constants in the form of Eqs. (2.29) and (2.30) are applicable to each of the two steps of the above mechanism.

The fluxes of A and B are given by

$$j_A = -k^0_{A/B} \exp\left[\frac{-\alpha_1 F}{RT}(E - E^0_f(A/B))\right][A]_0$$
$$+ k^0_{A/B} \exp\left[\frac{\beta_1 F}{RT}(E - E^0_f(A/B))\right][B]_0$$

$$j_B = k^0_{A/B} \exp\left[\frac{-\alpha_1 F}{RT}(E - E^0_f(A/B))\right][A]_0$$
$$- k^0_{A/B} \exp\left[\frac{\beta_1 F}{RT}(E - E^0_f(A/B))\right][B]_0$$
$$+ k^0_{B/C} \exp\left[\frac{\beta_2 F}{RT}(E - E^0_f(B/C))\right][C]_0$$
$$- k^0_{B/C} \exp\left[\frac{\alpha_2 F}{RT}(E - E^0_f(B/C))\right][B]_0$$

$$j_C = +k^0_{B/C} \exp\left[\frac{-\alpha_2 F}{RT}(E - E^0_f(B/C))\right][B]_0$$
$$- k^0_{B/C} \exp\left[\frac{\beta_2 F}{RT}(E - E^0_f(B/C))\right][C]_0,$$

where α_1 and β_1 relate to the one-electron reduction of A to B ($\alpha_1 + \beta_1 = 1$), whilst α_2 and β_2 relate to the one-electrode reduction of B to C ($\alpha_2 + \beta_2 = 1$). Note that $j_A + j_B + j_C = 0$ signifying conservation of matter.

We now consider the 2 cases identified above.

Case 1. The first step is rate-determining:

$$A + e^-(m) \longrightarrow B$$
$$B + e^-(m) \xrightarrow{fast} C.$$

In this case the concentration of B is tiny, so that

$$j_A = k^0_{A/B} \exp\left[-\frac{\alpha_1 F}{RT}(E - E^0_f(A/B))\right][A]_0 \tag{2.37}$$

and so, under the conditions of the Tafel analysis described in the previous section,

$$\ln|I_{red}| = -\frac{\alpha_1 FE}{RT} + constant. \tag{2.38}$$

Case 2. The second step is rate-limiting, so that the first step constitutes a pre-equilibrium:

$$A + e^-(m) \rightleftarrows B$$
$$B + e^-(m) \xrightarrow{slow} C.$$

In this case,

$$I_{red} = FA(j_A + j_B) = -FAj_C,$$

where A is the electrode area. It follows that, under Tafel conditions where the oxidation of C can be neglected,

$$j_C = -k^0_{B/C} \exp\left(\frac{-\alpha_2 F}{RT}(E - E^0(B/C))\right)[B]_0,$$

but because of the pre-equilibrium

$$\frac{[B]_0}{[A]_0} = \exp\left[\frac{-F}{RT}(E - E^0(A/B))\right],$$

it follows that

$$j_C = -k^0_{B/C} \exp\left(\frac{-\alpha_2 F}{RT}(E - E^0(B/C))\right) \exp\left(\frac{-F}{RT}(E - E^0(A/B))\right)[A]_0. \tag{2.39}$$

Again assuming $[A]_0$ remains close to its bulk value,

$$\ln|I_{ox}| = -\frac{(1+\alpha_2)FE}{RT} + constant. \qquad (2.40)$$

Comparison of Eqs. (2.38) and (2.40) shows that Tafel analysis locates the position of the transition state. Given that often

$$\alpha_1 \sim \beta_1, \sim \alpha_2 \sim \beta_2 \sim 0.5,$$

a Tafel slope of *ca.* 0.5 will indicate that case 1 operates and that the first electron transfer is rate-determining. In contrast, a Tafel slope of *ca.* 1.5 indicates case 2 and that the second electrode transfer is rate-determining. Figure. 2.6 shows the reaction profiles in the two cases.

A similar analysis applied to the two-step oxidation

$$Z - e^- \rightleftharpoons Y$$
$$Y - e^- \rightleftharpoons X$$

shows again two cases, as follows.

Case 1. The first step is rate-determining:

$$Z - e^- \xrightarrow{slow} Y$$
$$Y - e^- \xrightarrow{fast} X.$$

Tafel analysis gives

$$\ln|I_{ox}| = +\frac{\beta_1 F}{RT}E + constant, \qquad (2.41)$$

where β_1 is a transfer coefficient for the Y/Z couple.

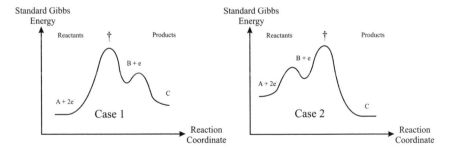

Fig. 2.6 Reaction profiles for a two-electron reduction.

Case 2. There is a pre-equilibrium involving Y and Z followed by a rate-determining step in which Y forms X:

$$Z - e^- \rightleftarrows Y$$

$$Y - e^- \xrightarrow{slow} X.$$

In this case, Tafel analysis shows that

$$\ln |I_{ox}| = +\frac{(1 + \beta_2)F}{RT}E + constant. \tag{2.42}$$

2.9 Tafel Analysis and the Hydrogen Evolution Reaction

Here, we illustrate the results of Section 2.8 and consider the hydrogen evolution reaction in acidic aqueous solution, where H^+ is reduced to form dihydrogen, H_2:

$$H^+(aq) + e^-(m) \rightarrow 1/2H_2(g).$$

The reaction proceeds with the intermediacy of adsorbed hydrogen atoms and this feature of the mechanism leads to huge variation in the rate of the reaction from one electrode material to another, spanning almost 10 orders of magnitude from platinum (fast) to lead (slow). The mechanism is also electrode sensitive and three limiting cases can be identified according to the values of the measured transfer coefficient.

Case A. This is characterized by transfer coefficients (α) of, or close to, 0.5 which is the case for electrodes made of Hg, Pb or Ni. The mechanism is as follows:

$$H^+(aq) + e^-(m) \xrightarrow[rds]{slow} H^\bullet(ads),$$

followed by either

$$2H(ads) \xrightarrow{fast} H_2(g)$$

or

$$H^+(aq) + H^\bullet(ads) + e^-(m) \xrightarrow{fast} H_2(g).$$

The fact that $\alpha \sim 0.5$ indicates it is the first electron transfer which is the rate-determining step (rds).

Case B. This applies to metal electrodes of Pb, Au or W, where transfer coefficients (α) of 1.5 are observed. The mechanism is thought to be

$$H^+(aq) + e^-(m) \xrightarrow{fast} H^\bullet(ads)$$

$$H^+(aq) + H(ads) + e^-(m) \xrightarrow[rds]{slow} H_2(g).$$

The observed transfer coefficient of $\alpha \sim 1.5$ suggests that it is the second electron transfer which is rate-determining.

Case C. This applies specifically to palladium electrodes where Tafel slopes giving $\alpha \sim 2$ are observed. This is consistent with the following mechanism in which the second step is rate-determining

$$H^+(aq) + e^-(m) \overset{fast}{\rightleftharpoons} H^\bullet(ads)$$

$$2H^\bullet(ads) \overset{slow}{\longrightarrow} H_2(g).$$

This is an example of a second order electrochemical process where

$$rate \propto [H^\bullet(ads)]^2$$

and $[H^\bullet(ads)]$ is the *surface* coverage of hydrogen atoms (mol cm^{-2}) on the electrode surface. Because of the pre-equilibrium step the electrode potential, E, will control this quantity via the Nernst equation,

$$E = E_f^0(H^+/H^\bullet(ads)) + \frac{RT}{F} \ln \left(\frac{[H^+(aq)]}{H^\bullet(ads)} \right),$$

so that

$$[H^\bullet(ads)] = [H^+(aq)] \exp \left(-\frac{F}{RT}(E - E_f^0)(H^+/H^\bullet(ads)) \right),$$

giving

$$rate \propto [H^\bullet(ads)]^2 \propto [H^+(aq)]^2 \exp \left(-\frac{2F}{RT}(E - E_f^0(H^+/H^\bullet(ads)) \right),$$

leading to an apparent transfer coefficient of 2, since

$$\ln |I_{red}| = -\frac{2F}{RT} E + constant. \tag{2.43}$$

Further insight into the mechanism of hydrogen evolution and, in particular, the transition between Case A and Case B, can be understood if the standard electrochemical rate constant, k^0, measured on different electrode materials is considered as a function of the estimated enthalpy of adsorption of H on the different electrode surfaces, $-\Delta H(ads)$. This is shown in Fig. 2.7; note the logarithmic scale on the y-axis.

The plot shows a 'volcano curve'; as the enthalpy of adsorption becomes more negative the plot first rises showing an increase of k^0 with the strength (exothermicity) of adsorption of H on the electrode surface. A maximum value of the rate

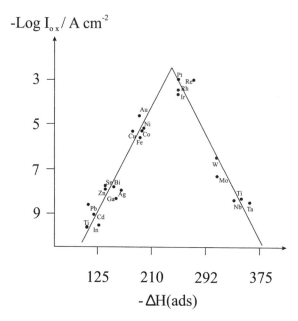

Fig. 2.7 Current density for electrolytic hydrogen evolution versus strength of intermediate metal–hydrogen bond formed during electrochemical reaction. Reprinted from Ref. [3] with permission from Elsevier.

constant is then attained before there is a steady decrease of k^0 with increasing exothermicity of adsorption. All of the metal electrodes on the ascending (left) side of the plot show transfer coefficients, $\alpha \sim 1/2$, whereas those on the descending part of the plot have $\alpha \sim 3/2$. This can be understood in terms of the general mechanism

$$H^+(aq) + e^-(m) \underset{}{\overset{\dagger_A}{\rightleftarrows}} H^\bullet(ads)$$

$$H^+(ads) + e^-(m) \underset{}{\overset{\dagger_B}{\rightleftarrows}} H_2(g),$$

where the transition state of the reaction is either $\dagger_A$ or $\dagger_B$ depending on whether the first or second step is rate-determining. For weak enthalpies of adsorption, the rate constant increases with the adsorption exothermicity since in the first step hydrogen atoms become adsorbed to the electrode surface. Since this adsorption is likely to be reflected in the transition state leading to reaction, the activation barrier to the process is likely to decrease as the exothermicity of the adsorption increases. Accordingly, the reaction rate rises from left to right in Fig. 2.7. Ultimately, at the curve maximum, the second step becomes rate-determining.

This can be understood since the second step undergoes removing $H^\bullet(ads)$ from the electrode surface, so as the adsorption becomes progressively stronger, then ultimately the second step must become rate-determining. Once this is the case, increased exothermicity of adsorption will progressively slow the reaction down, since the adsorption of $H^\bullet$ will likely be increasingly stronger in transition state $\dagger_B$. Accordingly, the energy barrier to loss of $H^\bullet(ads)$ gets increasingly larger, and the rate constant for hydrogen evolution slower, as the $\Delta H^\theta(ads)$ value becomes more negative as $H^\bullet$ forms stronger bonds with the metal atoms on the surface.

Finally, we briefly comment on the unusual case of hydrogen evolution at palladium electrodes. All electrochemists should be aware of the 1989 electrochemical 'cold fusion' paper of Fleischman and Pons[4] in which the reduction of alkaline heavy water at palladium electrodes was considered to generate 'excess heat' as a result of nuclear fusion events accompanying electrolysis. This was thought to be possible since, in the case of palladium electrodes, the formation of $H^\bullet(ads)$ is accompanied by transfer of the hydrogen atoms (or D atoms in the case of the electrolysis of heavy water, D_2O) into the metal lattice of the palladium electrode. Indeed, palladium metals shows a remarkable affinity for hydrogen, with atom ratios (H : Pd) of 1 : 1 being possible; simply inserting a palladium wire into hydrogen gas (with due care and attention) will lead to the wire glowing dull red as a result of the strongly favorable uptake of hydrogen atoms. In the electrochemical context Fleischman and Pons brought about the following electrochemical processes

$$D_2O + e^-(m) \rightarrow D^\bullet(ads) + OD^-$$
$$D^\bullet(ads) + D_2O + e^- \rightarrow D_2(g) + OD^-$$
$$2D^\bullet(ads) \rightarrow D_2(g)$$
$$D^\bullet(ads) \rightarrow D(lattice),$$

where $D(lattice)$ denotes a deuterium atom in the bulk palladium lattice. The high affinity of the palladium lattice for deuterium/hydrogen atoms, coupled with the fact that the atoms were known to be mobile, led Fleischman and Pons to speculate as follows.

'In view of the very high compression and mobility of the dissolved species there must therefore be a significant number of close collisions and one can ask the question: would nuclear fusion of D^+ such as

$$^2D + {}^2D \rightarrow {}^3T(1.01\,\text{MeV}) + {}^1H(3.02\,\text{MeV})$$

or

$$^2D + {}^2D \rightarrow {}^3He(0.82\,\text{MeV}) + n(2.45\,\text{MeV})$$

be feasible under these conditions?'[4]

Fleischman and Pons performed calorimetry and thought much more heat was generated than could be accounted for on the basis of the known electrochemical processes. They concluded that the energy had to be due to nuclear processes. The announcement of this work caused huge excitement. J. R. Huizenga, in his excellent book, describes a meeting of the American Chemical Society held on April 12, 1989 just two days after the publication of their paper in the *Journal of Electroanalytical Chemistry*.[5]

> 'Clayton Wallis, president of the AC in 1989, opened the session, which came to be referred to by some as the "Woodstock of Chemistry". He excited the seven thousand chemists, gathered in a large arena at the Dallas Convention Center, to an extremely high pitch by his introductory remarks, in which he hailed the tremendous potential of cold fusion as an energy source, and claimed it might be the discovery of the century. He then went on to detail the many problems physicists were having in achieving controlled nuclear fusion. "Now it appears that chemists may have come to the rescue" he said, and the thousands of chemists in the arena broke in loud applause and laughter.'

Huizenga describes Pons' lecture at Dallas as emphasizing 'the large amount of energy (up to 50 mega joules of heat) produced in his cells' and that

> '[Pons] pleased the large audience by showing his simple cell in a dish pan, that he claimed produced a sustained energy-producing fusion reaction at room temperature — "This is", he dead-panned, "the U-1 Utah Tokamak." The chemists went wild.'

The claims of cold fusion and excess heat are not generally accepted today. Where did Pons and Fleischman possibly go wrong; the latter, an eminent professor, a Fellow of the Royal Society (the website of which claims is comprised of 'the most distinguished scientists of the UK, of the Commonwealth countries and of the Republic of Ireland')?

Huizenga's analysis is as follows:

> 'Pons, in his lecture, made the same mistake as was made in the Fleischman — Pons publication. Based on his interpretation of the Nernst equations (taught in college freshman chemistry courses), Pons concluded that the deuterium pressure in the palladium cathode was equivalent to a hydrostatic pressure of approximately 10^{27}

atmospheres! It seems that it was this incorrect conclusion which led Fleischman and Pons to believe that, in a palladium cathode, the deuterium media would be forced together close enough to fuse. The Nernst equation, applicable under equilibrium conditions, was used to relate the overpotential in an electrochemical cell to the deuterium fugacity.[a] If this simple, but erroneous procedure is followed, a large overpotential does give a high deuterium fugacity. The use of the Nernst equation, however, for the large deuterium evaporation reaction under conditions of large values of the overpotential for estimation of the pressure of deuterium is inappropriate.'

The general message is clear: in using our pre-equilibrium arguments to establish Eq. (2.43), we must ensure that the pre-equilibrium is actually operative if the equation is to be meaningful. If large currents are being drawn, this may not be the case. Accordingly, the reader might note the need for caution in applying the analysis derived in Section 2.8. He/she might also wish to review the discussion of the Nernst equation in Chapter 1 of this book.

2.10 B. Stanley Pons

B. (Bobby) Stanley Pons was born in 1943 in the small town of Valdese in the North Carolina foothills, which, as Mallove[6] points out, is rather ironic since on the day of the cold fusion announcement, the huge oil tanker Exxon Valdez (same pronunciation as Valdese) was coming to grief on the rocky Alaskan coast. There soon appeared a cartoon (MacNelly, Chicago Tribune) connecting cold fusion with the oil spill; an oil-soaked bird adrift on a buoy was remarking to a similarly blackened seal or sea lion, 'Any more word on how those fusion experiments are going?'

Pons graduated in 1965 from Wake Forest University, North Carolina, and consequently studied for a doctorate at the University of Michigan at Ann Arbor. However, toward the end of his doctorate in 1967, he left to work in his father's textile mills and to manage a family restaurant. Pons' love for chemistry drew him back to science and he enrolled at the University of Southampton (UK) and received his Ph.D. in 1978. Martin Fleischmann was one of his professors at Southampton, where the pair meet. After being on the faculty at Oakland University in Rochester, Michigan, and the University of Alberta in Edmonton, Pons went to the University

---

[a] Fugacity approximates to pressure.

of Utah in 1983 as an associate professor, becoming a full professor in 1986, and Chairman of the Department in 1988.

2.11 Cold Fusion — The Musical!

The cold fusion fiasco even produced a musical! The following gives a glimpse of the action.

Fig. 2.8

Taken from Act One:
MUSIC starts.

> NARRATOR (CONT'D):
> Everybody knows about Adam and Eve
> Barney & Fred, Sonny and Cher
> Anthony and Cleopatra, Los Vegas and Sinatra
> And BJ and the Bear.
> Oh couples throughout history
> Leave their mark be they pros or cons
> But none have left such a legacy
> As Doctors Martin Fleischmann and B. Stanley Pons
> Two chemists and their tiny lab
> Plagued with doubt and ridicule
> Shocked the scientific world
> By untold kilo joules

'It works!' cried Pons, 'There's excess heat.'
'The people must be told.'
And on March 23 they gave the word
about fusion... and it was cold

FLEISCHMANN stands centre and sings Gilbert & Sullivenesquely.

FLEISCHMANN:
(singing)
My name is Doctor Martin Fleischmann
I have my degree
In the scientific study
Of electrochemistry
In the late 1960's
I saw an anomaly
In palladium, a by-product
Of titanium refinery

Slide: '1969'

The music gets quieter. FLEISCHMANN picks up a test tube or a microscope or whatever and ponders.

FLEISCHMANN:
Hey, man, grok what happens when these
Deuterium ions interact with this groovy palladium
lattice! Trippy!

PONS enters, singing and dancing.

PONS :
(singing)
My name is Dr. Stanley Pons
I have my degree
In the scientific study
Of electrochemistry
In the late 1970's
You taught me
When I went to South Hampton's
University

The play was written by Wes Borg and Paul Mather, and was produced by 'Three Dead Trolls' and 'Atomic Improv' in 1995.[7]

2.12 Why Are Some Standard Electrochemical Rate Constants Large but Others Slow? The Marcus Theory of Electron Transfer: An Introduction

To consider the basic principles underlying electron transfer at the electrode–solution interface, let us focus on the one-electron oxidation of aqueous ferrous ions, $Fe^{2+}(aq)$, to form ferric ions, $Fe^{3+}(aq)$, noting that in aqueous solution both ions have a primary solvation shell consisting of six water molecules:

$$Fe(H_2O)_6^{2+}(aq) \rightarrow Fe(H_2O)_6^{3+}(aq) + e^-(m).$$

Figure 2.9 illustrates the potential energy curves of the reactants and products as a function of a 'reaction coordinate', which in this example might be the $Fe-O$ bond distance within the octahedral complex. This is longer in the ferrous species (2.21 Å) than in the ferric species (2.05 Å), accounting for the relative location of the two potential energy curves in Fig. 2.9 We next pose the question as to why should there be an activation barrier to electron tunneling of an electron from the $Fe(II)$ species into the electrode?

The origin of the barrier resides in the Frank–Condon principle which tells us that the transfer of an electron occurs over a much shorter timescale ($\sim 10^{-15}$ s) than the timescale of molecular nuclear vibrations ($\sim 10^{-13}$ s). It follows that, when electron tunneling takes place, the 'jump' from the reactant potential energy curve (R) to that of the products (P) occurs as a vertical transition on the diagram corresponding to nuclei 'frozen' on the timescale of electron transfer. Consequently, if tunneling of the electron takes place from the ground state (lowest vibrational energy level with vibrational quantum number, $v = 0$, as shown in Fig. 2.9), then $Fe(H_2O)_6^{3+}$ must be formed in a highly vibrationally excited state (Fig. 2.9). Clearly, this must violate the principle of conservation of energy since this excitation cannot simply appear in the system (although of course under irradiation the transition might occur photochemically with the absorption of a photon). It follows that reaction proceeds via thermal activation of the reactant to the extent that the $Fe(H_2O)_6^{2+}$ ions have an energy corresponding to the cross-over point of the two potential energy curves. When this energy is attained and the molecular coordinates of the $Fe(H_2O)_6^{2+}$ ion reached those of the transition state (†, as given in Fig. 2.10) during vibrations of the thermally excited ions, electron tunneling can take place leading to the formation of a vibrationally excited $Fe(H_2O)_6^{3+}$ ion with an energy matching that of the $Fe(H_2O)_6^{2+}$ ion before tunneling.

Rapid thermal deactivation of this vibrationally excited $Fe(H_2O)_6^{3+}$ then takes place via collision with solvent molecules so producing a ground state $Fe(H_2O)_6^{3+}$ ion. It is evident that the barrier for electron transfer resides in the need to 'match'

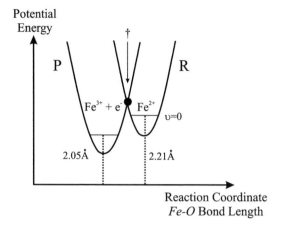

Fig. 2.9 Potential energy curves for the oxidation of Fe^{2+} (aq). ϕ_M and ϕ_S are fixed.

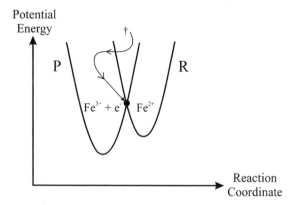

Fig. 2.10 The transition state is the cross-over point of the potential energy curves for R (reactants) and P (products).

the energy of the reactant and products in the transition state, †, requiring that the reactant $Fe(H_2O)_6^{2+}$ becomes thermally activated before electron transfer can occur. By 'thermal activation', we mean that the bond lengths and angles within the reactant species become stretched, compressed and/or distorted, and the solvation shells of the species may become similarly distorted. This is the basis of the Marcus theory for heterogeneous electron transfer. As we shall see later, it is possible to use a knowledge of the potential energy curves such as those shown in Fig. 2.9 and *quantitatively* predict electron transfer rate constants.

We focus on the standard electrochemical rate constants shown in Table 2.1. Note that the range of values covered spans many orders of magnitude.

A simple qualitative rule can be devised for deciding whether an electron transfer will be fast or slow based on the two situations shown in Fig. 2.11, which correspond to (A) a small or (B) a larger difference in reaction coordinate between reactants and products. It can be readily appreciated from Fig. 2.11 that the greater

Table 2.1. Some standard heterogeneous rate constants, $k^0/\text{cm s}^{-1}$ (in aqueous solution unless stated, 25°C).

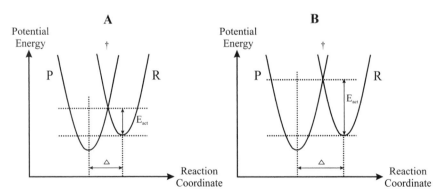

	4^a
$MnO_4^- + e^- \rightleftarrows MnO_4^{2-}$	0.2
$Fe(CN)_6^{3-} + e^- \rightleftarrows Fe(CN)_6^{4-}$	0.1
$Fe(H_2O)_6^{3+} + e^- \rightleftarrows Fe(H_2O)_6^{2+}$	7×10^{-3}
$V^{3+} + e^- \rightleftarrows V^{2+}$	4×10^{-3}
$Eu^{3+} + e^- \rightleftarrows Eu^{2+}$	3×10^{-4}
$Co(NH_3)_6^{3+} + e^- \rightleftarrows Co(NH_3)_6^{2+}$	5×10^{-8}

a In dimethylformamide.

Fig. 2.11 The greater the change in the reaction coordinate (Δ) between the reactant and products, then the higher in energy is the transition state, all other factors remaining equal. Note that a small Δ results in a small activation barrier, E_{act} shown above (curves A), while when Δ is large, a greater activation barrier occurs as shown above via curves B.

the difference in reaction coordinate between the ground state of the reactant and products then the higher in energy is the transition state †, since this is defined by the cross-over point of the two potential energy curves.

Since these changes in 'reaction coordinate' imply a change of bond length and/or angle, it follows that for the general electrode reaction

$$Ox + e \xrightarrow{k_0} Red$$

- If *Ox* and *Red* are close in molecular geometry (bond lengths, angles), then k^0 is large corresponding to a low activation barrier for reaction.
- If *Ox* and *Red* are structurally dissimilar, the k^0 is small and the activation barrier large.

These principles can be usefully applied to some of the entries in Table 2.1, as follows.

(i) The anthracene/anthracene radical anion couple

in dimethylforamide solution shows the very fast standard electrochemical rate constant of $4 \, \text{cm s}^{-1}$ (25°C). This reflects the fact that the bond angles are unchanged in the reactants and products and, since the electron in the anthracene radical anion is delocalised, the bond lengths are virtually unchanged. Accordingly, the barrier to electron transfer is very low and most probably derives from the slight changes of solvation between anthracene and its radical anion.

(ii) The couple

$$MnO_4^-(aq) + e^- \rightleftharpoons MnO_4^{2-}(aq)$$

shows fast electrode kinetics. Both anions are tetrahedral in geometry so that the bond angles are the same in both species. The $Mn-O$ bond length is 1.63 Å in the mono-anion and 1.66 Å in the di-anion. This is only a tiny change so that the simple principles elucidated above predict fast electrode kinetics, as is indeed observed.

(iii) The ferrous/ferric couple

$$Fe(H_2O)_6^{3+}(aq) + e^- \rightleftharpoons Fe(H_2O)_6^{2+}(aq)$$

in aqueous solution shows relatively sluggish kinetics with a k^0 of $7 \times 10^{-3} \, \text{cm s}^{-1}$. In this case, although both ions have an octahedral geometry, they differ somewhat

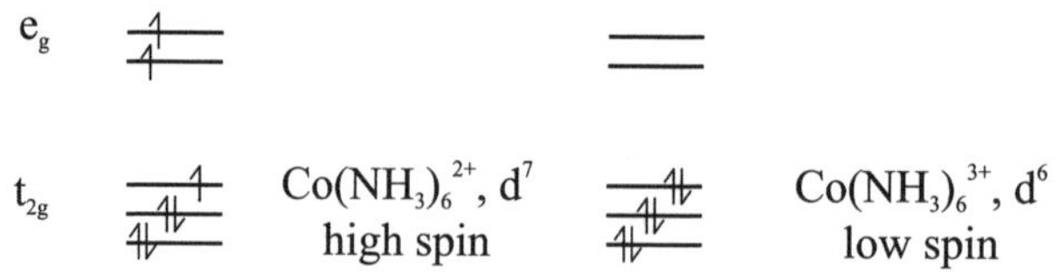

Fig. 2.12 The spin states for $Co(NH_3)_6^{2+}$ and $Co(NH_3)_6^{3+}$.

in the $Fe-O$ bond lengths, which are 2.21 Å in the $Fe(II)$ species and 2.05 Å in the Fe (III) species.

(iv) The couple

$$Co(NH_3)_6^{3+} + e^- \rightleftharpoons Co(NH_3)_6^{2+}$$

shows pathologically slow electron transfer kinetics. The standard electrochemical rate constant is reported as 5×10^{-8} cm s^{-1}. This is an interesting case where the slowness of the reactions is not thought to represent a huge barrier to electron transfer, but possibly rather reflects the fact that the $Co(II)$ couple is an octahedral d^7 species of high spin, whereas the $Co(III)$ is a low spin d^6 molecules, as shown in Fig. 2.12.

It follows that even when $Co(NH_3)_6^{2+}$ is thermally excited to an energy corresponding to the cross-over point of the potential energy curves where $Co(NH_3)_6^{2+}$ the activated molecule cannot 'jump' onto the $Co(NH_3)_6^{3+}$ curve without a simultaneous 'flip' of electron spins, as shown in Fig. 2.13.

Such a spin flip is unlikely and so the slowness of the electron transfer kinetics in this case does not reflect the activation barrier size, but rather the probability of jumping from one potential energy level to another *and* changing electron spin. The process is largely 'spin forbidden'.

Last, we examine three examples from organic electrochemistry in the light of the principles elucidated above.

(a) The reduction of cyclooctatetraene, *COT* has been studied in the non-aqueous solvent, dimethylformamide.[8] This molecule is tub-shaped as shown in Fig. 2.14. The molecule undergoes two separate one-electron reduction processes at two different potentials to form first the mono-anion, $COT^{\bullet-}$, and second the di-anion, COT^{2-}.

Both of these molecules are almost (but not exactly) planar, in contrast to the tub-shaped COT:

$$COT \underset{-e^-}{\overset{+e^-}{\rightleftharpoons}} COT^{\bullet-} \underset{-e^-}{\overset{+e^-}{\rightleftharpoons}} COT^{2-}$$

The standard electrochemical rate constant for the first reduction is $\sim 10^{-3}$ cm s^{-1}, whereas that for the second is $\sim 10^{-1}$ cm s^{-1}. The relative slowness of the first can

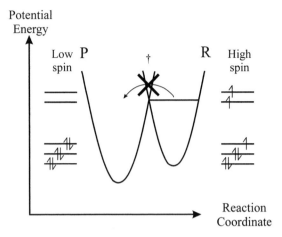

Fig. 2.13 The probability of the system 'jumping' from the reactant (R, $Co(NH_3)_6^{2+}$) potential energy curves to that of the product (P, $Co(NH_3)_6^{3+}$) is low since it requires a spin change.

Fig. 2.14 Structure of COT, which is tub-shaped.

be rationalised in terms of the significant change in structure in going from *COT* to *COT•*, i.e. from tub-shaped to almost flat. On the other hand, the reduction of *COT•* to *COT^{2-}* is relatively faster since both anions are almost flat and so little change of molecular geometries is required.

(b) The reduction of the molecule tetraphenylethylene, TPE, has been studied in dimethylformamide. The structure of this molecule is shown in Fig. 2.15.[9]

TPE undergoes the following electrochemical reduction

$$TPE \underset{-e^-}{\overset{+e^-}{\rightleftharpoons}} TPE^{\bullet} \underset{-e^-}{\overset{+e^-}{\rightleftharpoons}} TPE^{2-}.$$

The standard electrochemical rate constant for the first step is 0.10 cm s^{-1}, whilst that for the second step is slower: 8×10^{-3} cm s^{-1}. Both *TPE* and *TPE•* are thought to be planar, whereas in *TPE^{2-}* the two ends of the molecules rotate relative to each other so the di-anion is twisted about what was the double bond of the TPE before two electrons were added to the π^* orbital of the TPE to make *TPE^{2-}*. The relative

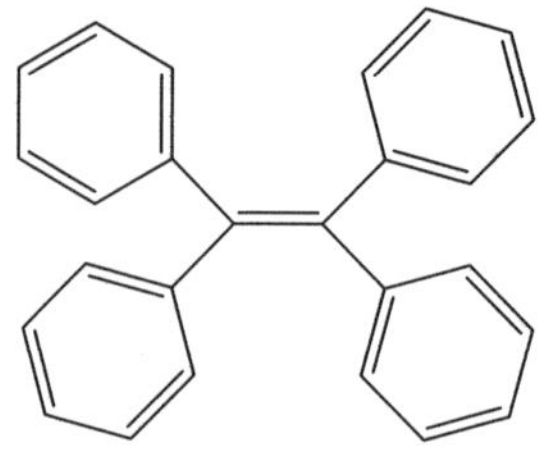

Fig. 2.15 The structure of TPE.

sizes of the two rate constants reflect first the similarity in structure between *TPE* and *TPE*• and second, the dissimilarity between *TPE*• and *TPE*²⁻.

(C) The oxidation of carboxylate anions in aqueous solution is synthetically useful for the formation of alkanes via the Kolbe oxidation

$$RCO_2^- \xrightarrow[Pt\ electrode]{-e^-} R^\bullet + CO_2 \longrightarrow 1/2R_2$$

The huge changes of bond lengths and bond angles are shown in Fig. 2.16. Accordingly, the reaction is very, very slow. Thermodynamically, the reaction should proceed at potentials positive of *ca* 0 volts on the standard hydrogen electrode scale, whereas in fact potentials of *ca* 1.0 volt typically have to be applied because of the slow kinetics. We return to the concept of 'overpotential' to drive slow processes later in this book.

2.13 Marcus Theory: Taking it Further. Inner and Outer Sphere Electron Transfer

The difference between inner and outer sphere electron transfer in the context of homogeneous chemistry is familiar largely from the pioneering work of H. Taube

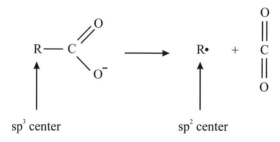

Fig. 2.16 The Kolbe oxidation shows very slow electrode kinetics: major changes in bond angles and bond lengths are required.

(winner of the 1983 Nobel Prize for chemistry). The outer sphere mechanism is characterized by weak interaction of the reactive species, with the inner coordination spheres remaining intact during the electrode transfer. A typical example is

$$^*Fe(CN)_6^{3-}(aq) + Fe(CN)_6^{4-}(aq) \rightleftharpoons {}^*Fe(CN)_6^{4-}(aq) + Fe(CN)_6^{3-}(aq).$$

The inner sphere mechanism involves reaction occurring through a common ligand shared by the metallic centres, as in

$$(NH_3)_5 CoCl^{2+}(aq) + Cr^{2+}(aq) \rightleftharpoons [(NH_3)_5 Co - Cl - Cr(H_2O)_5]^{4+}$$
$$[(NH_3)_5 Co - Cl - Cr(H_2O)_5]^{4+} \rightarrow Co(NH_3)_5^{2+} + ClCr(H_2O)_5^{2+}$$

and

$$CrCl_6^{2-}(aq) + Cr^{2+}(aq) \rightleftharpoons [Cl_5 Cr - Cl - Cr(H_2O)_5]$$
$$[Cl_5 Cr - Cl - Cr(H_2O)_5] \rightarrow CrCl_6^{3-}(aq) + Cr^{3+}(aq),$$

where there may or may not be transfer of the bridging ligand (Cl^- in the examples above) between the reactants and products.

The two types of mechanism are also observed in electrochemical processes. For an outer sphere process, Weaver and Anson[10] locate the reactant centre or ion in the 'Outer Helmholtz Plane' — the plane of closest approach for reactant where coordination spheres do not penetrate the layer of solvent molecules that are 'specifically adsorbed' or coordinated directly to the electrode surface. In contrast, inner sphere processes proceed through a common ligand and correspond to specifically adsorbed reactants and as such show kinetics strongly dependant of the chemical nature of the electrode surface. Fig. 2.17 shows the distinction between the outer and inner pathways.

2.14 Marcus Theory: Taking it Further. Adiabatic and Non-Adiabatic Reactions

Figure 2.18 shows the potential energy surfaces for a system moving from reactants, R, to products, P. Notice that where the two curves overlap there is a quantum mechanical splitting between the two surfaces; the energy gap is a *resonance energy*. If this splitting is strong, then a significant perturbation of the isolated R and P curves occurs and reaction from R to P takes place along the lower curve with a probability close to unity. This is the *adiabatic* limit. On the other hand, if resonance energy is small, there is only a tiny perturbation in the potential energy

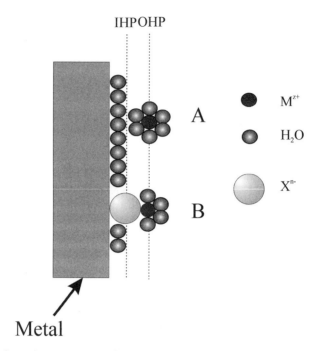

Fig. 2.17 Schematic representation for outer sphere (A) and inner sphere (B) redox reaction paths at electrodes.

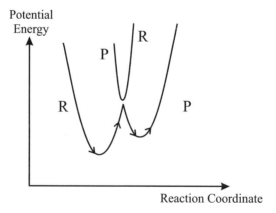

Fig. 2.18 Potential energy surface for a system going from reactants (curve R) to products (curve P). Adapted from Ref. [11]. Note the curves P and R are dependant on ϕ_M and ϕ_S.

curves of R and P, leading to a *non-adiabatic* process and a probability of transfer from R and P of much less than unity.

The power and relative simplicity of Marcus theory reside in the fact that for most electron transfer reaction (even inner sphere ones) the resonance energy is typically just a few kilojoules per mole or less. This is enough to ensure the reaction is adiabatic but not enough to significantly alter the R and P potential energy curves. It is thus possible to write for the electrochemical rate constant:

$$k = KZe^{-\frac{\Delta G(\dagger)}{RT}}, \qquad (2.44)$$

where K is the transition probability such that $K \sim 1$ for an adiabatic process, whereas $K \ll 1$ for a non-adiabatic reaction. $\Delta G(\dagger)$ is the Gibbs energy of activation and Z is a pre-exponential term.

The attraction of Marcus theory is that $\Delta G(\dagger)$ can be calculated with accuracy from knowledge of the potential energy curves of R and P neglecting the small resonance energy involved in the electron transfer reaction. Accordingly, the location of the transition state can be approximated to the cross-over potential of the R and P curves as already implicitly assumed in Section 2.8.

Marcus theory identifies two major contributions to the Gibbs energy of activation:

$$\Delta G(\dagger) = \Delta G_i(\dagger) + \Delta G_0(\dagger) = \frac{1}{4}(\lambda_i + \lambda_0), \qquad (2.45)$$

where $\Delta G_i(\dagger)$ is the activation energy arising from distortion of the inner coordination shell in the transition state geometry, whilst $\Delta G_0(\dagger)$ is that arising from rearrangement of the solvent dipoles between the reactants and the transition state, as shown in Fig. 2.19. The terms λ_i and λ_o are used below and are defined by Eq. (2.45).

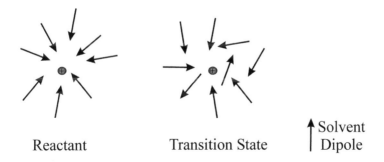

Fig. 2.19 Solvent dipole rearrangement contributes to λ_o.

2.15 Marcus Theory: Taking it Further. Calculating the Gibbs Energy of Activation

We consider the Gibbs energy versus reaction profile shown in Fig. 2.20. We shall see below that this is equally applicable for the consideration of $\Delta G_i(\dagger)$ and $\Delta G_o(\dagger)$.

We assume that the potential energy curves describing the reactants (R) and the products (P) are parabolae such as would apply if movement in the direction of the reaction were that of a harmonic oscillation. Accordingly,

$$G_R = G_R(X = X_R) + \frac{1}{2}k(X - X_R)^2$$

$$G_P = G_P(X = X_P) + \frac{1}{2}k(X - X_P)^2,$$

where k is the force constant describing this harmonic vibration. At the transition state,

$$G_\dagger = G_R(X = X_R) + \frac{1}{2}k(X_\dagger - X_R)^2$$

$$G_\dagger = G_P(X = X_P) + \frac{1}{2}k(X_\dagger - X_P)^2.$$

Solving these expression for $X_\dagger$ in terms of X_R and X_P, we obtain

$$X_\dagger = \frac{1}{2}(X_R + X_P) - \frac{G_P(X = X_P) - G_R(X = X_R)}{k(X_R - X_P)}.$$

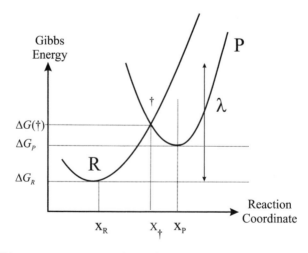

Fig. 2.20 Gibbs energy reaction coordinate diagram for an electron transfer process $R + e^- \rightarrow P$. Note the values of ϕ_M and ϕ_S are fixed.

Substituting this, we find an expression for the Gibbs energy of activation

$$\Delta G(\dagger) = G_\dagger - G_R(X = X_R) = \frac{1}{8}k(X_R - X_P)^2 + \frac{1}{2}[G_p(X = X_P)$$
$$- G_R(X = X_R)] + \frac{[G_P(X = X_P) - G_R(X = X_R)]^2}{2k(X_R - X_P)^2}.$$

Simplifying,

$$\Delta G(\dagger) = \frac{(\lambda + \Delta G)^2}{4\lambda}, \tag{2.46}$$

where

$$\lambda = \frac{1}{2}k(X_R - X_P)^2$$

and

$$\Delta G = G_P(X = X_P) - G_R(X = X_R).$$

Equation (2.46) is the basis of the simple principles elucidated in Section 2.9 relating the size of k^0 to changes in geometry between reactants and products (R and P): Equation (2.46) shows that a larger λ gives a larger $\Delta G(\dagger)$.

We next turn to consider the two contributions to λ, the so-called reorganisation energy; it is the energy of R at the coordinate X_P corresponding to the equilibrium geometry of P as shown in Fig. 2.20:

$$\lambda = \lambda_i + \lambda_o. \tag{2.47}$$

For the case of λ_i, it can be shown that

$$\lambda_i = \sum_j \frac{k_j^R k_j^P}{k_j^R + k_j^P}(X_R^j - X_P^j)^2,$$

where k_j^R and k_j^P are the normal mode force constants for the jth vibrational coordinate in the reactants (R) and products (P) respectively, whilst the terms $(X_R^j - X_P^j)$ are the changes in bond lengths and bond angles between the reactants and the products. To illustrate this, consider the electrode process

$$NO_2 \overset{k}{\rightleftarrows} NO_2^+ + e^-(m).$$

This involves a change of bond lengths and bond angles as shown in Fig. 2.18.

If it is assumed that only the $N-O$ stretches (symmetric and antisymmetric) and the $O-N-O$ bending modes contribute to the activation energy, then

$$\lambda_i = \frac{2k_{str}^R k_{str}^P}{k_{str}^R + k_{str}^P}(r_R - r_R)^2 + \frac{k_{bend}^R k_{bend}^P}{k_{bend}^R + k_{bend}^P}(\upsilon_R - \upsilon_P)^2,$$

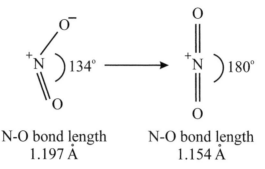

N-O bond length N-O bond length
1.197 Å   1.154 Å

Fig. 2.21 The bond length and angle changes in the oxidation of NO_2.

where v and r give the equilibrium bond angles and lengths (see Fig. 2.21). Eberson[11] lists the force constants as follows

$$k_{str}^{NO_2} = 11.04 \times 10^{-8} \text{ N Å}^{-1}$$

$$k_{str}^{NO_2^+} = 17.45 \times 10^{-8} \text{ N Å}^{-1}$$

$$k_{bend}^{NO_2} = 2.28 \times 10^{-8} \text{ N Å}^{-1} \text{ rad}^{-2}$$

$$k_{bend}^{NO_2^+} = 0.688 \times 10^{-8} \text{ N Å}^{-1} \text{ rad}^{-2}.$$

It is evident that the bending mode contributes most significantly to the activation energy and this is also true in the Kolbe oxidation considered in Section 2.8.

We next turn our attention to λ_0 corresponding to the solvent reorganisation. We assume that changes in solvation energy arise from chance fluctuation of the solvent dipoles around the species of interest and that the potential energy of the latter varies parabolically with displacement along the solvent coordinates. Under these conditions the following approximation has been derived,

$$\lambda_0 = \frac{e^2}{8\pi\varepsilon_0} \left(\frac{1}{r} - \frac{1}{2d} \right) \left(\frac{1}{\varepsilon_{op}} - \frac{1}{\varepsilon_S} \right), \tag{2.48}$$

where e is the electronic charge, d is the distance from the reactant to the electrode surface (which is often set to infinity), ε_{op} is the optical dielectric constant and ε_S is the static dielectric constant. ε_0 is the vacuum permittivity such that $4\pi\varepsilon_0 = 1.113 \times 10^{-10} \text{ J}^{-1} \text{ C}^2 \text{ m}^{-1}$. The term

$$\left(\frac{1}{\varepsilon_{op}} - \frac{1}{\varepsilon_S} \right)$$

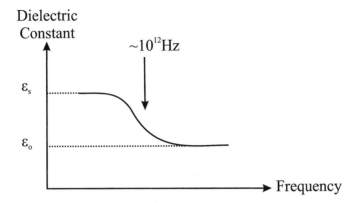

Fig. 2.22 The dielectric constant, ε_r, depends on the frequency.

arises, since dielectric constants (relative permittivities) are frequency dependent as sketched in Fig. 2.22.

The dielectric constant, ε_r, measures the reduction in force between two charges Z_+ and Z_- a distance r apart, when they are placed in a medium of dielectric constant, ε_r, as in Coulomb's law,

$$Force = \frac{Z_+ Z_-}{4\pi\varepsilon_o\varepsilon_r}\frac{1}{r^2}.$$

Two factors contribute to ε_r:

(a) Orientation of solvent dipoles in the medium, and
(b) Electronic polarization within the solvent molecules.

At the speeds pertinent to electron transfer the second contribution is always at equilibrium; the contribution from the solvent dipoles is therefore reflected in the difference between the reciprocals of ε_{op} and ε_S respectively as given in Eq. (2.48).

2.16 Relationship between Marcus Theory and Butler–Volmer Kinetics

In the previous section, we identified the following expression for the Gibbs energy of activation for an electrode process

$$\Delta G(\dagger) = \frac{\lambda}{4}\left(1 + \frac{\Delta G}{\lambda}\right)^2, \tag{2.49}$$

where for the electrode process

$$\Delta G = -nFE = G_P(X = P) - G_R(X = R),$$

as in Chapter 1, and

$$k = KZ \exp\left(-\frac{\Delta G(\dagger)}{RT}\right).$$

Furthermore, from our analysis of electrochemically irreversible Butler–Volmer kinetics,

$$\alpha = \frac{RT}{F} \frac{\partial \ln |I_{red}|}{\partial E} = \frac{RT}{F} \frac{\partial \ln |k|}{\partial E}$$

from which

$$\alpha = \frac{1}{2}\left(1 + \frac{\Delta G}{\lambda}\right). \tag{2.50}$$

It is clear from Eq. (2.50) that $\alpha \sim 1/2$ as noted earlier under the condition that $\lambda \gg \Delta G$. This insight provides a satisfactory link between the Marcus theory and Butler–Volmer kinetics.

It is noteworthy that in the expression for α,

$$\alpha = \frac{1}{2}\left(1 + \frac{\Delta G}{\lambda}\right),$$

the Gibbs energy term is *not* a pure standard Gibbs energy since the potential energy parabolae in Fig. 2.9 move relative to one another as the electrode potential changes since the chemical potential (Gibbs energy) of the electron in the metal energy changes. Accordingly, Eq. (2.50) does not predict α to be exactly constant at all potentials but to depend on the latter. At potentials close to E^0 where $\Delta G \sim 0$, then $\alpha \sim 1/2$. However, if $\Delta G \gg 0$ and the reaction is thermodynamically uphill, then since $\Delta G \sim \lambda$, under these conditions $\alpha \to 1$. Conversely, when $\Delta G \ll 0$ and the reaction is driven hard by the electrode potential, $\Delta G \sim -\lambda$ and $\alpha \to 0$. Implicit in this notion is the concept that Tafel plots can be curved, provided they are measured over a wide range of potentials due to the variation of α with potential. Such behavior has been reported for Cr^{3+}/Cr^{2+}, $Fe(CN)_6^{4-}/Fe(CN)_6^{3-}$ and for Fe^{2+}/Fe^{3+}. The latter is illustrated in Fig. 2.23.

2.17 Marcus Theory and Experiment. Success!

Table 2.2 lists experimental and calculated activation energies for various redox processes, as summarised by Albery[1] and Hale.[12] Given the simplicity of the model, the general level of agreement is highly satisfactory, encouraging belief in the concepts of Marcus Theory.

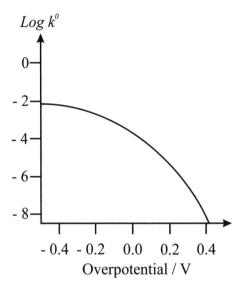

Fig. 2.23 The dependence of the rate constant of the reaction of $Fe^{3+} + e^- \rightarrow Fe^{2+}$ in 1 M $HClO_4$ on the overpotential.[13]

Table 2.2. Comparison of experimental and calculated standard* Gibbs energies of activation for $O + e^- \rightleftarrows R$.

	Experimental $\Delta G^0(\dagger)/\text{kJmol}^{-1}$	Calculated $\Delta G^0(\dagger)/\text{kJmol}^{-1}$
Tetracene (in DMF)	22	21
Naphthalene (in DMF)	23	24
$Fe(CN)_6^{3-}$	29	30
WO_4^{2-}	34	36
MnO_4^-	34	37
$Fe(H_2O)_6^{3+}$	36	38
$V(H_2O)_6^{3+}$	37	37
$Mn(H_2O)_6^{3+}$	41	45
$Ce(H_2O)_6^{4+}$	50	28
$Cr(H_2O)_6^{3+}$	52	42
$Co(H_2O)_6^{3+}$	56	38

DMF = N,N-dimethylformamide
Adapted from Refs. 1 and 12.
*Here, standard implies that the electrode potential matches the formal potential of the redox couple.

References

[1] W.J. Albery, *Electrode Kinetics*, Oxford University Press, 1975.

[2] Picture © Klaus Müller and kindly provided by Professor Evgeny Katz. Text adapted from Tafel's biography written by Klaus Müller: K. Müller, *J. Res. Inst. Catalysis, Hokkaido Univ.* **17** (1969) 54, with permission. Note that a comprehensive up-to-date website detailing many other electrochemical greats can be found at: http://chem.ch.huji.ac.il/~eugeniik/history/electrochemists.htm.
This impressive website was constructed by Professor Evgeny Katz, Department of Chemistry and Biomolecular Science, Clarkson University, Potsada, New York, USA.

[3] S. Trasatti, *J. Electroanalytical Chem.* **39** (1972) 163.

[4] M. Fleischman, S. Pons, *J. Electroanalytical Chem.* **261** (1989) 301.

[5] J.R. Huizenga, *Cold Fusion: The Scientific Fiasco of the Century*, Oxford University Press, 1993.

[6] E.F. Mallove, *Fire from Ice; Searching for Truth Behind the Cold Fusion Furror*, John Wiley & Sons, New York, 1991. The picture of Pons and the text are taken from Mallove's book. For more on Cold Fusion, the interested reader is directed to: F. David Peat, *Cold Fusion: The Making of a Scientific Controversy*, Contemporary Books, 1989; F. E. Close, *Too Hot to Handle: The Race for Cold Fusion*, Princeton University Press, 1991; G. Taubes, *Bad Science: The Short Life and Weird Times of Cold Fusion*, Random House, 1993.

[7] Text and picture ©copyright of Wes Borg and Paul Mather. See: http://www.subatomichumor.com/shows/coldfusion.html and www.deadtroll.com.

[8] R.D. Allendoerfer, P. H. Rieger *J. Am. Chem. Soc.* **87** (1965) 2336.

[9] M. Grzeszczuk, D.E. Smith, *J. Electroanalytical Chem.* **162** (1984) 189.

[10] M.J. Weaver, F.C. Anson, *Inorganic Chem.* **15** (1976) 1871.

[11] L. Eberson, (1987), *Electron Transfer Reactions in Organic Chemistry*, Springer, Berlin, 1987.

[12] J.M. Hale, in N.S. Hush (ed.), *Reactions of Molecules at Electrodes*, Wiley, 1971, p. 229.

[13] J. Koryta, J. Dvorak, L. Kavan, *Principles of Electrochemistry*, Wiley, 1993.

3

Diffusion

In this chapter we develop an understanding of the phenomenon of diffusion in liquids. This is based on the pioneering work of Adolf Fick who published the key papers in this area over 150 years ago.[1]

We start by examining Fick's law of diffusion in modern terms before briefly considering the experiments which led to its discovery. The rest of the chapter develops ideas key to the interplay of diffusion with the theory of voltammetry.

3.1 Fick's 1st Law of Diffusion

The diffusion as a phenomenon is readily perceived in the scent of flowers permeating a room or a drop of colourful dyestuff spreading throughout a liquid. For a more precise description, consider the concentration distribution shown in Fig. 3.1.

It is evident that diffusion will take place from high to low concentration, that is to say, down the concentration gradient as shown. At any point, x, there will be a diffusive flux quantified by Fick's 1st Law:

$$j = -D\frac{\partial c}{\partial x},\qquad(3.1)$$

where j is the flux/mol cm^{-2}s^{-1} corresponding to the number of moles passing through unit area in unit time, $\frac{\partial c}{\partial x}$ is the local concentration gradient at point x and D is known as the diffusion coefficient; as a rule of thumb the bigger the molecule, the smaller the diffusion coefficient. Note that the negative sign in Eq. (3.1) implies the flux is *down* the concentration gradient. The unit of D are cm^2 s^{-1} and typically the magnitude of D lies in the range $10^{-6} - 10^{-5}$ cm^2 s^{-1}

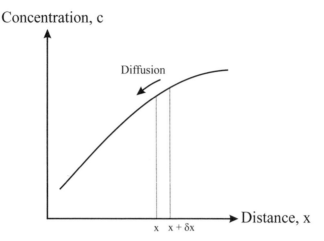

Concentration, c

Diffusion

Distance, x

x   x + δx

Fig. 3.1 An arbitrary plot of concentration against distance showing diffusion from high to low concentration.

around room temperature for almost all solvents encountered in voltammetric studies (water, acetonitrile, dimethylformamide,...), notable exceptions constituting some viscous room temperature ionic liquids.[2] Diffusion coefficients are strongly temperature sensitive, often following Arrhenius type relationships:

$$D = D_\infty \, \exp\left(\frac{-E_a}{RT}\right), \tag{3.2}$$

where E_a / kJ mol^{-1} is an activation energy for diffusion and D_∞ is the hypothetical value of D at an infinite temperature, T. This temperature sensitivity implies a need for thermostatically controlled experiments in voltammetry.

As an illustration of Fick's Law in practice, it is used to characterise diffusion though biological and other membranes. To make this application to a cell membrane of thickness, b, one must assume (i) that the membrane thickness is small so that steady-state diffusion prevails and (ii) that the solution phases either side of the membrane are well mixed. In this situation,

$$\frac{\partial c}{\partial x} \sim \frac{C_{outer} - C_{inner}}{b}, \tag{3.3}$$

where C_{outer} and C_{inner} are the concentrations outside and inside of the membrane. Accordingly, the rate (flux) of diffusion is given by $P(C_{outer} - C_{inner})$ where $P = D/b$ is refered to as a permeability.

Last, we emphasise that Fick's Law as stated in Eq. (3.1) presumes that the flux is driven solely by concentration differences within the solution and that there are no gradients of electrical potential (electric fields). In the case that the diffusion species is uncharged, the presence of the latter will not influence the flux, but for the case of ions there will be a significant effect. In normal voltammetric practice, however, the presence of sufficient quantities of supporting or background electrolytes as discussed in Chapter 2, Section 2.5, will eliminate the presence of significant electric fields in the solution except very close to the electrode surface; accordingly, material is brought to and from the electrode by diffusion (and sometimes convection). On the other hand, when a supporting electrolyte is absent, the diffusion of ions can lead to liquid junction potentials as described in Chapter 1, Section 1.5.

3.2 Fick's 2nd Law of Diffusion

Given that Fick's 1st law is operative and considering Fig. 3.1, the question arises as to how the concentration at point x varies with time. To this end consider the 'slab' of solution bounded by the planes at x and $x + \delta x$ (see Fig. 3.1), as shown in Fig. 3.2.

Over a time δt, the change in the number of moles, δn, of the diffusing species through the 'slab' is

$$\delta n = [J(x) - J(x + \delta x)]A\delta t.$$

It follows from Taylor expansion that

$$J(x + \delta x)] \approx J(x) + \delta x \left(\frac{\partial J}{\partial x}\right).$$

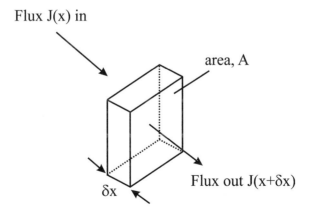

Flux J(x) in

area, A

Flux out J(x+δx)

δx

Fig. 3.2 Fick's 2nd Law can be derived by considering the fluxes in and out of the 'slab'.

Accordingly,

$$\delta n \sim -\delta x \left(\frac{\partial J}{\partial x}\right) A \cdot \delta t,$$

so that identifying the concentration change as

$$\delta c = \delta n / A \cdot \delta x$$

gives the rate of change of concentration at a point x, by

$$\frac{\delta c}{\delta t} \sim \frac{\partial c}{\partial t} = -\frac{\partial J}{\partial x} = D \frac{\partial^2 c}{\partial x^2},$$

so that

$$\frac{\partial c}{\partial t} = D \frac{d^2 c}{dx^2} \tag{3.4}$$

is a statement of Fick's 2nd Law in one dimension. In three dimensions, this becomes

$$\frac{\partial c}{\partial t} = D \left(\frac{\partial^2 c}{\partial x^2} + \frac{\partial^2 c}{\partial y^2} + \frac{\partial^2 c}{\partial z^2}\right) \tag{3.5}$$

or

$$\frac{\partial c}{\partial t} = D \nabla^2 c$$

for the Cartesian coordinates $(x, y$ and $z)$. The operator

$$\nabla^2 = \frac{\partial^2}{\partial x^2} + \frac{\partial^2}{\partial y^2} + \frac{\partial^2}{\partial z^2}.$$

In cylindrical coordinates, (r, ϕ, z), as appropriate say for a disc electrode, and assuming no angular variation (c is not a function of ϕ) then

$$\frac{\partial c}{\partial t} = D \left(\frac{\partial^2 c}{\partial r^2} + \frac{1}{r}\frac{\partial c}{\partial r} + \frac{\partial^2 c}{\partial z^2}\right) \tag{3.6}$$

where r is the radial and z is the normal coordinate (Fig. 3.3).

Last, the case of spherical coordinates is needed for dealing with spherical and hemispherical electrodes:

$$\frac{\partial c}{\partial t} = D \left(\frac{\partial^2 c}{\partial r^2} + \frac{2}{r}\frac{\partial c}{\partial r}\right) \tag{3.7}$$

where r is the radial coordinate from the centre of the sphere or hemisphere. Finally, we note that in the above D has been assumed constant, independent of c. This is usually an excellent approximation in voltammetric studies.

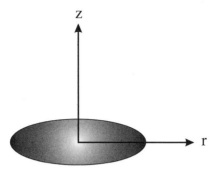

Fig. 3.3 Cylindrical coordinates.

3.3 The Molecular Basis of Fick's Laws

The physical basis of Eq. 3.1 was given independently by Einstein and Van Smoluckowskii.[3,4] In molecular terms, consider Fig. 3.4 which shows an arbitrary concentration–distance profile and highlights a 'box' of width $2\delta x$ at a general coordinate x such that half the box is on either side of the coordinate x. We assume that

- in each 'half box' molecules are as likely to be moving to the right as to the left, and that
- a particle, on the average, moves δx in time δt. It follows that the number (moles) travelling from left to right $= 1/2c_1 A\delta x$.

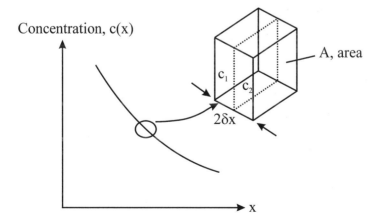

Fig. 3.4 The molecular basis of Fick's Laws.

Similarly the number (moles) travelling from right to left

$$= 1/2c_2 A\delta x.$$

It follows that the net rate of transfer through a plane positioned at x is,

$$rate = \frac{(c_1 - c_2)A\delta x}{2\delta t}.$$

But,

$$c_1 - c_2 \sim -\delta x \left(\frac{\partial c}{\partial x}\right),$$

so that

$$rate \sim -\frac{A(\delta x)^2}{2(\delta t)}\left(\frac{\partial c}{\partial x}\right).$$

Accordingly,

$$flux, j = -\frac{(\delta x)^2}{2\delta t}\left(\frac{\partial c}{\partial x}\right), \tag{3.8}$$

which is equivalent to Fick's 1st Law,

$$j = -D\left(\frac{\partial c}{\partial x}\right),$$

if

$$D = \frac{(\delta x)^2}{2\delta t}.$$

It follows that the diffusion coefficient, D, of a molecule provides a measure of how far the molecule travels (diffuses) in a certain time. Specifically, the root mean square displacement in time, t

$$\sqrt{\langle x^2 \rangle} = \sqrt{2Dt}. \tag{3.9}$$

This is an equation of huge significance in voltammetry: it enables us to estimate distances diffused in a time, t. Eq. (3.9) shows that the movement of the diffusing species dramatically decreases with the distance from its source (typically a zone of high concentration, or an electrode generating the species of interest). As Fig. 3.5 implies, a small solute can diffuse across a biological cell in around a second, whilst several years (sic) are required for diffusion to occur over one meter.

Albery[5] points to the slowness of diffusion as to the reason it is necessary to stir a cup of tea after the addition of sugar if the sweetness is to reach anywhere other than the bottom of the cup before it cools. We return to the idea of supplementing diffusion with convection in Chapter 8.

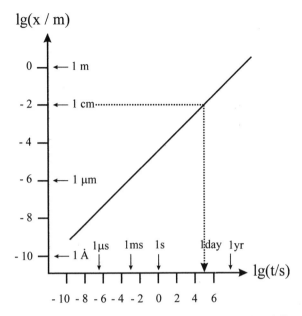

Fig. 3.5 The root mean square distance diffused by a molecule with a diffusion coefficient of 5×10^{-6} cm^{2}s^{-1} corresponding to a typical solution value.

3.4 How Did Fick Discover His Laws?

Adolph Fick was born in Kassel, Germany in 1829, the youngest son of the city's municipal architect. He studied first at Marburg, initially physics and chemistry but later, on the advice of his elder brother, turned to medicine. This mix of backgrounds enabled him to conduct cross-disciplinary research long before it was the hot and well-funded activity it is today! After Marburg he worked with his long-term mentor Carl Ludwig in Zurich for 16 years from 1852, after which he became professor of physiology in Würzburg until his retirement at the age of 70. He died in 1901 at the coastal town of Blankenberge, in Belgium.

Adolf Fick[a]

---

[a]The picture of Fick is thought to be in the public domain worldwide due to the date of death of its author (as the author has been dead for over 70 years) and due to its date of publication; the picture of Fick is therefore ineligible for copyright. That said, we will gladly amend accordingly if copyright can be established.

Fick's first paper dates from 1849 and analysed the musculoskeletal system of the pelvis in terms of mechanism, relating experimentally measured torques to the forces generated by the muscles and the geometry of the system. In 1870, he was the first to design a technique for measuring cardiac output, called the Fick principle.[6] In 1887, he constructed and fitted what was to be considered the first ever contact lens, although Leonardo da Vinci had created sketches encapsulating the concept as early as 1508. The first lens, tested naturally on rabbits and then on Fick himself, were made from thick glass and covered the entire eye. Needless to say they were not very comfortable to wear but were able to successfully correct some vision problems. Fick's book of 1856 entitled 'Medizinische Physik' covered a huge range of innovative contributions: the mixing of air in the lungs, the measurement of carbon dioxide output in humans, the heat economy of the body, the mechanism of limbs, bioelectricity, sound and its production, heat and its generation in organisms, the hydrodynamics of circulation, and of course diffusion.[6]

It is interesting to evaluate Fick's paper entitled 'On Liquid Diffusion'. Agutter, Malone and Wheatley[6] say

'like many classics of science, Fick's 1855 paper is far more widely acknowledged than read. It is reputed to have established the law of diffusion inductively from experimental data, but this imaginative reconstruction — inspired by old-fashioned empiricist beliefs about science — is far from reality. The paper has great scientific virtues, but it also has defects of reasoning ...'

What is the basis for this criticism? Primary it arises because Fick's own arguments derive from analogy rather than scientific deduction. In Fick's own words:[1]

'It was quite natural to suppose, that this law for the diffusion of a salt in its solvent must be identical with that, according to which the diffusion of heat in a conducting body takes place; upon this law Fourier founded his celebrated theory of heat, and it is the same which Ohm applied with such extraordinary success, to the diffusion of electricity in a conductor. According to this law, the transfer of salt and water occurring in a unit of time, between two elements of space filled with differently concentrated solutions of the same salt, must be, *ceteris paribus,* directly proportional to the difference of concentration, and inversely proportional to the distance of the elements from one another.'

Fick went on to establish via conservation of material, much as we did in Section 3.2, the second law in the form for one dimensional diffusion

$$\frac{\partial y}{\partial t} = k\frac{\partial^2 y}{\partial x^2}, \tag{3.10}$$

where y is concentration and x is distance. We recognise the constant k as the diffusion coefficient of Eq. (3.4). He appreciated that this equation would relate to diffusion in a cylinder for which steady-state diffusion could be experimentally studied.

'This is most easily attained by cementing the lower end of the vessel filled with solution, and in which the diffusion-current takes place, into a reservoir of salt, so that the section at the lower end is always maintained in a state of perfect saturation by immediate contact with solid salt; the whole being then sunk in a relatively infinitely large reservoir of pure water, the section at the upper end, which pass into pure water, always maintains a concentration $= 0$. Now for a cylindrical vessel, the condition $\frac{dy}{dt} = 0$ becomes by virtue of [Eq. (3.10)]

$$0 = \frac{d^2 y}{dx^2}.$$

The integral of this equation, $y = ax + b$ contains the following proposition: If, in a cylindrical vessel, dynamic equilibrium shall be produced, the differences of concentration of any two pairs of strata must be proportional to the distances of the strata in the two pairs or in other words, the decrease of concentration must diminish from below upwards as the ordinates of a straight line. Experiment fully confirms this proposition',[1]

The experiment is illustrated in Fig. 3.6.

Fick measured the specific gravity of the solution as a function of distance along the cylinder, related it to the concentration of dissolved salt and shows that, within his own assessment of experimental error, the predicted linear change of concentration took place. However, Fick made no reported attempts to change the depth of the cylinder and his data showed systematic deviation from linearity — which, Fick says, 'is easily explained by the consideration, that the stationary condition had not been perfectly attained.' He was not the last experimentalist to make this excuse!

Next Fick considered an experiment using a conical funnel, probably much as shown in Fig. 3.7, with the apex down and containing a saturated salt solution as above.

In this case the second law needs reformulating to allow for the fact that the cross-sectional of the funnel A, increases with the distance, x. If we adopt the

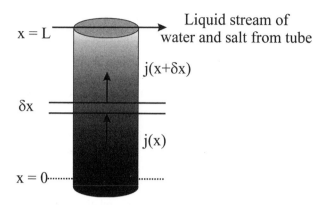

Fig. 3.6 Fick's diffusion experiment in a cylindrical tube with a salt reservoir at the bottom and flowing fresh water at the top. Adapted from Ref. [7].

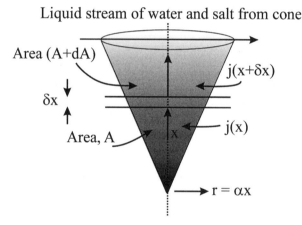

Fig. 3.7 Fick's conical funnel experiment. Adapted from Ref. [7].

discussion of Section 3.2 to consider the fluxes in and out of the cross sectional areas A and $A + dA$ (see Fig. 3.7), we find that

$$\frac{\delta n}{\delta t} = -(A + dA)j(x + \delta x) + Aj(x).$$

But

$$j = -D\frac{\partial c}{\partial x},$$

so

$$\frac{\delta n}{D\delta t} = \left(A + \frac{dA}{dx}dx\right)\left(\frac{dc}{dx} + \frac{d^2c}{dx^2}dx\right) - A\frac{dc}{dx}$$

$$\frac{\delta n}{D\delta t \cdot \delta x} \sim \frac{Ad^2c}{dx^2} + \frac{dA}{dx}\frac{dc}{dx} + O(dx)^2.$$

But

$$\frac{\delta n}{A\delta x} = \delta c,$$

so

$$\frac{dc}{dt} = D\frac{d^2c}{dx^2} + \frac{D}{A}\frac{dA}{dx} \cdot \frac{dc}{dx} \tag{3.11}$$

is the appropriate form of Fick's law assuming that at each point x the concentration is radially uniform. For a cone,

$$A = \pi\alpha^2 x^2$$

(see Fig. 3.7), and for steady-state diffusion

$$0 = \frac{d^2c}{dx^2} + \frac{2}{x}\frac{dc}{dx}.$$

The latter integrates to

$$c = -\frac{A}{x} + B,$$

where A and B are constants defined by the concentrations at the bottom and top of the conical funnel. Again, Fick presented experimental data consistent with this model albeit prefaced with the remark 'I here annex a short table of the *best* experiments with common salt' (our italics)!

The conical funnel experiment merits further analysis. Following Patzek[7]

'In an inverted conical funnel, and in the absence of gravity, the salt concentration contours would be section of concentric spheres centered on the salt reservoir at the funnel tip. One may argue that in the gravitational field the spherical concentration profiles of salt will be flattened vigorously by buoyancy force. The denser salt solution near the funnel axis will sink, while the less dense solution near the walls will be buoyed. The concentration profiles will hence become almost perfectly horizontal.'

Given that the experiment might reasonably be described via Eq. (3.11), there remains the issue of fitting Eq. (3.12) to the measured salt concentration data as a function of x. Fick fails to report the location of the funnel–salt reservoir relative to the tip apex. Patzek[7] has back calculated that this must have been a distance close to 5 cm.

So a referee of a modern scientific paper might report the following criticisms of Fick's paper.

- The theory is not rigorous, but simply based on analogy.
- The experimental details are insufficiently detailed as to permit the experiments to be reproduced.
- The data is reported selectively.
- The experimental data for the tubular cylinder experiment shows systematic deviations from theory.

It is debatable whether Fick's paper would be published in today's journals! Nevertheless, it is undisputable that Fick's insights were correct and that in referring to Eqs. (3.1) and (3.4) as Fick's laws of diffusion we are honouring the memory of a great intuitive scientist.

3.5 The Cottrell Equation: Solving Fick's 2nd Law

In the earlier part of this chapter, we established Fick's 2nd law,

$$\frac{\partial c}{\partial t} = D \frac{\partial^2 c}{\partial x^2},$$

as the equation describing the evolution of the concentration, c at a point x, in time t. To show this equation 'in action', we consider an electrode placed in a solution of a diffusing, electroactive species of concentration c^*. At the start of the experiment the electrode, located at $x = 0$, is passive; no potential is applied and no current drawn. Then at a time, $t = 0$, a large potential is applied such that the species is oxidised or reduced at a very fast rate (compared to diffusion) such that its concentration at the electrode surface is zero. This problem, in modern parlance, constitutes 'potential step chronoamperometry at a planar macroelectrode'. It was first considered by Frederick G. Cottrell (1877–1948) at the University of Leipzig and published in 1902.[8]

Mathematically, the problem involves solving Eq. (3.4) subject to the boundary conditions

$$t = 0, \; all \; x, \; c = c^*$$
$$t > 0, \; x = 0, \; c = 0$$
$$t > 0, \; x \to \infty \; c = c^*.$$

The trick to solving this problem requires us to introduce a new variable

$$\Gamma = \frac{x}{2\sqrt{Dt}}.$$

It follows that Fick's 2nd law becomes

$$\frac{d^2c}{d\Gamma^2} + 2\Gamma\frac{dc}{d\Gamma} = 0,$$

as can be verified by direct substitution. Integration of this shows

$$\frac{dc}{d\Gamma} = a\exp(-\Gamma^2),$$

where a is constant of integration.

$$\int_c^{c^*} dc = a\int_\Gamma^\infty \exp(-\Gamma^2)d\Gamma,$$

so that

$$c^* - c = a\left\{\int_0^\infty \exp(-\Gamma^2)d\Gamma - \int_0^\Gamma \exp(-\Gamma^2)d\Gamma\right\}.$$

Given that

$$\int_0^\infty \exp(-\Gamma^2)d\Gamma = \frac{\sqrt{\pi}}{2}$$

and applying the boundary conditions, we find that

$$c = c^* \frac{2}{\sqrt{\pi}}\int_0^\Gamma \exp(-\Gamma^2)d\Gamma \tag{3.12}$$

or

$$c = c^* erf\left(\frac{x}{2\sqrt{Dt}}\right), \tag{3.13}$$

where *erf* is the so-called error function and is defined by the integral in Eq. (3.12). Figure 3.8 shows how concentration, c, varies with x and t for two different values of D.

In experimental practice the current, I at the electrode is measured as a function of time; this is given via Eq. (2.2),

$$I = nFAj,$$

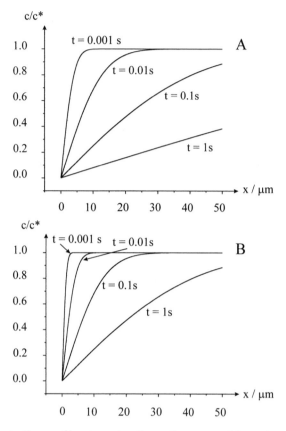

Fig. 3.8 Concentration profiles at varying times after a potential step into the diffusion-controlled region (A) $D = 5 \times 10^{-5}$ cm^{2}s^{-1} and (B) $D = 5 \times 10^{-6}$ cm^{2}s^{-1}.

where n is the number of electrons passed in the electrode reaction. Further the flux, j, is given by Ficks 1st Law:

$$j = D\frac{\partial c}{\partial x}|_{x=0} = \frac{D}{2\sqrt{Dt}} \frac{\partial c}{\partial \Gamma}|_{\Gamma=0}. \tag{3.14}$$

It follows that

$$I = \frac{nFA\sqrt{D}c^*}{\sqrt{\pi t}}, \tag{3.15}$$

a result known as the Cottrell equation, where A is the electrode area. It shows that the current resulting from a potential step decays to zero with a dependence which is inversely proportional to the square root of time. Figure 3.9 shows the form of

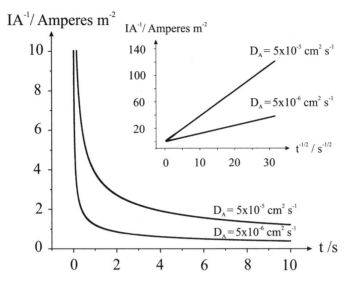

Fig. 3.9 The current density (I/A) transient resulting from a potential step at a planar macroelectrode.

the current transients. Further we note that Eq. (3.7) enables us to calculate the charge $Q(t)$ passed as a function of time:

$$Q = \int_0^t I dt$$

$$Q = 2nFA \frac{\sqrt{Dt}}{\sqrt{\pi}} c^*. \qquad (3.16)$$

This quantifies the extent of the electrolysis in terms of the number of moles of reactant, N, consumed given by

$$N = \frac{Q}{nF} = 2A \frac{\sqrt{Dt}}{\sqrt{\pi}} c^*. \qquad (3.17)$$

Figure 3.8 shows the depletion of material near the electrode surface as the timescale of the experiment progresses. The zone of depletion is known as the *diffusion layer* of the electrode. Electroactive material must diffuse across this in order to reach the electrode and react. As time progresses, the diffusion layer becomes thicker and the rate of diffusion falls, as does the current.

As the electroactive species becomes depleted near the electrode, there is a build up of the product of the electrode reaction near the electrode surface. In the case that the diffusion coefficients of the reactant and product are equal, the concentration profile of the product is given by

$$c(product) = c^* \left[1 - erf \left(\frac{x}{2\sqrt{Dt}} \right) \right]. \tag{3.18}$$

Note that the sum of the concentration of the reactant and the products is equal to c^* in the case of equal diffusion coefficients.

It is interesting to examine Fig. 3.8 in the light of the Einstein–Van Smoluchowskii equation established in Section 3.3:

$$\sqrt{\langle x^2 \rangle} = \sqrt{2Dt},$$

where the root mean square distance diffused is related to the diffusion coefficient and the time of the experiment. For the two different diffusion coefficients considered in drawing Fig. 3.8, we can calculate $\sqrt{\langle x^2 \rangle}$ as a function of time, t.

(a) $\quad D = 5 \times 10^{-6} \text{ cm}^2 \text{ s}^{-1}, \sqrt{\langle x^2 \rangle} = 32 \,\mu\text{m} \qquad (t = 1 \text{ s})$

$\quad\quad\ D = 5 \times 10^{-6} \text{ cm}^2 \text{ s}^{-1}, \sqrt{\langle x^2 \rangle} = 3.2 \,\mu\text{m} \qquad (t = 0.01 \text{ s})$

(b) $\quad D = 5 \times 10^{-5} \text{ cm}^2 \text{ s}^{-1}, \sqrt{\langle x^2 \rangle} = 100 \,\mu\text{m} \qquad (t = 1 \text{ s})$

$\quad\quad\ D = 5 \times 10^{-5} \text{ cm}^2 \text{ s}^{-1}, \sqrt{\langle x^2 \rangle} = 10 \,\mu\text{m} \qquad (t = 0.01 \text{ s})$

The distance calculates correspond roughly to the scale of the diffusion layer shown in Fig. 3.8. We again emphasise the power of Eq. (3.9) in estimating the scale (distance) of diffusion.

3.6 The Cottrell Problem: The Case of Unequal Diffusion Coefficients

It is illuminating to develop the problem posed in the last section for the general case where the reactants and products of the electrode reaction have unequal diffusion coefficients. Consider the reaction

$$A + e \longrightarrow B,$$

where the solution only contains A at a concentration of c^* for times, $t < 0$. The electrode potential is stepped from a value at which no current flows (c^* is uniform throughout the solution) to one where the above reaction is so strongly driven that the concentration of A is reduced to zero at the electrode surface.

Mathematically, the problem is formulated as follows. We need to solve the diffusion equations

$$\frac{\partial [A]}{\partial t} = D_A \frac{\partial^2 [A]}{\partial x^2} \tag{3.19}$$

and

$$\frac{\partial[B]}{\partial t} = D_B \frac{\partial^2[B]}{\partial x^2},$$ (3.20)

subject to the boundary conditions

$$t < 0, \quad all\ x: \quad [A] = c^*, \quad [B] = 0$$

$$t \geq 0, \quad x = 0: \quad [A] = 0, \quad D_A \frac{\partial[A]}{\partial x}\Big|_0 = -D_B \frac{\partial[B]}{\partial x}\Big|_0$$

$$t \geq 0, \quad x \to 0: \quad [A] \to c^*, \quad [B] \to 0.$$

Notice that the second boundary condition puts the flux of A into the electrode equal to the flux of B away, utilising Fick's 1st Law to quantify the fluxes. The solution of the problem in respect of $[A]$ is, of course exactly as given in the preceding section:

$$[A] = c^* \frac{2}{\sqrt{\pi}} \int_0^{\Gamma_A} \exp\left(-\Gamma^2\right) d\Gamma$$

$$[A] = c^* erf\left(\frac{x}{2\sqrt{D_A t}}\right).$$

Also,

$$\frac{\partial[A]}{\partial \Gamma_A} = c^* \frac{2}{\sqrt{\pi}} \exp\left(-\Gamma_A^2\right),$$ (3.21)

where

$$\Gamma_A = \frac{x}{2\sqrt{D_A t}}.$$

Similarly it can be deduced that

$$\frac{\partial^2[B]}{\partial \Gamma_B^2} + 2\Gamma_B \frac{\partial[B]}{\partial \Gamma_B} = 0,$$ (3.22)

where

$$\Gamma_B = \frac{x}{2\sqrt{D_B t}}.$$

It follows that

$$\frac{\partial[B]}{\partial \Gamma_B} = b \exp(-\Gamma_B^2),$$

where b is a constant. Rewriting the boundary conditions equating the fluxes of A to, and B from, the electrode gives

$$\sqrt{D_A} \frac{\partial[A]}{\partial \Gamma_A}\Big|_0 = -\sqrt{D_B} \frac{\partial[B]}{\partial \Gamma_B}\Big|_0$$

so that

$$\frac{\partial[B]}{\partial\Gamma_B} = -\sqrt{\frac{D_A}{D_B}}c^* \frac{2}{\sqrt{\pi}} \exp\left(-\Gamma_B^2\right).$$

Integrating and applying the other boundary conditions

$$[B] = c^*\sqrt{\frac{D_A}{D_B}}\left[1 - erf\left(\frac{x}{2\sqrt{D_B t}}\right)\right]. \tag{3.23}$$

It is clear that if $D_B < D_A$, then the concentration of B near the electrode is in excess of c^*, whereas if $D_B > D_A$, it is less than c^*.

Figure 3.10 shows plots of reactants and products for the following reaction

$$Fe(CN)_6^{3-}(aq) + e^-(m) \longrightarrow Fe(CN)_6^{4-}(aq),$$

where the diffusion coefficients are 0.76×10^{-5} cm^{2}s^{-1} ($Fe(CN)_6^{3-}$) and 0.63×10^{-5} cm^{2}s^{-1} ($Fe(CN)_6^{4-}$), relating to aqueous solutions of 1M KCl at 25°C.

3.7 The Nernst Diffusion Layer

When potential step experiments of the type described in the last two sections are carried out experimentally, the predicted dependence of the current of the reciprocal of the square root of time is typically observed for up to a few seconds, except at very short times (tens of milliseconds or less) when the current due to the electron transfer reaction at the electrode (to so-called 'Faradaic' processes) is obscured by 'charging current' due to movement near the electrode of the ions comprising the supporting electrolyte. At long time, however, rather than the current falling-off to zero as predicted by the Cottrell equation, the experimentally measured currents tend to an approximately steady value. This is consistent with the model proposed in Fig. 3.11 in which the bulk solution beyond a critical distance, δ, from the electrode is well mixed, so that the concentration of the electroactive species is maintained at a constant bulk value. Thus, mixing is due to 'natural convection' that is, movement of the solution induced by density differences. Consideration of Fig. 3.10 shows that such movements are intrinsic to electrolysis since the latter changes an initially uniform solution into one in which reactants are depleted but products built up near the electrode surface; if these species have different densities then natural convection is inevitable. In addition, slight temperature variation throughout the bulk of the solution as may arise from imperfect thermostating, can also provide a driving force for natural convection.

Close to the (solid) electrode, the natural convection dies away due to the rigidity of the electrode and to the operation of frictional forces. This is the zone

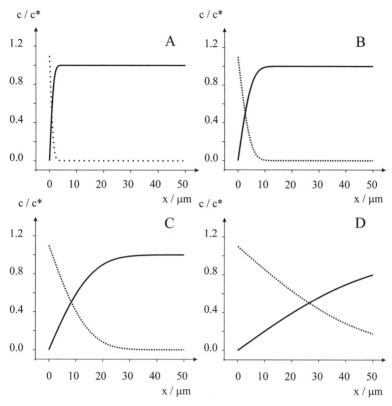

Fig. 3.10 For the electrode reaction $Fe(CN)_6^{3-} + e^-(m) \rightarrow Fe(CN)_6^{4-}$, the concentration of $Fe(CN)_6^{4-}$ local to the electrode can rise above the bulk value, c^*, since $Fe(CN)_6^{4-}$ diffuses more slowly than $Fe(CN)_6^{3-}$. Note the following potential step times are: $A = 0.001$ s, $B = 0.001$ s, $C = 0.1$ s, $D = 1$ s. The solid line shows $[Fe(CN)_6^{3-}]$ and the dashed line $[Fe(CN)_6^{4-}]$.

in Fig. 3.11 between $x = 0$ and $x = \delta$, the so-called *diffusion layer*. Since the concentration changes only in this zone, it is in this region that diffusional transport is operative. From Fick's 1st Law the steady-state diffusional flux, is

$$j = D\frac{\partial c}{\partial x} = \frac{DC^*}{\delta}. \tag{3.24}$$

It follows that the corresponding current is

$$I_{ss} = \frac{nFADc^*}{\delta}$$
$$I_{ss} = nFAm_T c^*, \tag{3.25}$$

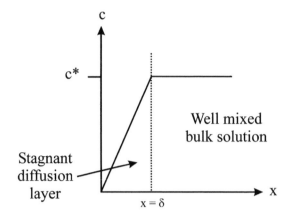

Fig. 3.11 The Nernst diffusion layer model.

where m_T is the so-called mass transport coefficient (unit cm s^{-1}) given by

$$m_T = \frac{D}{\delta}.$$

Note that his quantity has the same units as k^0, the standard electrochemical rate constant, so that a direct comparison of the two quantities is possible and can provide an indication of the relative speeds of electron transfer and mass transport, as will be addressed in the next section.

Returning to the Nernst diffusion layer, it is important to recognise that this is a simplified model, and that the zone of mixing via natural convection and the stagnant zone of diffusion in reality merge into one another. Nevertheless, the concept is helpful and insightful. Experiment shows that the size of the layer is of the order of tens to hundreds of microns. In developing voltammetric experiments which are to be theoretically interpreted via diffusion-only models, it is important that these are restricted to timescales such that the concentration changes are kept to a distance significantly closer to the electrode surface than δ.

Last, we introduce the concept of a *limiting current* — that which flows when the electrode potential drives the electrode reaction so 'hard' that the concentration of the electroactive species falls to zero at the electrode surface. The current predicted by the Cottrell equation is an example of such a current and in this case the limiting current falls with time. In contrast, in the case of the Nernst diffusion layer, the limiting current is attained when the concentration decays from the bulk value c^* to zero at the electrode surface over the distance of the diffusion layer, δ. The current given by Eq. (3.25) is thus the limiting current.

3.8 Mass Transfer vs. Electrode Kinetics: Steady-State Current-Voltage Waveshapes

Suppose we consider an electrode at which the following electrode process is studied

$$A(aq) + e^-(m) \underset{k_a}{\overset{k_c}{\rightleftharpoons}} B(aq).$$

We suppose that in bulk solution the concentration of A and B are $[A]_{bulk}$ and $[B]_{bulk}$ respectively. We also suppose that the electrode has a Nernst diffusion layer of thickness, δ, such that its mass transfer coefficient is

$$m_T = \frac{D}{\delta},$$

where D is the diffusion coefficient of the diffusing species, A or B; D_A or D_B assumed equal. Last, we assume that we can control the electrode potential, E, by means of a suitable three-electrode system linked to a potentiostat (see Chapter 2) and hence control the electrochemical rate constants

$$k_c = k^0 \exp\left(\frac{-\alpha F}{RT}[E - E_f^0(A/B)]\right)$$

$$k_a = k^0 \exp\left(\frac{\beta F}{RT}[E - E_f^0(A/B)]\right),$$

where $\alpha + \beta = 1$.

If the electrode is 'uniformly accessible', that is the flux (and current) is uniform over the electrode surface, the problem is one-dimensional and we consider the following expressions for the fluxes of A and B:

$$j_A = m_T([A]_0 - [A]_{bulk}) = -j_B, \tag{3.26}$$

$$j_B = m_T([B]_0 - [B]_{bulk}) \tag{3.27}$$

and

$$-j_A = k_c[A]_0 - k_a[B]_0,$$

where $[\ ]_0$ denotes a concentration at the electrode surface, see Fig. 3.12. These three equations can be solved by eliminating the unknown surface concentrations between the equations and introducing the transport limited currents for the electrolysis of each of the two species, A and B:

$$j_{A,lim} = -m_T[A]_{bulk} \tag{3.28}$$

$$j_{B,lim} = -m_T[B]_{bulk}. \tag{3.29}$$

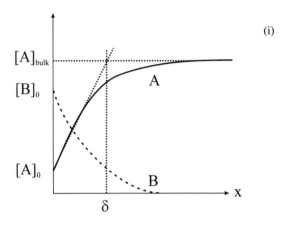

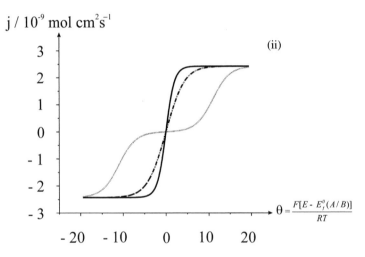

Fig. 3.12 (i) Concentration profiles of species A and B. (ii) Steady-state voltammograms for the one-electron reduction of A to B when $k^0 = 10^{-1}$, (solid line), 10^{-3} (dot-dash line), and 10^{-5} (dotted line) cm s^{-1}. Parameters: $[A]_0 = [B]_0 = 1$mM; $D_A = D_B = 10^{-5}$ cm^2 s^{-1}; $m_T = 10^{-3}$ cm s^{-1}.

It follows that flux,

$$j = -j_A = j_B \tag{3.30}$$

$$j = \frac{k_c j_{A,lim} - k_a j_{B,lim}}{m_T + k_c + k_a}. \tag{3.31}$$

We can consider three limiting cases of this equation.

Case (i)

$$k_c \gg m_T, k_a.$$

Here,

$$j \rightarrow j_{A,lim}$$

and a constant, potential independent current is seen corresponding to the case where the flux is controlled by the maximum rate of diffusion to the electrode; that is $[A]_0 = 0$ and

$$j_{A,lim} = \frac{-D[A]_{bulk}}{\delta}.$$

This will be achieved experimentally by applying a large negative potential to the electrode encouraging a rapid rate for the reaction

$$A + e \longrightarrow B,$$

such that the back reaction (controlled by k_a) is negligible.

Case (ii)

$$k_a \gg m_T, k_c.$$

Here,

$$j \rightarrow -j_{B,lim}$$

and again a transport controlled potential independent flux (current) is seen such that

$$j_{B,lim} = \frac{D}{\delta}[B]_{bulk},$$

and this will occur at large positive potentials which make the rate constant (k_a) for the process

$$B \longrightarrow A + e$$

large, and the rate constant (k_c) for the reverse process tiny. In both cases (i) and (ii) the process is, kinetically speaking, 'transport controlled'.

Case (iii)

$$m_T \gg k_a, k_c.$$

In this situation,

$$j = \frac{k_c J_{A,lim} - k_a J_{B,lim}}{m_T} \tag{3.32}$$

$$j = -k_c[A]_{bulk} + k_a[B]_{bulk} . \tag{3.33}$$

This corresponds to a situation which is under the control of 'electrode kinetics' and the current flowing is sensitive to the electrode potential. Note that Eq. (3.33) implies that the concentrations of A and B are, in this limit, essentially unperturbed from those of bulk solution.

Figure 3.12(ii) shows the dependence of flux on electrode potential (a 'voltammogram') for the cases where $k^0 = 10^{-1}, 10^{-3},$ and 10^{-5} cm s^{-1} with $\alpha = 0.5$ and $E_f^0 = 0$ V.

A mass transfer coefficient of $m_T = 10^{-3}$ cm s^{-1} has been assumed. Notice that in all three cases at both positive and negative extremes of potential a transport limited current is seen (cases (i) and (ii) above). In between these limits, different responses are seen depending on the relative size of k^0 and m_T. In particular, three distinct types of behavior are noted.

- Electrochemically 'reversible' voltammetry corresponding to the case $k^0 \gg m_T$. Here, a single voltammetric wave is centered around the formal potential for the A/B couple.
- Electrochemically 'irreversible' voltammetry corresponding to the limit where $k^0 \ll m_T$. Here, two distinct waves are seen for the processes

$$A + e \to B$$

and

$$B - e \to A.$$

Consequently, negligible current flows for potentials close to the formal potential, E_f^0. Rather 'overpotentials' in both the cathodic (negative potential) and anodic (positive potentials) direction must be applied in order to drive the electrode process because of the small value of k^0.
- Electrochemically 'quasi-reversible' voltammetry corresponds to the intermediate case, $k_0 \sim m_T$.

We return to these distinctions in the following section.

3.9 Mass Transport Corrected Tafel Relationships

We now develop further and more quantitatively the ideas of the last section, but focus on the experimentally more usual situation where the solution under study only contains one, not both, species comprising a redox couple. So, pursuing the example described in the previous section

$$A + e \to B,$$

we assume there is only A in bulk solution, so that $[B]_{bulk} = 0$. For the simple electrode process above,

$$j_A = D\left(\frac{[A]_0 - [A]_{bulk}}{\delta}\right). \tag{3.34}$$

For this process, the magnitude of the flux densities of A and B must be equal:

$$j_A = -j_B = D\frac{[B]_0}{\delta}, \tag{3.35}$$

where j_A and j_B are the flux densities of species A and B, respectively and D is a diffusion coefficient. It is assumed that the diffusion coefficients and diffusion layer thicknesses of the two species are equal. We next consider the two limits of electrochemical reversibility and irreversibility:

Case (a). Electrochemically irreversible electron transfer

Another expression for the flux density at the electrode comes from considering the rate of electron transfer, which is described by Butler–Volmer kinetics for an irreversible process:

$$j_A = -k^0 e^{-\alpha\theta}[A]_0, \tag{3.36}$$

where k^0 is the standard heterogeneous rate constant, α is the charge transfer coefficient and $\theta = \frac{F}{RT}(E - E_f^0)$ where E is the electrode potential and E_f^0 is the formal electrode potential. The unknown quantity $[A]_0$ may be eliminated from Eqs. (3.34) and (3.36):

$$\frac{1}{j_A} = -\frac{e^{\alpha\theta}}{k^0[A]_{bulk}} - \frac{\delta}{D[A]_{bulk}}. \tag{3.37}$$

At sufficiently negative potentials, the exponential terms approach zero and flux density at the electrode approaches its mass transport limiting value, j_{lim}:

$$\frac{1}{j_{lim}} = \frac{-\delta}{D[A]_{bulk}}. \tag{3.38}$$

The term δ may be eliminated from Eqs. (3.37) and (3.38) making the assumption that its value is independent of the potential:

$$\frac{1}{j_A} - \frac{1}{j_{lim}} = \frac{-e^{\alpha\theta}}{k^0[A]_{bulk}}. \tag{3.39}$$

The total current, I is related to the flux density by an integral over the electrode. For a uniformly accessible electrode the relationship between current and flux density is simply:

$$I = nFA_{elec}j_A, \tag{3.40}$$

where n is the number of electrons transferred per molecule of A and A_{elec} is the electrode area. Substituting Eq. (3.40) into (3.29) and rearranging gives

$$\ln\left(\frac{I_{lim}}{I} - 1\right) = \ln\left(\frac{-I_{lim}}{nFA_{elec}k^0[A]_{bulk}}\right) + \alpha\theta, \tag{3.41}$$

where I_{lim} is the mass transport limiting current. The collection of terms in parentheses on the right-hand side of Eq. (3.41) form a positive dimensionless constant which is independent of potential. Hence under conditions where the reverse process of an electron transfer may be considered negligible, a plot of θ vs. $\ln[(\frac{I_{lim}}{I}) - 1]$ is predicted to be linear with a gradient $1/\alpha$.

Case (b). Electrochemically reversible electron transfer

For reversible electron transfer the Butler–Volmer expression for the flux density at the electrode must contain terms for both anodic and cathodic processes:

$$j_A = -k_0 e^{-\alpha\theta}[A]_0 + k^0 e^{(1-\alpha)\theta}[B]_0. \tag{3.42}$$

The unknown quantities $[A]_0$ and $[B]_0$ are eliminated from Eq. (3.42) by substituting in Eqs. (3.34) and (3.35) and rearranging to give

$$j_A = \frac{-k^0 e^{-\alpha\theta}[A]_{bulk}}{1 + k^0 \frac{\delta}{D}(e^{-\alpha\theta} + e^{(1-\alpha)\theta})}. \tag{3.43}$$

Equation (3.43) may be simplified by considering a fast electron transfer for which $(k^0\delta/D) \gg 1$. The term $[A]_{bulk}$ is eliminated by substituting in the expression for the mass transport limiting flux density in Eq. (3.8). Further, rearranging gives

$$j_A = \frac{j_{lim}}{1 + e^\theta}. \tag{3.44}$$

Substituting in the expression for current at a uniformly accessible electrode in Eq. (3.40) leads to

$$\ln\left(\frac{I_{lim}}{I} - 1\right) = \theta. \tag{3.45}$$

Hence for sufficiently fast reversible electron transfer a plot of θ vs. $\ln[(\frac{I_{lim}}{I}) - 1]$ is predicted to be linear with a gradient of 1 (unity).

Last, we note that the two cases derived above strictly apply only to electrode reaction of the stoichiometry

$$A \pm e \longrightarrow B.$$

In experimental practice, other stoichiometries will often be encountered, for example,

$$3Br^- - 2e^- \longrightarrow Br_3^-$$

and

$$H^+ + e^- \longrightarrow 1/2H_2.$$

It is important to realise that in these cases modification of the expressions given above are required.[9]

Let us consider a reversible redox reaction taking place at the surface of a uniformly accessible hydrodynamic electrode

$$mA \pm e^- \rightleftarrows nB, \tag{3.46}$$

where m and n are rational numbers. Obviously, Eq. (3.46) represents all possible stoichiometries of electrode processes since any stoichiometric equation with integer coefficients can be divided by the number of electrons transferred to obtain (3.46). The use of the generalised Nernst equation gives the following relation for the surface concentrations of reacting species denoted as $[\ ]_0$.

$$\frac{([A]_0/[A]^o)^m}{([B]_0/[B]^o)^n} = \exp(\pm \theta), \tag{3.47}$$

where $\theta = (F/RT)(E - E_f^0)$, E^0 is the formal potential of the A/B redox couple, symbols F, R, T and E have their usual significance and $[A]^o$ and $[B]^o$ are the concentrations of the standard thermodynamic states of A and B, respectively. The upper sign $(+)$ within the exponential on the right-hand side of (3.47) corresponds to a reduction process while the lower one $(-)$ to an oxidation in (3.46). Assuming x to be a coordinate normal to the electrode we can express the flux conservation property at its surface as

$$nD_A \left.\frac{\partial[A]}{\partial x}\right|_{x=0} = -mD_B \left.\frac{\partial[B]}{\partial x}\right|_{x=0}, \tag{3.48}$$

where D_A and D_B are the diffusion coefficients of the corresponding species. Under steady-state conditions the mass transport of species occurs within a diffusion layer of thickness, δ, adjacent to the surface of the electrode. Presuming that initially the solution contained exclusively species A at a bulk concentration, $[A]_{bulk}$, we can write

$$nD_A \frac{[A]_{bulk} - [A]_0}{\delta} = mD_B \frac{[B]_0}{\delta}. \tag{3.49}$$

Therefore, the current flowing at the surface of the electrode can be expressed as

$$I = \pm \frac{1}{n} \frac{FD_B[B]_0}{\delta} A, \tag{3.50}$$

where A denotes the area of the electrode. Similarly the limiting current flowing through the electrode is

$$I = \pm \frac{1}{m} \frac{FD_A[A]_{bulk}}{\delta} A. \tag{3.51}$$

Utilising Eqs. (3.47), (3.50) and (3.51) gives

$$\left(\frac{I}{I_{lim}}\right)^{n/m} = \left(\frac{m}{n}\frac{D_B[B]^o}{D_A[A]_{bulk}}\right)^{n/m}\frac{1}{[A]^o}\exp\left(\pm\frac{\theta}{m}\right)[A]_0. \qquad (3.52)$$

The current can be written as

$$I = \pm\frac{1}{m}\frac{FD_A}{\delta}([A]_{bulk} - [A]_0)A = I_{lim}\left(1 - \frac{[A]_0}{[A]_{bulk}}\right), \qquad (3.53)$$

which leads us to the final result

$$-\ln\left[\left(\frac{I_{lim}}{I}\right)^{n/m} - \left(\frac{1}{I_{lim}}\right)^{1-n/m}\right] = \mp\frac{\theta}{m} + \frac{n}{m}\ln\left(\frac{m}{n}\frac{D_B}{D_A}\frac{[B]^o}{([A]^o)^{m/n}}[A]_{bulk}^{m/n-1}\right). \qquad (3.54)$$

Equation (3.54) allows for the current response of a reversible system of any stoichiometry to be linearised as a plot of the left-hand term against θ. The slope of the ensuing line is $\pm m^{-1}$ and the intercept is given by the right most term of Eq. (3.54), which depends on the ratios of the stoichiometric coefficients of reactant species and their diffusion coefficients as well as the bulk concentration of A.

Figure 3.13 shows typical simulated voltammograms for these three oxidation reactions:

$$A - e^- \rightleftarrows B \quad m = 1, n = 1 \qquad (3.55)$$

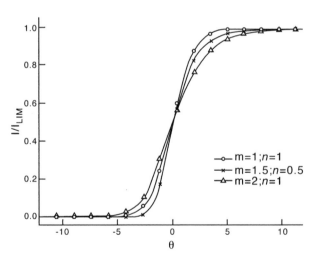

Fig. 3.13 Simulated steady-state voltammograms for three different stoichiometries of electrode process. Reprinted from Ref. [9] with permission from Elsevier.

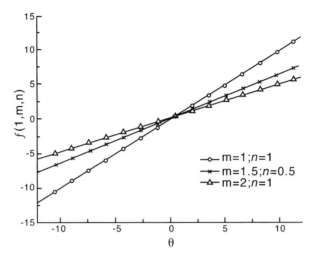

Fig. 3.14 Tafel plots of the voltammograms shown in Fig. 3.13 (see text). Reprinted from Ref. [9] with permission from Elsevier.

$$3A - 2e^- \rightleftarrows B \quad m = 3/2, n = 1/2 \quad (3.56)$$

$$2A - e^- \rightleftarrows B \quad m = 2, n = 1 \quad (3.57)$$

The waveshapes for these three cases with $D_A = D_B = 10^{-5} \ \text{cm}^2 \ \text{s}^{-1}$ and $[A]_{bulk} = 1 \ \text{M}$ are shown in Fig. 3.13.

It is clear from the figure that difference in stoichiometry leads to different curvature of the voltammograms at the foot and top of the wave. The Tafel analysis of the voltammograms is presented in Fig. 3.14, where

$$f(I, m, n) = -\ln\left[\left(\frac{I_{lim}}{I}\right)^{n/m} - \left(\frac{I}{I_{lim}}\right)^{1-n/m}\right]. \quad (3.58)$$

The slopes of the straight lines obtained are as predicted by Eq. (3.54): slope = 1 for (3.55), 2/3 for (3.56) and 1/2 for (3.57). The intercepts of the straight lines in Fig. 3.14 are also in excellent agreement with Eq. (3.54).

References

[1] A. Fick, Uber Diffusion, *Poggendorff's Annel. Physik.* **94** (1855) 59, in German. In English translation: *The London, Edinburgh and Dublin Philosophical Magazine* **10** (1855) 30 and *Journal of Science* **16** (1855) 30.
[2] M.C. Buzzeo, R.G. Evans, R.G. Compton, *Chem. Phys. Chem.* **5** (2004) 1106.

[3] A. Einstein, *Annalen de Physik* **17** (1905) 549.

[4] M. Van Smoluckowskii, *Annalen de Physik* **21** (1906) 756.

[5] W.J. Albery, *Electrode Kinetics*, Oxford University Press, 1975

[6] P.S. Agutter, P.C. Malone, D.N. Wheatley, *Journal of the History of Biology* **33** (2000) 71.

[7] T.W. Patzek, 'Fick's Diffusion Experiments Revisited', personal communication. Can be found at: http://petroleum.berkeley.edu/papers/patzek/Fick%20Revisited% 20V2.pdf.

[8] F.G. Cottrell, *Z. Physik. Chem.* **44** (1902) 385.

[9] O.V. Klymenko, R.G. Compton, *J. Electroanal. Chem.* **571** (2004) 571.

4

Cyclic Voltammetry at Macroelectrodes

This chapter seeks to develop the interplay of electrode kinetics and diffusion as initiated towards the end of the preceding chapter. We start by introducing cyclic voltammetry — the most important and most widely employed piece of methodology in the entire field of voltammetry and the study of electrode kinetics.

4.1 Cyclic Voltammetry: The Experiment

The cyclic voltammetry experiment involves applying a potential to the working electrode which changes with time as shown in Fig. 4.1.

The experiment records the current flowing through the working electrode as a function of the applied potential and a plot of current versus potential is constructed, which is known as a 'voltammogram'. The potential of the working electrode starts at a value, E_1, typically but not obligatory chosen to correspond to negligible current flow. That is to say, the starting voltage is usually selected so that the chemical species under investigation are not initially oxidised or reduced. The potential is then swept in a linear manner to a voltage, E_2, at which point the direction of scan is reversed and the working electrode potential reversed usually to is original value. The potential E_2 is usually selected so that the potential interval $(E_2 - E_1)$ contains an oxidation or reduction process of interest. If this results in chemical reaction and the formation of new chemical species, then the reverse scan may be extended beyond E_1 so as to allow their characterisation and/or second triangular potential sweeps employed to learn more about the system under study

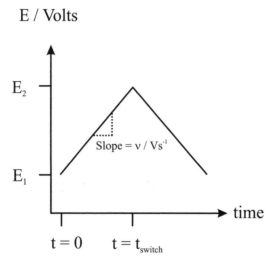

Fig. 4.1 The potential waveform applied to the working electrode in the cyclic voltammetry experiment.

and its electrochemical reactivity. We will address the issue of 'follow up' chemistry in Chapter 7. The present chapter focuses exclusively on simple electrode processes of the form

$$A \pm e \rightleftarrows B,$$

where both A and B are solution phase species. We next seek to predict the form of the voltammogram for this process, that is to say, how the current varies with the potential applied to the working electrode assuming that the potential varies as shown in Fig. 4.1. We can anticipate that the observed voltammogram will depend on the following:

- The standard electrochemical rate constant, k^0, and the formal potential of the A/B couple.
- The diffusion coefficients of A and B, and
- the voltage sweep rate, $\upsilon/\,\mathrm{Vs}^{-1}$, together with the voltages E_1 and E_2.

The problem can be formulated in general terms as requiring the solution of the following diffusion equations describing the concentrations of A and B as a function of distance normal to the electrode, x, and time t.

$$\frac{\partial[A]}{\partial t} = D_A \frac{\partial^2[A]}{\partial x^2} \tag{4.1}$$

$$\frac{\partial[B]}{\partial t} = D_B \frac{\partial^2[B]}{\partial x^2}. \tag{4.2}$$

The equations are coupled via the following boundary conditions corresponding to a case where there is only A in bulk solution:

$$t < 0, \quad \text{all } x, \quad [A] = [A]_{bulk}, \quad [B] = 0$$

$$t > 0, \quad x \to \infty, \quad [A] = [A]_{bulk}, \quad [B] = 0$$

$$t > 0, \quad x = 0, \quad D_A \frac{\partial [A]}{\partial x}|_{x=0} = -D_B \frac{\partial [B]}{\partial x}|_{x=0}$$

$$t > 0, \quad x = 0, \quad D_A \frac{\partial [A]}{\partial x}|_{x=0} = +k_c[A]_{x=0} - k_a[B]_{x=0},$$

where the last equation refers to a reduction only. The electrochemical rate constants are

$$k_c = k^0 \exp\left[\frac{-\alpha F}{RT}(E - E_f^0(A/B)) \right]$$

$$k_a = k^0 \exp\left[\frac{\beta F}{RT}(E - E_f^0(A/B)) \right],$$

where $E_f^0(A/B)$ is the formal potential of the A/B couple and E is the potential applied to the working electrode. The latter is shown in Fig. 4.1 and can be written algebraically as

$$0 < t \le t_{switch} \quad E = E_1 + vt \tag{4.3}$$

$$t_{switch} \le t \quad E = E_1 + vt_{switch} - v(t - t_{switch})$$

$$E = E_1 + 2vt_{switch} - vt, \tag{4.4}$$

noting that the scan rate, v, can be positive or negative in sign corresponding to the likely study of an oxidation or a reduction, respectively.

4.2 Cyclic Voltammetry: Solving the Transport Equations

Equations (4.1) and (4.2) together with the boundary conditions specified at the end of the previous section are not easily solved. The development of cyclic voltammetry was facilitated by the work of Nicholson and Shain[1,2] in solving the problems of interest and other related ones by means of integral equation methods. These allow the formulation of 'answers' to the problems but typically only after the numerical evaluation of integrals and series. Nevertheless, tables of data have been presented which allow the analysis of voltammetric data. That said, and all students of voltammetry should at least glance at the classic Nicholson and Shain papers at some point in their careers, it is the case that the contemporary approach is to employ 'simulation' software, which allows the immediate solution of the diffusion equations of interest. Such programs, widely available through suppliers of commercial electrochemical equipment, will 'simulate' the voltammogram for any specified electrochemical mechanism, subject to the voltammetry occurring at a macroelectrode so that the diffusion equations involve one spatial dimension only.[a] Moreover, whilst such software will typically allow a problem of almost any complexity to be interrogated, there is an upper limit to that which can be unambiguously understood. In the context of the present chapter, we intend to solve the cyclic voltammetry problem for the process

$$A \pm e \rightleftarrows B$$

and explore the impact of the different parameters identified in the previous section on the resulting voltammogram. For such purposes, 'simulation software' is perfect and provides almost immediate answers to problems that only a few decades ago posed real applied mathematical difficulty. By exploiting the power of modern computers, the electrochemist can explore the basics and the nuances of cyclic voltammetry at the touch of a keyboard. The following sections detail the results of such explorations. We encourage the reader to undertake 'computation experiments' using the software of the type described if it is available in their own laboratory or classroom; it is usually available in all competitive research laboratories.

The basis of using 'simulation software' is the solution of the relevant diffusion equations and boundary conditions via so-called finite difference and, increasingly, finite element methods. The state of art of these methods has been authoritatively reviewed by Fisher.[3] A very brief outline of finite difference modelling is given in the appendix to this book. However, those seeking to begin to work with their own code should consult the truly excellent work by D. Britz entitled 'Digital Simulation in Electrochemistry'.[4]

We end this section with a comment. Many simulation programs allow the exploration of very complex electrochemical mechanisms: this is fine for generating understanding but it is quite a different matter to 'fit' experimental data using a multiplicity of parameters. It is always required that the simulator demonstrates uniqueness of fit under these latter conditions.

---

[a] Note that in the next chapter we will discuss 'microelectrodes' where the diffusion equations involve 2 or 3 spatial coordinates. Current software is less easily able to cope with these problems and one often must write one's own code.

4.3 Cyclic Voltammetry: Reversible and Irreversible Kinetics

We saw in Chapters 2 and 3 that electron transfer processes showed different limiting behaviours corresponding to 'large' or 'small' values[b] of k^0 the standard electrochemical rate constant. These are characterised by the labels 'electrochemically reversible' and 'electrochemically irreversible' corresponding to 'fast' and 'slow' electrode kinetics respectively.

Figure 4.2 shows three voltammograms simulated for all the same conditions other than a difference in the standard electrochemical rate constant. The common parameters were

$$[A]_{bulk} = 10^{-3} \text{M} \quad [B]_{bulk} = 0$$
$$D_A = D_B = 10^{-5} \text{ cm}^2\text{s}^{-1}$$
$$\text{Electrode area} = 1 \text{ cm}^2$$
$$E_f^0(A/B) = 0 \text{V}$$
$$\text{Voltage sweep rate, } \nu = 1 \text{ } Vs^{-1}$$
$$E_1 = +0.5\text{V } E_2 = -0.5\text{V}$$

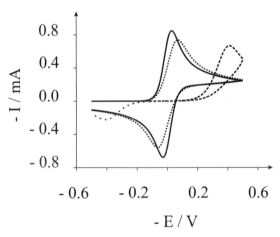

Fig. 4.2 Cyclic voltammograms for the reduction of A to B. Parameter: $E_f^0 = 0\text{V}; \alpha = 0.5$, $\upsilon = 1\text{V s}^{-1}; A = 1 \text{ cm}^2; [A]_0 = 1 \text{ mM}; D_A = D_B = 10^{-5} \text{ cm}^2\text{s}^{-1}$. The values of the standard electrochemical rate constant were 1 cm s^{-1} (solid line), $10^{-2} \text{ cm s}^{-1}$ (dotted line) and $10^{-5} \text{ cm s}^{-1}$ (dashed line).

---

[b] We shall see later that 'large' and 'small' are, of course, relative terms. More precisely we mean 'large' or 'small' relative to the prevailing rate of mass transport.

and

$$\alpha = 0.5 = \beta.$$

It is assumed that A is reduced to B

$$A + e \rightleftarrows B.$$

The three standard electrochemical rate constants used were 1, 10^{-2} and 10^{-5} cm s^{-1}.

The first point to consider when examining a voltammogram is to be clear as to what potential range is being swept on the forward going and on the reverse scan. In the present case the potential applied to the working electrode changes with time as shown in Fig. 4.3, so that the potential in the voltammogram in Fig. 4.2 starts at the left ($+0.5$ V) moves through 0 V to -0.5 V and then sweeps back to the starting potential.

The current shown is a reduction current and so the y-axis is labelled as $-I$/mA. This is in accordance with the International Union of Pure and Applied Chemistry (IUPAC) convention[5] as elucidated by a panel of no less than 21 IUPAC delegates under the chairmanship of R.G. Bates (USA) and secretaryship of J.F. Coetzee (also USA):

> 'The fundamental convention will consist of assigning positive values to anodic current and negative ones to cathodic current. Anodic and

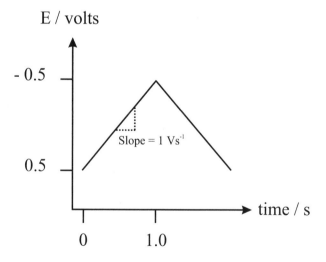

Fig. 4.3 The potential cycle applied in Fig. 4.2.

cathodic currents will continue to be defined as corresponding to net oxidation and net reduction, respectively, at the indicator or working electrode ...

... Any reasonable choice of coordinates is appropriate in plotting any such [voltammetric] curve, provided that the abscissa[c] and ordinate axes are clearly labelled. Most of the polarographic and other voltammetric curves in the existing literature are plotted with cathodic current above the abscissa axis and negative values of the applied emf to the rig hot the ordinate axis. Those who wish to follow the new convention and also to facilitate comparison of their curves with those in the prior literature may achieve both aims by choosing — i as the positive ordinate and — E as the positive abscissa.'[5]

The reader will note that we have ensured that Fig. 4.2 conforms to the dictated convection. Whether the panel of 21[d] really believed that their convention would be systematically applied must be doubtful since they recognise that

'Conformity to this convention will require many chemists who work with polarographic waves, chronopotentiograms, and other electrochemical response curves to reformulate some of the equations associate with them and adjust related procedures.'

The reality is that many (perhaps even most) voltammograms are published with scant regard for the IUPAC convention. It follows that those learning the

---

[c] Abscissa is the x-axis and ordinate the y-axis in plain English.

[d] One name stands out in the 21 names: Marco Branica of the then Yugoslavia who pioneered the application of electroanalytical methods in marine research. The authors recall with pleasure meeting him in his laboratories at Zagreb, Croatia in September 2004 shortly before his death. He was a man of great vision and depth of thought.

Shown left is Marco Branica (1931–2004). In their obituary (Croatia Chemica Acta, 79, 2006, xiii–xxiii), Goran Kniewald and Milivoj Lovrié wrote, 'Branica's life-long motto was that the (Adriatic) sea was of paramount importance to Croatia and its future, and he understood that an effort must be made to provide appropriate education to coming generations of scientists. As early as 1971, he organised a postgraduate course in Oceanography at the University of Zagreb. Marco was course director from the beginning and in 35 years of its existence more than 200 students received their master degrees by taking this course.' Picture and text copyright of Croatia Chemica Acta and used with permission.

subject of voltammetry need to become adept at making sure that their first act on encountering a new voltammogram is to clarify what potential sweep has been applied, that is, to figure out the relevant analogy of Fig. 4.3 and to ensure that they are au fait with which currents are anodic and which are cathodic. Returning to Fig. 4.2, the relevant features to note are shown in Fig. 4.4.

It can be seen that for all three voltammograms there are three zones on the forward going sweep from E_1 to E_2:

- At relatively positive potentials, no current flows since the electrode is insufficiently negative to reduce A to B.
- At more negative potentials, as the electrochemical rate constants, k_a, becomes suitably large, the current rises as the potential becomes more negative.
- At yet more negative potentials, the current passes through a maximum and decreases.

At the end of Chapter 3, we considered the current–voltage curves resulting under steady-state conditions with a fixed (constant) diffusion layer thickness. Under those conditions, the current–voltage curves showed limiting currents at extremes of potential in contrast to the peak shown in Fig. 4.2. The reason for the latter behaviour is that under the cyclic voltammetry conditions the diffusion layer is constantly expanding, so that once the electrode potential has reached a value at which the surface ($x = 0$) concentration of A is close to zero the current decays, approximately as $1/\sqrt{t}$ as in the potential step experiments described by the Cottrell equation.

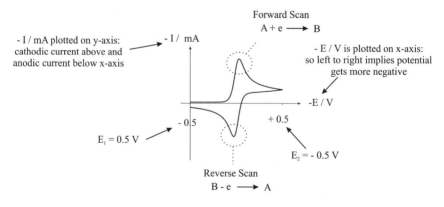

Fig. 4.4

The three curves shown in Fig. 4.2 correspond to the limits of electrochemical irreversibility, reversibility, and quasi-reversibility, respectively. We next consider the two extreme cases in more detail.

Case (i): Electrochemically irreversible behaviour
Figure 4.5 shows the voltammogram calculated for $k^0 = 10^{-5} \text{cm s}^{-1}$ and the other parameters specified above, together with concentration 'profiles' for A and B at six different locations on the voltammetric wave. The six points (A) to (F) to note are as follows:

(A) This point is before the start of the peak corresponding to the reduction of A. Accordingly, only a small amount of A has been consumed at the electrode surface and only a small layer of B has been built up. The spatial extent of the diffusion layer is relatively small, of the order of 0.01 mm (10 μm). Note that at the surface

$$\frac{\partial[A]}{\partial x}\bigg|_{x=0} = -\frac{\partial[B]}{\partial x}\bigg|_{x=0},$$

fulfilling one of the boundary conditions specified above. Also as $x \to \infty$, $[A] \to [A]_{bulk}$ and $[B] = 0$.

(B) This point corresponds to the maximum reduction current in the voltammetric wave. Figure 4.5(B) shows a greater depletion of A and a larger build up of B as compared to Fig. 4.5(A). The diffusion layer has thickened.

(C) This location corresponds to a point on the reduction peak where the current is decreasing with increasing potential. The concentration profile plot shows the concentration of A at the electrode surface to be close to zero, so that this part of the voltammogram is under diffusion control, whereas at (A) it was the electrode kinetics which controlled the response. The diffusion layer has reached a *ca* 40 μm thickness. Part (C) corresponds to the point at which the direction of potential scan is reversed.

(D) This point is where the working electrode potential has the value of 0 V corresponding to the formal potential of the A/B couple. At this point the electrode potential is insufficient to noticeably reduce A or oxidise B. Accordingly, at the electrode surface,

$$\frac{\partial[A]}{\partial x}\bigg|_{x=0} = -\frac{\partial[B]}{\partial x}\bigg|_{x=0} \sim 0.$$

In going from (C) to (D), the diffusion layer has thickened: B has continued to diffusion into bulk solution and the zone of depletion of A has extended further into solution, although the concentration of A at the electrode surface has been partially replenished.

(E) This location corresponds to the peak in the reverse scan due to the re-conversion of B to A. The concentration profiles show the build up of A and

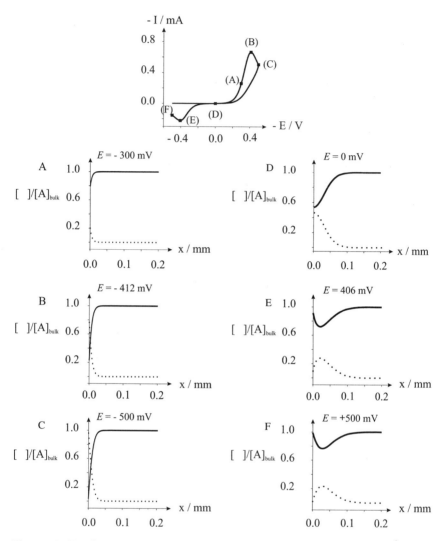

Fig. 4.5 Cyclic voltammogram for the irreversible reduction of A to B. Parameters: $E_f^0 = 0\,\mathrm{V}$; $\alpha = 0.5; k^0 = 10^{-5}\,\mathrm{cm\,s^{-1}}; \upsilon = 1\,\mathrm{Vs^{-1}}; A = 1\,\mathrm{cm^2}; [A]_0 = 1\mathrm{mM}; D_A = D_B = 10^{-5}\,\mathrm{cm^2}$ $\mathrm{s^{-1}}$. The six concentration profiles shown (A–F) correspond to the dotted potentials marked on the voltammograms. The solid line represents the concentration of A, whilst the dotted line shows that of B.

depletion of B relative to point (D). Again, conservation of matter dictates equality of fluxes:

$$\frac{\partial [A]}{\partial x}\bigg|_{x=0} = -\frac{\partial [B]}{\partial x}\bigg|_{x=0}.$$

The profile of [A] has a minimum, whilst the concentration–distance variation of [B] shows a maximum.

(F) This corresponds to a point on the reverse peak beyond the maximum (E) and shows that the concentration of B is very close to zero at the electrode surface, whilst that of A has returned to almost its original value nearly that in the bulk solution.

Case (ii): Electrochemically reversible behaviour
Figure 4.6 shows the voltammogram simulated for $k^o = 1\ \text{cm s}^{-1}$ and the parameters specified above again along with the concentration–distance plots at eight different parts on the voltammetric wave. These correspond to similar points on the current peaks for the irreversible case, except that the potential necessary to bring about current flow are markedly different.

In the irreversible case, a significant potential over and above that thermodynamically required was needed to bring about the reduction of A to B in the forward-going scan and the oxidation of B to A in the reverse scan. In contrast, significant current flows at potentials around the formal potential of the A−B couple in the reversible case (Fig. 4.6).

In the 'reversible' limit, the electrode kinetics are so 'fast' (relative to the rate of mass transport — see below) that Nernstian equilibrium is attained at the electrode surface throughout the voltammogram:

$$A + e \rightleftarrows B.$$

The implication is that the concentrations of A and B at the electrode surface obey the Nernst equation:

$$E = E_f^0(A/B) - \frac{RT}{F} \ln \frac{[B]_0}{[A]_0},$$

where E is now the applied potential which defines the ratio of the surface concentrations $[A]_0$ and $[B]_0$ once $E_f^0(A/B)$ is specified. Examination of Fig. 4.6 shows how the concentration profiles and the surface concentrations change during the voltammogram. Noting that $\left(\frac{RT}{F}\right) = 25.7\text{mV}$ at 298 K, Fig. 4.6 shows the concentration profiles on the forward going sweep but at a potential $\left(\frac{RT}{F}\right)$ positive of the formal potential (A), then at the formal potential (B) and third at $\left(\frac{RT}{F}\right)$ negative of the formal potential (C). Note that as required by the Nernst equation the surface

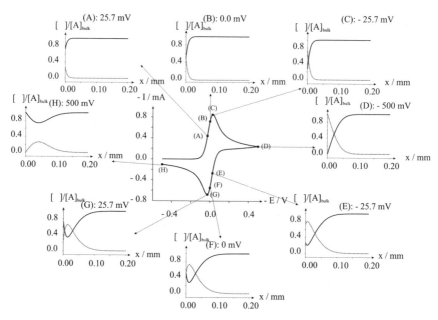

Fig. 4.6 Cyclic voltammogram for the reversible reduction of A to B. Parameters: $E^0 = 0\,V$; $\alpha = 0.5$; $k^0 = 1\,cm\,s^{-1}$; $v = 1\,Vs^{-1}$; $A = 1\,cm^2$; $[A]_0 = 1mM$; $D_A = D_B = 10^{-5}\,cm^2\,s^{-1}$. The concentration profiles show the distributions of A (solid line) and B (dashed line) at eight locations, A–H, on the voltammogram.

concentration ratios are as follows:

Points (A and G)

$$E = E_f^0 + \frac{RT}{F}$$

$$\frac{[B]_0}{[A]_0} = \frac{1}{e} = \frac{1}{2.7183},$$

Points (B and F)

$$E = E_f^0$$

$$\frac{[B]_0}{[A]_0} = 1,$$

Points (C and E)

$$E = E_f^0 - \frac{RT}{F}$$

$$\frac{[B]_0}{[A]_0} = e = 2.7183.$$

The current peak occurs at a potential *ca.* 29 mV negative of the formal potential. Fig. 4.6(C) shows the concentration profiles near this potential. Note that as with the irreversible case the increase in the thickness of the diffusion layer throughout the potential scan from -0.5 V through (A) to (B), to (E), to (D). On the reverse scan, the surface concentrations continue to obey the Nernst equation and so the calculation predicted above for parts (A), (B) and (C) also applies to (G), (F) and (E), respectively.

4.4 What Dictates 'Reversible' and 'Irreversible' Behaviour?

We stated above that reversible voltammetry was seen for 'fast' electrode kinetics, irreversible for 'slow'. However, 'fast' and 'slow' are relative terms and we must pose the question as to what they are fast or slow in relation to? The answer is the rate of mass transport to the electrode.

The rate of the electron transfer kinetics is measured by the standard electrochemical rate constant, k^0, whilst the rate of mass transport is measured by the mass transport coefficient,

$$m_T = \frac{D}{\delta},$$

where δ is the diffusion layer thickness. The latter depends on time, t, according to

$$\delta \sim \sqrt{Dt},$$

where we are interested in order of magnitude estimated, and so the factor of $\sqrt{\pi}$ seen in Chapter 3 in the context of the Cottrell equation has not been included. It is evident from the discussion at the end of the previous section that the 'time' taken to scan the voltammogram embracing the potentials at which there is current flow is similarly of an order of magnitude,

$$t \sim \frac{RT}{Fv},$$

where v is the voltage scan rate. It follows that, again as an order of magnitude estimate,

$$m_T \sim \sqrt{\frac{D}{(RT/Fv)}}$$

for the cyclic voltammetric experiment.

The distinction between fast and slow electrode kinetics, relates to the prevailing rate of mass transport given by

$$k^0 \gg m_T \quad \text{(reversible)}$$

or

$$k^0 \ll m_T \quad \text{(irreversible)}.$$

It follows that the transition between the reversible and irreversible limits can be followed by means of the parameter, Λ, introduced first in a classical paper by Matsuda and Ayabe:[6]

$$\Lambda = \frac{k^0}{\left(\frac{FDv}{RT}\right)^{1/2}}. \tag{4.5}$$

Matsuda and Ayabe suggested the following ranges for the three different classification at stationary macroelectrodes:

Reversible

$$\Lambda \geq 15 \quad \Lambda \geq 0.3v^{1/2} \text{cm s}^{-1}$$

Quasi-reversible

$$15 > \Lambda > 10^{-3} \quad 0.3v^{1/2} > k^0 > 2 \times 10^{-5}v^{1/2} \text{ cm s}^{-1}$$

Irreversible

$$\Lambda \leq 10^{-3} \quad k^0 \leq 2 \times 10^{-5}v^{1/2},$$

where the numerical values relate to 298 K and we have assumed $\alpha \sim 0.5$.

4.5 Reversible and Irreversible Behaviour: The Effect of Voltage Scan Rate

The Matuda–Ayabe conditions given in Section 4.4 show that for a given electrochemical rate constant the observed reversible or irreversible behaviour depends on the voltage scan rate and that for a sufficiently fast scan rate, at least in principle, all processes can appear to be electrochemically irreversible. Why is this?

We have seen in Figs. 4.5 and 4.6 that as the voltammogram is swept from E_1 to E_2 (Fig. 4.1), the thickness of the diffusion layer surrounding the electrode increases. The longer the time taken to scan the voltammogram, the thicker the diffusion layer. Conversely, the faster the voltage sweep rate, the thinner the diffusion layer. The thickness of the diffusion layer, as noted in the previous section, controls the rate of mass transport to the electrode, as parameterised by the mass transport coefficient, but ultimately reflects Ficks 1st law which predicts greater fluxes for a fixed concentration drop over a thinner diffusion layer. Since the 'reversible versus irreversible' distinction reflects competition between the electrode kinetics and the mass transport, it follows that faster scan rates will encourage greater electrochemical irreversibility. Consequently, it is unwise to refer to any redox couple as being electrochemically 'reversible' or 'irreversible', since all will tend to be irreversible if the voltammetry can be undertaken at sufficiently fast scan rates. Indeed, we will see in the next chapter that with microelectrodes scan rates in excess of $10^6 \, \text{V s}^{-1}$ can be realised.

The fact that increased scan rates lead to enhanced fluxes is also apparent from the variation of the peak current, I_p, with voltage scan rate. For both reversible and irreversible processes, a square root dependence is seen for the simple one-electron reduction of A to B:

Reversible $\qquad I_p = 0.446 FA[A]_{bulk} \sqrt{\frac{FD\upsilon}{RT}}$

Leading to $\qquad I_p = 2.69x10^5 AD^{1/2}[A]_{bulk}\upsilon^{1/2}$ $\qquad$ at 298 K

Irreversible $\qquad I_p = 0.496\sqrt{\alpha}FA[A]_{bulk}\sqrt{\frac{FD\upsilon}{RT}}$

Leading to $\qquad I_p = 2.99x10^5 \sqrt{\alpha}D^{1/2}[A]_{bulk}A\upsilon^{1/2}$ $\qquad$ at 298 K.

It is illuminating to consider the concentration profiles associated with voltammograms run at two different voltage scan rates. Figure 4.7 shows the concentration profiles for an electrochemically reversible system run at 100 mV s^{-1} and at 10 V s^{-1}, all other parameters kept constant.

The greater thickness of the diffusion layer is evident at the lower scan rate and this is manifested in a peak current which is ten times smaller than that seen at the faster scan rates.

It was pointed out above that the reversibility or irreversibility of a redox couple was a function of voltage scan rate with a switch over for $\Lambda \sim 1$. It follows that the equations given above for I_p will only apply in the correct domains. Figure 4.8 shows how the peak current increases with scan rate for a system with $A = 1 \, \text{cm}^2$, $D_A = D_B = 10^{-5} \, \text{cm}^2 \, \text{s}^{-1}$, $[A]_{bulk} = 10^{-3}$M and $k^o = 10^{-2} \, \text{cm s}^{-1}$.

The data is plotted as a log–log graph and the expected slopes of (1/2) are obtained in the reversible (low scan rate) and irreversible (fast scan rate) limits as predicted by the equations above. Note that the scan rate where $\Lambda = 1$ is marked and that this corresponds to a zone of switch over between the two limits. Here, the peak current does not exactly scale with the square root of the scan rate.

Returning to Fig. 4.2 we see that the major difference between a reversible and an irreversible voltammogram is the potential separation between the two peaks, ΔE_{pp}, where

$$\Delta E_{pp} = |E_p(anodic) - E_p(cathodic)|$$

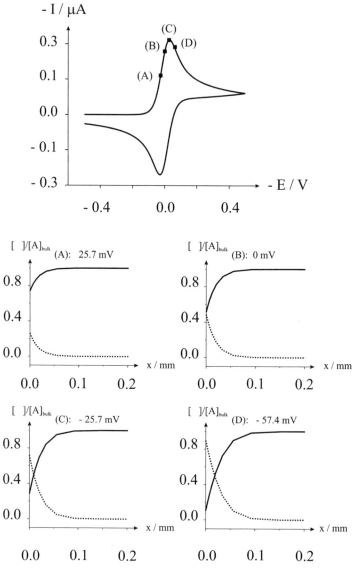

Fig. 4.7(A) Cyclic voltammogram for the reversible reduction of A to B. Parameters: $E^0 = 0\,V$; $\alpha = 0.5$; $k^0 = 1\,cm\,s^{-1}$; $\upsilon = 0.1\,Vs^{-1}$; $A = 1\,cm^2$; $[A]_0 = 1\,mM$; $D = 10^{-5}\,cm^2\,s^{-1}$. The concentration profiles show the distributions of A (solid line) and B (dashed line) at four locations, A–D, on the voltammogram.

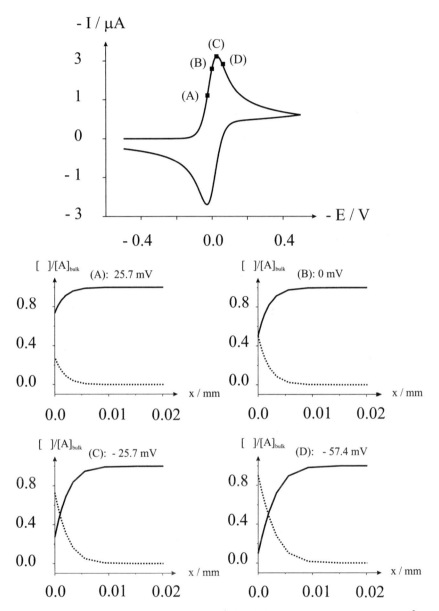

Fig. 4.7(B) Cyclic voltammogram for the reversible reduction of A to B. Parameters: $E^0 = 0\,\text{V}$; $\alpha = 0.5$; $k^0 = 1\,\text{cm s}^{-1}$; $\upsilon = 10\,\text{Vs}^{-1}$; $A = 1\,\text{cm}^2$; $[A]_0 = 1\,\text{mM}$; $D = 10^{-5}\,\text{cm}^2\,\text{s}^{-1}$. The concentration profiles show the distributions of A (solid line) and B (dashed line) at four locations, A–D, on the voltammogram.

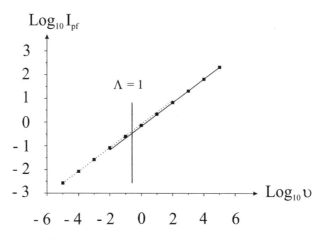

Fig. 4.8 Forward peak current (i_{pf}) versus scan rate for the simple one-electron reduction of A to B. Parameters: $E^0 = 0\,V$; $\alpha = 0.5$; $k^0 = 10^{-2}\mathrm{cm\ s^{-1}}$; $\upsilon = 1\,\mathrm{Vs^{-1}}$; $A = 1\,\mathrm{cm^2}$; $[A]_0 = 1\,\mathrm{mM}$; $D_A = D_B = 10^{-5}\ \mathrm{cm^2\ s^{-1}}$. The dotted line shows the reversible limit and the solid line the irreversible limit.

and $E_p(anodic)$ and $E_p(cathodic)$ are the potential of the peaks for the oxidation of B and the reduction of A, respectively. As can be anticipated from the earlier discussion, ΔE_{pp} is a function of the parameter Λ and also the transfer coefficient $\alpha\ (= 1 - \beta)$, although as pointed out in Chapter 2 the latter is often close to 1/2.

Figure 4.9 shows the simulation of some voltammograms with different values of Λ and assuming $\alpha = \beta = 1/2$. Note that the value of E_1 and E_2 are such that they do not influence the shape of the voltammogram (see later in this chapter). It is clear that the lower the value of Λ, the greater the peak-to-peak separation.

Moreover, by fitting experimental voltammograms of varying scan rates so as to give different Λ values, it is possible to evaluate k^0, the standard electrochemical rate constant. Figure 4.10 shows simulated data for $k^0 = 10^{-2}\mathrm{cm\ s^{-1}}$ showing how ΔE_{pp} changes, at low scan rates, from the reversible limit where

$$\Delta E_{pp} = 2.218\frac{RT}{F}$$
$$= 57\,\mathrm{mV}\quad(298\ \mathrm{K}),$$

to, at high scan rates, the irreversible limit where

$$\Delta E_{pp} = \frac{RT}{\alpha F}\ln \upsilon + constant$$

$$\Delta E_{pp} = \frac{59.4}{\alpha}\log_{10}\upsilon + constant\quad(\mathrm{mV}, 298\ \mathrm{K}).$$

The relative nature of terms such as reversible and irreversible is again evident as is the scope for measuring the values of k^0.

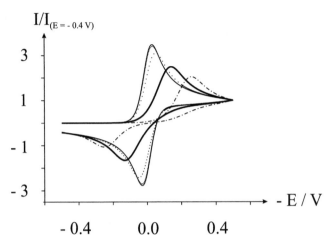

Fig. 4.9 Overlay of voltammograms at five values of Λ (100, 10, 1, 0.1 and 0.01) for the reduction of A to B. Parameters: $E_f^0 = 0\,\mathrm{V}$; $\alpha = 0.5$; $A = 1\,\mathrm{cm}^2$; $[A]_0 = 1\,\mathrm{mM}$; $D_A = D_B = 10^{-5}\,\mathrm{cm}^2\,\mathrm{s}^{-1}$. Note that the current has been normalised to its value at the limit of the potential scan.

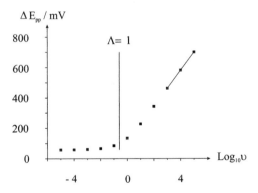

Fig. 4.10 Peak-to-peak potential versus scan rate for the reduction of A to B. Parameters: $E_f^0 = 0\,\mathrm{V}$; $\alpha = 0.5$; $k^0 = 10^{-2}\,\mathrm{cm}\,\mathrm{s}^{-1}$; $A = 1\,\mathrm{cm}^2$; $[A]_0 = 1\,\mathrm{mM}$; $D_A = D_B = 10^{-5}\,\mathrm{cm}^2\,\mathrm{s}^{-1}$.

4.6 Reversible versus Irreversible Voltammetry: A Summary

It is helpful to summarise some diagnostics which can indicate whether a particular voltammogram corresponds to the reversible or irreversible limit. Note that under quasi-reversible conditions neither will apply, but typically if a system is studied under a range of scan rates as wide as possible one or other limit may be reached. In any case, it is essential experimental practice to record voltammograms over a wide scan rate range. Three diagnostics can be usefully identified.

a) Peak-to-peak separation, ΔE_{pp}
In the reversible limit $\Delta E_{pp} \approx 57$ mV (at 298 K) and is independent of scan rate. Under quasi- and irreversible conditions, ΔE_{pp} depends on the voltage scan rate.

b) Peak current, I_p
In both limits, the peak current varies with the square root of the voltage scan rates but with a different coefficient of proportionality. The dependence does not hold in the quasi-reversible limit.

c) Waveshape of the forward peak
This can be usefully characterised by means of the difference in potential between the potential, E_p, corresponding to the peak current, and that, $E_{1/2}$, corresponding to half the peak current. For a reversible system,

$$|E_p - E_{1/2}| = 2.218\frac{RT}{F},\tag{4.6}$$

whereas for an irreversible reduction

$$|E_p - E_{1/2}| = 1.857\frac{RT}{\alpha F}\tag{4.7}$$

or

$$|E_p - E_{1/2}| = \frac{47.7}{\alpha}\text{ mV}\quad\text{(at 298 K)}$$

and for an irreversible oxidation

$$|E_p - E_{1/2}| = 1.857\frac{RT}{\beta F}.\tag{4.8}$$

The formal potential $E_f^0(A/B)$ may, in all cases, be found as the potential midway between the two peaks provided $D_A = D_B$. The case of unequal diffusion coefficients is considered later in this chapter.

First, however, we consider issues relating to the recording of cyclic voltammograms suitable for quantitative analysis.

4.7 The Measurement of Cyclic Voltammograms: Three Practical Considerations

In making experimental measurements and recording data, it is essential to be wary that the voltammograms recorded will be compatible with ready quantitative analysis. Accordingly, voltammetric experiments require a large amount of operator knowledge, carefulness and, in particular, intervention in order to obtain good or even satisfactory results. It is this feature that typically discriminates the professional researcher from the naive. This section addresses three issues that are not unknown to arise.

The first concerns the selection of the interval $|E_1 - E_2|$ over which the voltammogram is recorded. It is essential that this is wide in comparison with the voltammetric features recorded for the purpose of quantitative analysis. It is preferable that E_1 and E_2 are selected by trial and error, if necessary, so that their values do not influence the shape of the voltammogram recorded. Figure 4.11 shows reversible voltammograms recorded with a suitable value of E_1, such as to have no impact on the voltammetric waveshape, but with different values of E_2.

It can be seen that if E_2 is not selected sufficiently negative, in the case of a reduction, to the forward peak, the peak potential of the reverse (oxidative) peak becomes a function of E_2. Of course, this can be simulated with appro-

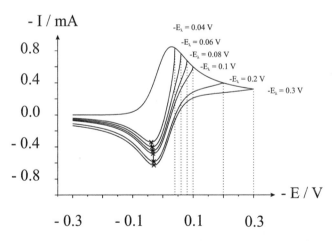

Fig. 4.11 Cyclic voltammograms for the reduction of A to B at varying switching potentials, E_2. Parameters: $E_f^0 = 0\,\text{V}$; $\alpha = 0.5$; $k^0 = 10^{-2}\,\text{cm s}^{-1}$; $\upsilon = 1\,\text{V s}^{-1}$; $A = 1\,\text{cm}^2$; $[A]_0 = 1\text{mM}$; $D_A = D_B = 10^{-5}\,\text{cm}^2\,\text{s}^{-1}$.

priate software but the simple criteria described in the previous section are compromised.

Second, it is essential to note that the criteria established in the previous section relate to the first voltammetric scan. If the triangular sweep is cycled several times to allow the voltammogram to approach steady-state, then the final voltammogram is different from the first. Provided E_1 and E_2 have been set correctly (see above), this is a small but significant difference, and the diagnostics given in the previous section should not be expected to apply. Figures 4.12 and 4.13 show the voltammograms resulting from second cycles — notice both the peak currents (for the forward and reverse scans) as well as the peak potential vary.

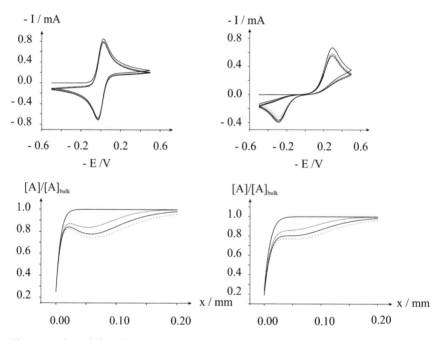

Fig. 4.12 (top left) The first four voltammetric scans for the reversible reduction of A to B. Parameters: $E_f^0 = 0\,\text{V}$; $\alpha = 0.5$; $k^0 = 1\,\text{cm s}^{-1}$; $\upsilon = 1\,\text{V s}^{-1}$; $A = 1\,\text{cm}^2$; $[A]_0 = 1\,\text{mM}$; $D_A = D_B = 10^{-5}\,\text{cm}^2\,\text{s}^{-1}$. Concentration profiles (bottom left) at $E = -29\,\text{mV}$ for the reversible reduction of A to B.

Fig. 4.13 (top right) The first four voltammetric scans for the irreversible reduction of A to B. Parameters: $E_f^0 = 0\,\text{V}$; $\alpha = 0.5$; $k^0 = 10^{-4}\,\text{cm s}^{-1}$; $\upsilon = 1\,\text{V s}^{-1}$; $A = 1\,\text{cm}^2$; $[A]_0 = 1\,\text{mM}$; $D_A = D_B = 10^{-5}\,\text{cm}^2\,\text{s}^{-1}$. Concentration profiles (bottom right) at $E = -293\,\text{mV}$ for the irreversible reduction of A to B.

Also shown are the concentration–distance plots for the reactive species A at the formal peak potential. The very significant difference over just four cycles emphasises the need that quantitative data should be obtained from a solution in which the concentration is initially uniform. An experimental implication is that if several voltammograms are first run as 'sighters', then those finally recorded for quantitative analysis should be measured in a freshly stirred solutions.

The third issue to be addressed concerns the following apparently simple question: 'How big is my back peak?' At the obvious level, it is straightforwardly the current measured on the voltammogram at the potential of the reverse peak, and this is, of course, entirely appropriate for study via simulation. However, in the early pre-simulation literature, much use is made of the ratio of the forward and back peaks in order to establish the stability or otherwise of the electro-generated species B. This criterion required the extrapolation of the forward scan in time as shown in Fig. 4.14.

With this extrapolation, the ratio of the cathodic and anodic peak current

$$\frac{I_{p,c}}{I_{p,a}} = 1,$$

if B is stable, but less than unity otherwise. Historically, these ratios were used extensively in the determination of homogeneous coupled chemistry (see Chapter 8) but are unnecessary in the era of simulation-based voltammetry. Needless to say, the non-linear extrapolation required in the old method was typically not unambiguous!

4.8 The Effect of Unequal Diffusion Coefficients, $D_A \neq D_B$

The analysis hitherto, and in particular the diagnostics presented in Section 4.5 have presumed that both A and B have equal diffusion coefficients. One consequence of equality of diffusion coefficient is that, in the absence of any coupled chemical reactions, the local concentrations at any point sum to the bulk concentrations of A (assuming that of B is zero):

$$[A] + [B] = [A]_{bulk} \qquad (4.9)$$

thus follows, since if

$$\frac{\partial [A]}{\partial t} = D_A \frac{\partial^2 [A]}{\partial x^2}$$

and

$$\frac{\partial [B]}{\partial t} = D_B \frac{\partial^2 [B]}{\partial x^2},$$

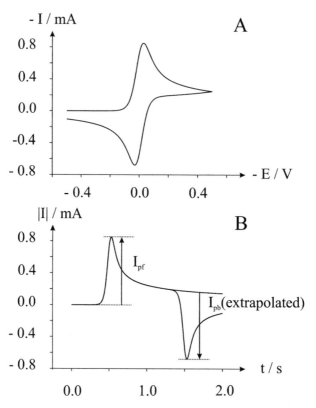

Fig. 4.14 (A) cyclic voltammogram for the reversible reduction of A to B. (B) Cyclic voltammograms in the form of current plotted against time rather than potential for the reversible reduction of A to B. Parameters: $E_f^0 = 0\,V$; $\alpha = 0.5$; $k^0 = 10^{-2}\,cm\,s^{-1}$; $v = 1\,Vs^{-1}$; $A = 1\,cm^2$; $[A]_0 = 1mM$; $D_A = D_B = 10^{-5}\,cm^2\,s^{-1}$.

then

$$\frac{\partial\{[A] + [B]\}}{\partial t} = D\frac{\partial^2\{[A] + [B]\}}{\partial x^2}, \tag{4.10}$$

if $D = D_A = D_B$. The boundary conditions for the cyclic voltammetric experiment include

$$x \to \infty, \quad \text{all } t, \quad [A] = [A]_{bulk},$$
$$[B] = 0$$
$$x = 0, \quad \text{all } t, \quad D\frac{\partial[A]}{\partial x}\Big|_0 = -D\frac{\partial[B]}{\partial x}, \quad \text{or} \quad \frac{\partial\{[A]+[B]\}}{\partial x}\Big|_0 = 0.$$

Equation (4.9) follows on solution of Eq. (4.10) subject to these boundary conditions.

When $D_A \neq D_B$, then

$$[A] + [B] \neq [A]_{bulk}.$$

An extreme case of inequality of diffusion coefficients in a simple electrode process arises in the reduction of molecular oxygen

$$O_2 + e^- \rightarrow O_2^{\bullet-}$$

in the room temperature ionic liquid (see Section 5.4) hexyltriethylammonium bis(trifluoromethyl)sulfonyl)imide[7] where at 25°C

$$D_{O_2} = 1.48 \times 10^{-10} \, m^2 s^{-1}$$

$$D_{O_2^{\bullet-}} = 4.66 \times 10^{-12} \, m^2 s^{-1}$$

corresponding to difference of a factor of over 30! The contrast arises from the relative speeds of diffusion of the charged and neutral species in the entirely ionic media. Figure 4.15 shows the concentration profiles of O_2 and $O_2^{\bullet-}$ at different points on the electrochemically reversible cyclic voltammogram for a scan rate of $97.8 \, mV \, s^{-1}$.

Notice that the slower moving superoxide ion builds up significantly near the electrode surface so that

$$[O_2^{\bullet-}] \gg [O_2]_{bulk}.$$

Moreover, because of the slow loss of $O_2^{\bullet-}$ from the interface, significant memory effects can be expected: Fig. 4.15(F) shows a 'layer' of superoxide *ca.* 10 μm from the electrode after the potential sweep has returned to its initial value.

In practice, for many electrode processes the assumption of equal diffusion coefficients is a reasonable approximation; in the following chapter the use of potential step chronoamperometry for the measurement of D_A and D_B is described, and this allows an assessment of the validity of this assumption. The most important circumstances when inequality of diffusion coefficients needs to be recognised is in the calculation of formal potentials from cyclic voltammograms. When $D_A = D_B$, the formal potential of the A/B couple is the potential corresponding to the mid point between the two peaks in the voltammogram corresponding to the conversion of A to B and its reverse. Inequality of diffusion coefficient has different effects in the electrochemically reversible and irreversible limits.

In the reversible limit for the process

$$A + e \rightleftarrows B,$$

the mid point potential corresponds to

$$E_{mid} = E_f^0 + \frac{RT}{2F} \ln \frac{D_B}{D_A},$$

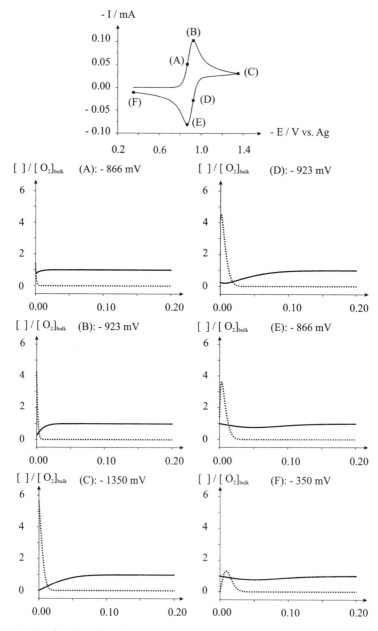

Fig. 4.15 Simulated cyclic voltammetry and concentration profiles for the reversible reduction of O_2 to $O_2^{\bullet-}$ in $[N_{6222}][N(Tf)_2]$. Parameters: $E_f^0 = -0.85\,V$ versus Ag; $\alpha = 0.5$; $\Lambda = 1000$; $A = 1\,cm^2$; $[O_2]_{bulk} = 1\,mM$; $D_{O_2} = 1.48 \times 10^{-6}\,cm^2\,s^{-1}$; $D_{O_2^{\bullet-}} = 4.66 \times 10^{-8}\,cm^2\,s^{-1}$. In the concentration profiles the solid line represent O_2 and the dotted line $O_2^{\bullet-}$. The x-axis scale is measured in mm.

so that if $D_B \gg D_A$ the reductive process in the cyclic voltammogram occurs at less negative potential then when $D_A = D_B$. One way of appreciating this is to see the enhanced diffusion of B relative to A as 'pulling' the electrochemical equilibrium over in favour of the products following Le Chatelier's principle.

In the irreversible limit, provided the potential limits of the voltammogram are sufficiently wide as not to influence the peak potentials, if

$$\alpha = \beta = 1/2,$$

the midpoint potential is

$$E_{mid} = \frac{E_{p,forward} + E_{p,reverse}}{2} = E_f^0 + \frac{RT}{F} \ln \frac{D_B}{D_A}.$$

4.9 Multiple Electron Transfer: Reversible Electrode Kinetics

We will consider the following two-step reduction before generalising to the case of n-electron transfer:

$$A + e \rightleftharpoons B \quad E_f^0(A/B)$$
$$B + e \rightleftharpoons C \quad E_f^0(B/C).$$

In the case that the electrode kinetics of both couples are fast compared to mass transport and so the electrochemically reversible limit operates, then what is seen in the voltammetry of the solution containing only A in bulk solution depends on the relative magnitude of the formal potentials $E_f^0(A/B)$ and $E_f^0(B/C)$.

If B is more easily reduced than A so that

$$E_f^0(B/C) \gg E_f^0(A/B),$$

it can be appreciated that the voltammetric reduction of A leads to the formation of B at the surface of an electrode which is at a potential well negative of that required to reduce B to C. Accordingly, a single voltammetric wave is seen, corresponding to the net conversion of A to C, an overall two-electron process. On the other hand, when

$$E_f^0(A/B) \gg E_f^0(B/C),$$

two voltammetric waves will be seen, the first at relatively positive potentials corresponding to the reduction of A to B and the second at relatively negative potentials due to the conversion of B to C. This arises since the potentials at which A is converted to B are insufficiently negative to reduce B; only after the potential has been

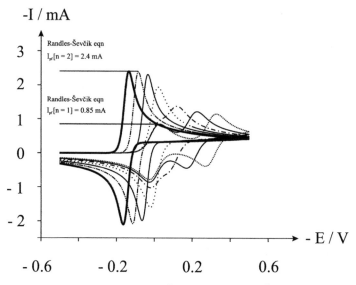

Fig. 4.16 The observed voltammetry for $E_f^0(A/B) = 0\,\text{V}$ and $E_f^0(B/C) = +0.3\,\text{V}, +0.2\,\text{V},$ $+0.1\,\text{V}, 0.0, -0.1\,\text{V}, -0.2\,\text{V}, -0.3\,\text{V}$ assuming both redox couples display electrochemically reversible behaviour with $A = 1\,\text{cm}^2$, $[A]_{bulk} = 10^{-3}\,\text{M}$, $D_A = D_B = 10^{-5}\,\text{cm}^2\,\text{s}^{-1}$ with scan rate of $1\,\text{V}\,\text{s}^{-1}$.

scanned sufficiently negative to values in the vicinity of $E_f^0(B/C)$ will the second wave be seen. Figure 4.16 shows the development of two waves from one merged wave as the difference in formal potentials changes sign.

Examples of separate waves include the following three examples in aprotic solvents such as dichloromethane and acetonitrile.

Example (1)

Example (2)

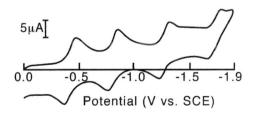

Example (3)

$$C_{60} + e \rightleftarrows C_{60}^{\bullet -}$$

$$C_{60}^{\bullet -} + e \rightleftarrows C_{60}^{2-}$$

$$C_{60}^{2-} + e \rightleftarrows C_{60}^{3-}$$

$$C_{60}^{3-} + e \rightleftarrows C_{60}^{4-}$$

The corresponding voltammograms are shown in Fig. 4.17.

Fig. 4.17(A) Cyclic voltammogram of C_{60}.

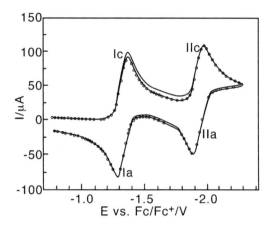

Fig. 4.17(B) Cyclic voltammograms of 1.07 mM anthraquinone at $1\,V\,s^{-1}$ in 0.1 M Bu$_4$NPF$_6$ in acetonitrile (full curve) on a glassy carbon electron. Simulation of 70 Ω of resistance included (symbols). 140 Ω was compensated electronically by positive feedback. Reprinted from Ref. [8] with permission from Elsevier.

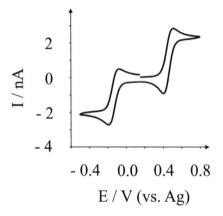

Fig. 4.17(C) Cyclic voltammogram of the radical cation of tetramethyl-p-phenylene diamine in [EMIM][N(Tf)2]. Scan rate $200\,mV\,s^{-1}$. The scan starts from $+0.2\,V$ in an oxidising direction forming first the di-cation ($\sim +0.5\,V$) and then the neutral molecule ($\sim -0.2\,V$). Reprinted from Ref. [9] with permission from Wiley.

In contrast, the reduction of anthraquinone in aqueous solution shows a two-electron reduction for all pHs below *ca.* 10:

Similar considerations in respect of relative values of formal potentials apply to the observation of merged or separate voltammetric waves in the case of n-electron process. Where a single n-electron electrochemically reversible reduction is observed

$$A + ne^- \rightleftharpoons products,$$

the voltammetric wave can be characterised as follows:

(a) The peak current is given by

$$I_{peak} = -0.446nFA[A]_{bulk}\sqrt{\frac{nFvD}{RT}}.$$

Note the dependence on $n^{3/2}$, where n is the total number of electrons transferred.

(b) The peak-to-peak separation between the formal and reverse peaks is

$$\Delta E_{pp} = \frac{2.218RT}{nF}$$

$$\Delta E_{pp} = \frac{57}{n}\text{mV} \quad \text{at 298 K.}$$

Note that the quantity is independent of the voltage scan rate.

(c) The voltammetric waveshape can be characterised by the difference in the peak potential, E_p, and the potential at which the current is one half of the peak current, $E_{p/2}$:

$$|E_{p/2} - E_{1/2}| = 2.218\frac{RT}{nF} = 2|E_p - E_{mid}|,$$

where E_{mid} is the potential midway between the two peaks, forward (f) and reverse (r),

$$E_{mid} = \frac{E_{peak,f} + E_{peak,r}}{2} = E_f^0(A/B) + \frac{RT}{2F}\ln\frac{D_B}{D_A}.$$

Last, we return to the case of separate electron reversible waves and consider the case of $n = 2$;

$$A + e^- \rightleftharpoons B$$
$$B + e^- \rightleftharpoons C,$$

where, since two waves are by hypothesis present,

$$E_f^0(A/B) \gg E_f^0(B/C).$$

It follows from this that the reaction

$$A + C \rightleftarrows 2B$$

is thermodynamically downhill, since

$$\Delta G^o \simeq -F\{E_f^0(A/B) - E_f^0(B/C)\}$$
$$\ll 0.$$

Accordingly, when C is produced in the second voltammetric wave it can, at least in principle, react with A to form B. The question arises as to how this is apparent in the voltammetry? The surprising answer is that, provided the waves are electrochemically reversible and $D_A = D_B = D$, then the voltammetry is blind to whether or not disproportionation occurs! The relevant analysis was given in a classic paper by Andrieux and Savéant [10]. We follow their approach here.

Let us suppose the following arbitrary second order reaction constituting

$$B + B \xrightarrow{k_1} A + C$$

and

$$A + C \xrightarrow{k_2} 2B.$$

The equilibrium constant,

$$K = \frac{[A][C]}{[B]^2} = \exp\left\{\frac{F}{RT}(E_f^0(B/C) - E_f^0(A/B))\right\} = \frac{k_1}{k_2}.$$

In this case, Fick's laws of diffusion must be modified to allow for the homogeneous kinetics (see Chapters 3 and 7):

$$\frac{\partial[A]}{\partial t} = D_A \frac{\partial^2[A]}{\partial x^2} + k_1[B]^2 - k_2[A][C] \tag{4.11}$$

$$\frac{\partial[B]}{\partial t} = D_B \frac{\partial^2[B]}{\partial x^2} - 2k_1[B]^2 + 2k_2[A][C] \tag{4.12}$$

$$\frac{\partial[C]}{\partial t} = D_C \frac{\partial^2[C]}{\partial x^2} + k_1[B]^2 - k_2[A][C] \tag{4.13}$$

The boundary conditions are as follows

$$t = 0, \text{all} x \qquad [A] = [A]_{bulk}, \quad [B] = [C] = 0$$

$$\text{all } t, x \to \infty \qquad [A] = [A]_{bulk}, \quad [B] = [C] = 0$$

$$x = 0, \text{ all } t \qquad D_A \frac{\partial[A]}{\partial x} + D_B \frac{\partial[B]}{\partial x} + D_C \frac{\partial[C]}{\partial x} = 0$$

$$\text{and} \qquad \frac{[A]_{x=0}}{[B]_{x=0}} = \exp\left[\frac{F}{RT}\{E - E_f^0(A/B)\}\right]$$

$$\frac{[A]_{x=0}}{[C]_{x=0}} = \exp\left[\frac{F}{RT}\{E - E_f^0(B/C)\}\right],$$

where E is the potential applied to the working electrode.

The current is

$$i = FA \left[2D_A \left. \frac{\partial [A]}{\partial x} \right|_{x=0} + D_B \left. \frac{\partial [B]}{\partial x} \right|_{x=0} \right].$$

In the case that $D_A = D_B = D$ we can write

$$\frac{\partial (2[A] + [B])}{\partial t} = D \frac{\partial^2}{\partial x^2} (2[A] + [B]),$$

from Eqs. (4.11) and (4.12). This equation can be solved using the boundary conditions

$t = 0$, all x $2[A] + [B] = 2[A]_{bulk}$

all t, $x \to \infty$ $2[A] + [B] = 2[A]_{bulk}$

$x = 0$, $t \geq 0$:

$2[A] + [B]$

$$= \frac{[A]_{bulk} \{ 2 + \exp(-F/RT)[E - E_f^0(A/B)] \}}{1 + \exp(-F/RT)[E - E_f^0(A/B)] + \exp(-2F/RT)[E - 1/2(E_f^0(A/B) + E_f^0(B/C)]}.$$

Since the current is given by

$$I = FAD \left[\frac{2\partial [A]}{\partial x} + \frac{\partial [B]}{\partial x} \right]_{x=0},$$

this expression cannot depend on k_2 (or k_1). It follows that the voltammetric response (I versus E) is completely independent of the existence and kinetics of the disproportionation reaction in solution: exactly the same voltammograms are seen irrespective of whether k_2 is zero or fast, provided $D_A = D_B$ and the two-electrode processes (A/B and B/C) are both electrochemically reversible. Experimental methods other than voltammetric must be deployed in order to probe the magnitude of k_2; of particular inportance in this context are spectroscopic methods coupled to voltammetry. Andrieux and Savéant deployed electron spin resonance in their classic study for this purpose.

Simulation allows the exploration of the voltammetry when disproportionation occurs, and when it does not. Figure 4.18 shows the concentration profiles for those values of k_2 corresponding to zero, intermediate and fast homogenous kinetics; specifically, second order rate constants of 0 and 10^8 cm^{3}mol^{-1} s^{-1} were used to generate the data for a sweep rate of $1\,\text{V}\,\text{s}^{-1}$ along with the parameters: electrode area = 1 cm^2, $D_A = D_B = 10^{-5}$ cm^2 s^{-1}, $[A]_{bulk} = 10^{-3}$ M.

The electrochemical rate constant for both couples was set in the electrochemically reversible range. It can be seen for the two cases that whilst *exactly* the same

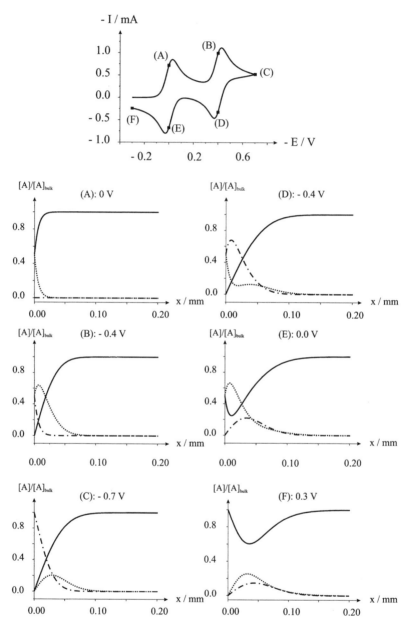

Fig. 4.18(A) Cyclic voltammogram for the two-electron reversible reduction of A to B to C with $k_1 = 0\,cm^3\,mol^{-1}\,s^{-1}$. Parameters: $E_f^0(A/B) = 0\,V$, $E_f^0(B/C) = -0.4\,V$; $\alpha_1 = \alpha_2 = 0.5$; $k_1^0 = k_2^0 = 1\,cm\,s^{-1}$; $A = 1\,cm^2$, $[A]_{bulk} = 10^{-3}\,M$, $D_A = D_B = D_c = 10^{-5}\,cm^2\,s^{-1}$ with a scan rate of $1\,V\,s^{-1}$. The solid line shows the concentration of A, the dotted line that of B, and the dash-dot line that of C.

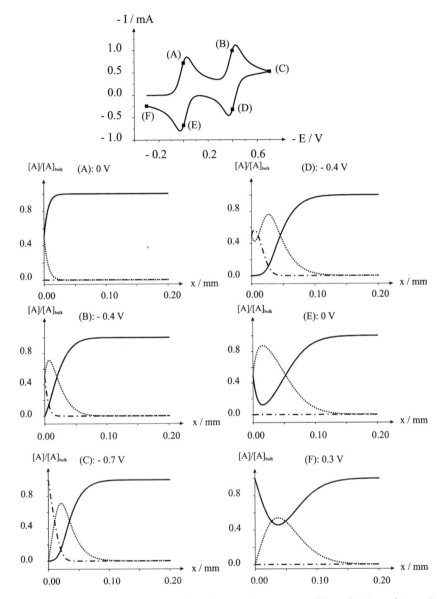

Fig. 4.18(B) Cyclic voltammogram for the two-electron reversible reduction of A to B to C with $k_1 = 10^8 \text{ cm}^3 \text{ mol}^{-1} \text{ s}^{-1}$. Parameters: $E_f^0(A/B) = 0 \text{ V}$, $E_f^0(B/C) = -0.4 \text{ V}$; $\alpha_1 = \alpha_2 = 0.5$; $k_1^0 = k_2^0 = 1 \text{ cm s}^{-1}$; $A = 1 \text{ cm}^2$, $[A]_{bulk} = 10^{-3} \text{M}$, $D_A = D_B = D_c = 10^{-5} \text{ cm}^2 \text{ s}^{-1}$ with a scan rate of 1 V s^{-1}. The solid line shows the concentration of A, the dotted line that of B, and the dash-dot line that of C.

voltammogram is generated, the concentration profiles for A, B and C are very significantly different.

Last, we point out that if the diffusion coefficients differ, $D_A \neq D_B$ and/or one or both of the redox couples A/B and B/C are not fully reversible then the presence or absence of the disproportionation can become evident in the voltammetry as described by Rongfeng and Evans.[11]

4.10 Multiple Electron Transfer: Irreversible Electrode Kinetics

Let us consider the following scheme for the two-electron reduction of A to C:

$$A + e \underset{k_{1,a}}{\overset{k_{1,c}}{\rightleftarrows}} B$$

$$B + e \underset{k_{2,a}}{\overset{k_{2,c}}{\rightleftarrows}} C,$$

where the electrochemical rate constants are given by the Butler–Volmer equation

$$k_{1,c} = k_1^0 \exp\left[-\frac{\alpha_1 F}{RT}\{E - E_f^0(A/B)\}\right]$$

$$k_{1,a} = k_1^0 \exp\left[\frac{(1 - \alpha_1)F}{RT}\{E - E_f^0(A/B)\}\right]$$

$$k_{2,c} = k_2^0 \exp\left[\frac{-\alpha_2 F}{RT}\{E - E_f^0(B/C)\}\right]$$

$$k_{2,a} = k_2^0 \exp\left[\frac{(1 - \alpha_2)F}{RT}\{E - E_f^0(B/C)\}\right],$$

where E is the applied potential.

As in the previous section one merged or two separated voltammetric waves can be seen depending on the relative values of the electrode kinetics. If $k_{2,c} \gg k_{1,a}$ then the electrochemically irreversible reduction of A to C occurs, provided $k_{2,c}$ is slow compared to the rate of mass transport, either as two separate waves if

$$k_{1,c} > k_{2,a}$$

or as a single wave if

$$k_{2,c} > k_{1,c}.$$

When two waves are seen then each is characterised by its own value of α_1 or α_2, should Tafel analysis (see Chapter 3) be attempted on the rising part of the voltammetric peak. In the case that one merged wave is seen, Tafel analysis will give a value of α_1 consistent with the mechanism

$$A + e \xrightarrow{slow} B$$
$$B + e \xrightarrow{fast} C$$

with the first step rate determining.

Figure 4.19 shows the voltammograms obtained for a fixed value of $k_1^0 = 10^{-4} \, cm \, s^{-1}$ and $E_f^0(A/B) = -0.1 \, V$ with different values of k_2^0 but with $E_f^0(B/C)$ fixed at $+0.1 \, V$. When k_2^0 is tiny, two irreversible waves are seen but as it increases the expected merging of the two waves occurs. Note that the merged wave, when subjected to Tafel analysis [of the part of the voltammogram before the peak; more strictly ca the middle 50% of the rising curve], shows a Tafel slope reflecting a transfer coefficient of α_1 $(= 0.5$ in Fig. 4.19) consistent with the mechanism

$$A + e \xrightarrow{slow} B$$
$$B + e \longrightarrow C.$$

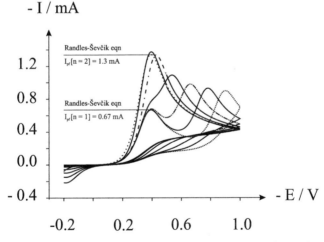

Fig. 4.19 Overlay of cyclic voltammograms for varying values of k_2^0 (10^{-4}, 10^{-5}, 10^{-6}, 10^{-7}, 10^{-8}, 10^{-9}, $10^{-10} \, cm \, s^{-1}$), for the two-electron reversible reduction of A to B to C. Parameters: $E_1 = 0.1 \, V$, $E_2 = 0.1 \, V$; $\alpha_1 = \alpha_2 = 0.5$; $k_1^0 = 10^{-4} \, cm \, s^{-1}$; $A = 1 \, cm^2$, $[A]_{bulk} = 10^{-3} \, M$, $D_A = D_B = D_c = 10^{-5} \, cm^2 \, s^{-1}$ with scan rate of $1 \, V \, s^{-1}$.

We next consider a second example. Figure 4.20 shows simulation for the case where a fast ($k_1^0 = 1\,\mathrm{cm\,s^{-1}}$, $\alpha_1 = 0.5$, $E_f^0(A/B) = -0.1\,\mathrm{V}$) electron transfer for the A to B conversion is followed by an electrochemically irreversible process ($k_2^0 = 10^{-5}\,\mathrm{cm\,s^{-1}}$, $\alpha_2 = 0.5$) for different values of $E_f^0(B/C)$. For very low values of $k_{2,c}$, two waves are seen: the first corresponds to the electrochemically reversible reduction of A to B; the second to the electrochemically irreversible reduction of B to C. If Tafel analysis is made of the two values — again taking the mid 50% of the rising part of the voltammogram, to avoid, near the peak, diffusional effects and, near the foot, measuring currents too close to the baseline — then following the discussion at the end of Chapter 3 in the case of the first wave an apparent transfer coefficient of unity will be obtained reflecting the fast electrode kinetics, whereas similar analysis of the second wave will record α_2 ($= 0.5$ in Fig. 4.20).

Referring again to Fig. 4.18, as the value of $E_f^o(B/C)$ is made more and more positive the two waves merge to produce one. First, a sharp single wave is produced (for $E_f^o(B/C)$ around $+0.6\,\mathrm{V}$) which on Tafel analysis shows that

$$\frac{\partial \ln i}{\partial E} = \frac{(1+\alpha_2)F}{RT},$$

where in the case illustrated in Fig. 4.20 corresponds to

$$1 + \alpha_2 \sim 3/2.$$

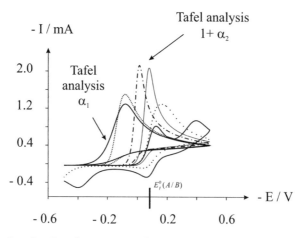

Fig. 4.20 Overlay of cyclic voltammograms for varying values of E_2 ($0, 0.2, 0.4, 0.6, 1.0$, and $1.5\,\mathrm{V}$) for the two-electron reversible reduction of A to B to C. Parameters: $E_f^0(A/B) = -0.1\,\mathrm{V}$; $\alpha_1 = \alpha_2 = 0.5$; $k_1 = 1\,\mathrm{cm\,s^{-1}}$; $k_2 = 10^{-5}\,\mathrm{cm\,s^{-1}}$ $A = 1\,\mathrm{cm^2}$, $[A]_{bulk} = 10^{-3}\,\mathrm{M}$, $D_A = D_B = D_c = 10^{-5}\,\mathrm{cm^2\,s^{-1}}$ with scan rate of $1\,\mathrm{V\,s^{-1}}$.

From this, the mechanism can be concluded to be

$$A + e \leftrightarrows B$$
$$B + e \xrightarrow{slow} C,$$

with the second step rate determining. On making $E_f^o(B/C)$ yet more positive the single wave shifts anodically and changes shape consistent with Tafel analysis, showing

$$\frac{\partial \ln I}{\partial E} = \frac{\alpha_1 F}{RT},$$

so that, for the system shown in Fig. 4.18,

$$\alpha_1 \sim 0.5.$$

This shows that the mechanism is

$$A + e \xrightarrow{slow} B$$
$$B + e \longrightarrow C,$$

where the first step is now rendered rate-determining and electrochemically irreversible by the increased rate constant for the removal of B.

It follows from the above that the observed voltammetry when two or more electrons can be transferred, is a subtle interaction of thermodynamic (E_f^o) and kinetic (k^0, v) effects and that simulation is essential to analyse all but some limiting cases. That said we conclude this section by briefly examining a general n-electron, electrochemically irreversible process. This can be written as follows:

$$
\left.
\begin{aligned}
&\left.
\begin{aligned}
A + e &\rightleftarrows B \\
B + e &\rightleftarrows C \\
&\vdots \\
&\vdots \\
&\vdots \\
L' + e &\rightleftarrows M' \\
M' + e &\rightleftarrows N'
\end{aligned}
\right\} n'\ electrons\ transferred \\
&N' + e \xrightarrow{slow} O' \\
&\qquad \vdots \\
&\qquad \vdots \\
&\qquad \vdots \\
&X + e \longrightarrow N
\end{aligned}
\right\} n\ electrons\ transferred,
$$

where the slow, rate-determining step is shown as the addition of the $(n' + 1)$th electron of the total n transferred. It follows that the rate of the overall process is

$$rate \propto k^0_{n'+1}[N']_0 \exp\left(-\frac{\alpha_{n'+1}F}{RT}\{E - E^0_f(N'/O')\}\right),$$

where $[N']_0$ is the concentration of N' at the electrode surface, $\alpha_{n'+1}$ is the transfer coefficient for the rate determining step and $k^0_{n'+1}$ the corresponding standard electrochemical rate constant. Since the steps preceding the rate determining step form pre-equilibria,

$$[N']_0 = [M']_0 \exp\left[-\frac{F}{RT}(E - E^0_f(M'/N')\right]$$

$$[M']_0 = [L']_0 \exp\left[-\frac{F}{RT}(E - E^0_f(L'/M')\right]$$

$$\vdots$$

$$\vdots$$

$$\vdots$$

$$\vdots$$

$$[B]_0 = [A]_0 \exp\left[-\frac{F}{RT}(E - E^0_f(A/B)\right].$$

It follows that

$$rate \propto [A]_o \exp\left(-\frac{(n' + \alpha_{n'+1})FE}{RT}\right), \tag{4.14}$$

so that, in a form appropriate for Tafel analysis (see Chapters 2 and 3),

$$\frac{\partial \ln I}{\partial E} = \frac{(n' + \alpha_{n'+1})F}{RT}$$

and the Tafel slopes gives a value of $(n' + \alpha_{n'+1})$, the sum of the number of electrons, n', transferred before the rate determining electron transfer and $\alpha_{n'+1}$ the transfer coefficient of this slow step.

The use of Eq. (4.14) as a boundary condition in the solution of the cyclic voltammetry problem leads to the following expression for the peak current,

$$I_{peak} = -0.496\sqrt{n' + \alpha_{n'+1}}\, nFA[A]_{bulk}\sqrt{\frac{FvD}{RT}}$$

for a totally irreversible n-electron wave, where the symbols have their usual significance. Similarly, the waveshape can be characterised by the difference between the peak potential, $E_{p/2}$ and the potential, $E_{1/2}$ corresponding to one half the maximum current:

$$|E_p - E_{p/2}| = \frac{1.857RT}{(n' + \alpha_{n'+1})F}$$

$$|E_p - E_{1/2}| = \frac{47}{(n' + \alpha_{n'+1})}\,mV \quad \text{at 298 K.}$$

Note that this expression is helpful for making preliminary estimates of $(n' + \alpha_{n'+1})$ prior to simulation.

Finally, a comment: in the case of a one-electron transfer process, it is true that

$$\alpha + \beta = 1,$$

so that if the transfer coefficients for reduction (α) is known, that for oxidation (β) is fixed. In the case of a multi-step mechanism with slow electron kinetics, the reductive and oxidative peaks occur at quite different potentials. So whilst Tafel analysis might, for example, indicate for the reduction waves a value of $(n' + \alpha_{n'+1})$, it does *NOT* follow that the transfer coefficient for the oxidation wave is $(n - n' - \alpha_{n'+1})$, because the rate-determining step of the mechanism can easily change with potential. The need for caution in approaching the analysis of multi-step processes is evident. In the next and final section, we consider the role of protons in such reactions when conducted in aqueous solutions.

4.11 The Influence of pH on Cyclic Voltammetry

Whilst the topic of coupled homogeneous chemistry will be addressed in Chapter 8 of this book, it is appropriate at this point to emphasise that voltammetric measurements carried out in aqueous solution are likely to be pH dependant since the addition or removal of an electron from an organic molecule in particular may induce the uptake or loss of a proton. In this section, we address the issue as to how the position of a chemically reversible voltammetric wave moves with pH as measured by the peak potential or by the potential midway between the two peaks in the voltammogram which would, in the case of the simple electrode process

$$A \pm e^- \rightleftarrows B,$$

correspond to the formal potential of the A/B couple assuming equality of diffusion coefficients D_A and D_B.

For generality, we consider a reduction involving the uptake of m-protons and consumption of n-electrons:

$$A + mH^+ + ne^- \rightleftarrows B.$$

The limiting cases correspond to those of electrochemical reversibility and irreversibility.

Case (a) In the case of the electrode process being fully electrochemically reversible, we can write for the relevant Nernst equation

$$E = E_f^0(A/B) - \frac{RT}{nF} \ln \frac{[B]}{[A][H^+]^m}$$

$$E = E_f^0(A/B) + \frac{RT}{nF} \ln[H^+]^m - \frac{RT}{nF} \ln \frac{[B]}{[A]}$$

$$E = E_f^0(A/B) - 2.303\frac{mRT}{nF}pH - \frac{RT}{nF} \ln \frac{[B]}{[A]}.$$

It can be seen that the quantity

$$E_{f,eff}^0 = E_f^0(A/B) - 2.303\frac{mRT}{nF}pH$$

acts as an 'effective' formal potential. Accordingly, the pH dependence of reversible system can be obtained by replacing $E_f^0(A/B)$ by $E_{f,eff}^0$ in the theoretical analysis given earlier in this chapter. It follows that provided $D_A = D_B$ the potential midway between the peaks for the reduction of A and the oxidation of B corresponds to $E_{f,eff}^0$ with the shape of the voltammogram being otherwise unaffected. Accordingly, the midpoint potential varies by an amount $2.303\frac{mRT}{nF}$ per pH unit. In the commonly seen case where $m = n$, this corresponds to *ca.* 59 mV per pH unit at 25°C. For example, in the case of the reduction of p-benzoquinone (BQ) to hydroquinone (HQ)

(BQ)   (HQ)

$m = 2, n = 2$ at pH values below *ca.* 9. Note that the upper limit corresponds to the appropriate pK_a (acid dissociation constant) of hydroquinone.

Case (b) In analysing the electrochemically irreversible case, we assume that the protons and electrons are transferred in separate steps:

$$A + mH^+ \rightleftharpoons AH_m^{m+}$$
$$AH_m^{m+} + ne \to B,$$

where n is the number of electrons transferred in the second step. In this case, we focus on the location of peak potential for the reduction of A.

The pre-equilibrium can be described by the following equilibrium constant

$$K = \frac{[A][H^+]^m}{[AH_m^{m+}]}.$$

If the total concentration of both A and AH_m^{n+} present is $[A]_{Total}$, then

$$[A]_{Total} = [A] + [AH_m^{m+}],$$

so that

$$[AH_m^{m+}] = \frac{[H^+]^m[A]_{Total}}{K + [H^+]^m}.$$

Under electrochemically irreversible conditions, we expect that the current is given by

$$I \propto [AH_m^{n+}]_o \exp\left(-\frac{(n' + \alpha)F}{RT}\eta\right), \qquad (4.15)$$

where α is the transfer coefficient and $n'(\leq n)$ is the number of electrons transferred before the rate-determining electron transfer step (see previous section). The overpotential is given by

$$\eta = E - E_f^0(AH_m^{n+}/B),$$

where the formal potential reflects the standard Gibbs energy difference between AH_m^{n+} and B, not between A and B.

We can rewrite Eq. (4.15) as

$$I \propto [A]_{Total}\frac{[H^+]^m}{K + [H^+]^m}\exp\left(-\frac{(n' + \alpha)F}{RT}\eta\right)$$

$$I \propto [A]_{Total}\exp\left(-\frac{(n' + \alpha)F}{RT}\eta + \ln\frac{[H^+]^m}{K + [H^+]^m}\right)$$

$$I \propto [A]_{Total}\exp\left(-\frac{(n' + \alpha)F}{RT}\eta'\right),$$

where

$$\eta' = E - E_f^0(AH_m^{n+}/B) + \frac{RT}{(n' + \alpha)F}\ln\left(\frac{[H^+]^m}{K + [H^+]^m}\right).$$

It follows that the peak potential E_p will show a pH dependence given by

$$E_p = const + \frac{RT}{(n' + \alpha)F}\ln\frac{[H^+]^m}{K + [H^+]^m}.$$

It is instructive to consider two limits of the equation.

Limit (i) $[H^+] \ll K$. In this situation,

$$E_p \sim const + \frac{RT}{(n'+\alpha)F} \ln \frac{[H^+]^m}{K}$$

$$E_p \sim const - \frac{2.303RTm}{(n'+\alpha)F}pH + \frac{2.303RT}{(n'+\alpha)F} \log_{10} K,$$

so that as the pH increases (and $[H^+]$ decreases) the peak potential for the reduction of A shifts to more negative potentials.

Limit (ii) $[H^+] \gg K$. In this limit,

$$E_p \sim const$$

and there is no variation of the peak potential with pH since all A in the solution is in the form of AH_m^{n+}.

The switchs over between the two limits has been observed in the reduction of phenacyl sulfonium salts:[12]

$$PhCOCH^- S^+ R_1 R_2 + H^+ \rightleftarrows PhCOCH_2 S^+ R_1 R_2$$

$$PhCOCH_2 S^+ R_1 R_2 + 2e \rightleftarrows PhCOCH_2^- + R_1 SR_2.$$

Below *ca* pH 8 the reduction potential is pH independent whereas above it the variation with pH suggests that $n' + \alpha \sim 0.5$ implying that it is the first electron transfer of the two in the step following the protonation that is rate-determining. Note that the intersection of the data for the two limits can be used to provide information about the pKa for the acid dissociation, as shown in Fig. 4.21.

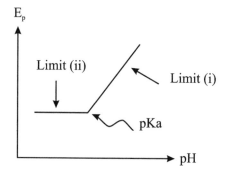

Fig. 4.21

4.12 The Scheme of Squares

The arguments of the previous section can be developed by means of the scheme of squares first proposed by Jacq[13] and best illustrated by means of a $2H^+$, $2e^-$ system such as might apply to the reduction of a quinone:

The model is based on the assumption that electron transfer is the rate limiting step and that all protonations are at equilibrium. A general scheme is presented below:

The reduction route from Q to QH_2 is heavily dependent on the pK_a values associated with the various intermediates, the pH of the environment local to the electrode, and the various E^0 values.

AQMS   AQDS

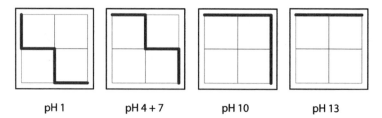

| pH 1 | pH 4 + 7 | pH 10 | pH 13 |

Fig. 4.22 Scheme showing the major mechanistic pathways for both AQMS and AQDS at varying pH. Horizontal movement represents electron transfer and vertical movements represent proton transfer.

For the cases of anthraquinone-2-sulfonate (AQMS) and anthraquinone-2,6-disulfonate (AQDS) the route has been mapped out[14] across the full pH range in aqueous solution and shown to vary as in Fig. 4.22.

Note that in Fig. 4.22 horizontal movement represents electron transfer whilst the vertical movement depicts proton transfer. Thus at pH 1 the mechanism is:

$$Q + H^+ \rightleftarrows QH^+$$

$$QH^+ + e^- \rightleftarrows QH^\bullet$$

$$QH^\bullet + H^+ \rightleftarrows QH_2^{\bullet+}$$

$$QH_2^{\bullet+} + e^- \rightleftarrows QH_2$$

where Q is either AQMS or AQDS.

The scheme of square is invaluable in mechanistically deciphering complex sequences of electron and proton transfer.

4.13 Simultaneous Two-Electron Transfer in Electrode Kinetics?

Our analysis of multiple electron transfer in, for example, Sections 2.8, 4.10 and 4.12 has implicitly assumed that electrons must be transferred one at a time. This, of course, necessitates the assumption that various intermediates, sometimes of high energy and hence instability, must be present during the course of the reaction. Gileadi[15] has reflected as follows:

> 'In a seminal paper by Bockris *et al.* in 1961[16], the mechanism of iron deposition was discussed in detail. It was concluded that the two

electrons were transferred one at a time, and an intermediate mono-valent iron species was formed

$$Fe^{2+} + e_M^- + OH^- \rightarrow [FeOH]_{ads}.$$

The authors were, of course, fully aware of the fact that monovalent iron is not stable, and overcame this problem by assuming that it is adsorbed on the surface, (a valid common practice in the study of electrode kinetics).'

And further:

'The approach taken by the above authors[16] to discuss the mechanism of iron deposition has been widely accepted since then, not only for the case of iron, but also for nickel and cobalt deposition, although in all these cases the monovalent intermediates that must be assumed, if electrons are to be transferred one at a time, are not known in chemistry, and have not been detected in the study of deposition of the above metals. Indeed it may be stated that, in the study of the mechanism of electrode reactions, simultaneous $2e^-$ electron transfer is excluded almost axiomatically (to say nothing of simultaneous $3e^-$ transfer!), and the results are almost invariably analysed on the assumption that the data must be fitted to a sequence of single-electron transfer steps that add up to the overall reaction. In some of the earlier literature on electrode kinetics it was common to use the term αn_a in equations describing the potential dependence of the electrochemical rate constant where n_a was said to be 'the number of electrons transferred in the rate determining step'. However, in the second edition of *Electrochemical Methods* by Bard and Faulkner, the authors rejected this practice, stating that a 'widely held concept in electrochemistry is that truly elementary electron-transfer reactions always involve the exchange of one electron'.[17]'

The use of $(n' + \alpha_{n'})$ in Section 4.10 rather than αn_a is consistent with the general outlook described. That said the possibility of concerted two-electron transfers should be considered.

In Section 4.9 we saw that for the two-electron reduction of A to C via B that if

$$E_f^0(B/C) \gg E_f^0(A/B)$$

corresponding to so-called 'potential inversion'[18] then a single voltammetric wave is seen and this will occur for reversible electrode kinetics near the standard potential for the overall process, viz.

$$1/2[E_f^0(A/B) + E_f^0(B/C)]$$

For the case of a concerted two-electron reduction

$$A + 2e^- \longrightarrow C$$

then the intermediate B would have to be of too high an energy to be accessed. Gileadi[15] and Evans[18] compare the energy of this intermediate with an activation barrier calculated using an expression in the form of Eq. (2.49) but for a two-electron transfer, noting that the Marcus expression for the outer sphere reorganisation energy depends on the square of the charge transferred (Eq. (2.48)). Fore sufficiently unstable intermediates (B) the concerted two-electron process is found to have a lower energy of activation so it was concluded that this type of process was not necessarily implausible. The situation, with respect to real systems, was summarised by Evans[18] as follows:

'What are the requirements for the concerted two-electron reaction to have a lower barrier than that of sequential transfer of two electrons? First, it is important to realise that the reactions must be electron transfer reactions and not reactions that are coupled to chemical steps. Thus several examples ... do not qualify as simple electron transfer reactions. For example, the reduction of Ni^{2+}(aq) to metallic nickel involves, in addition to electron transfer, steps of deaquation and adsorption/crystallisation. Similarly the reduction of CrO_4^{2-}(aq) to metallic chromium involves, in addition to electron transfer, conversion to the tetrahedrally coordinated oxygen atoms in CrO_4^{2-} to OH^- (in the neutral to alkaline medium needed to stabilise CrO_4^{2-}). The overall reaction, $CrO_4^{2-} + 4H_2O + 6e^- \rightleftarrows Cr(s) + 8OH^-$, obviously involves much more than electron transfer. The same can be said for the possibility of simultaneous transfer of three electrons to chromate, $CrO_4^{2-} + 4H_2O + 3e^- \rightleftarrows Cr(OH)_3(s) + 5OH^-$.

One example ... does seem to qualify as a two-electron reaction without coupled chemical steps. That reaction is Tl^{3+}(aq) $+ 2e^- \rightleftarrows Tl^+$(aq) which is studied in acidic media owing to the insolubility of thallium (III) hydroxide. Both of these ions (as well as the hypothetical intermediate, Tl^{2+}(aq)) have filled d and f orbitals so that water molecules are not strongly coordinated indicating that the ions are probably hydrated in a facile reversible fashion.'

It is possible that Tl(III) is reduced to Tl(I) in a concerted two-electron process.

References

[1] R.S. Nicholson, I. Shain, *Anal. Chem.* **36** (1964) 706.

[2] R.S. Nicholson, I. Shain, *Anal. Chem.* **37** (1965) 179.

[3] A.C. Fisher, *Encyclopaedia of Electrochemistry* **2** (2003) 122.

[4] D. Britz, *Digital Simulation in Electrochemistry*, 3rd edn, Springer, Berlin, 2005.

[5] *Pure Appl. Chem.* **45** (1976) 131.

[6] H. Matsuda, Y. Ayabe, *Z. Elecktrochem* **59** (1955) 494.

[7] M.C. Buzzeo, O.V. Klymenko, J.D. Wadhawan, C. Hardacre, K.R. Seddon, R.G. Compton, *J. Phys. Chem.* A **107** (2003) 8872.

[8] M.W. Lehmann, D.H. Evans, *J. Electroanal. Chem.* **500** (2001) 12.

[9] R.G. Evans, O.V. Klymenko, P.D. Price, S.G. Davis, C. Hardacre, R.G. Compton, *Chem. Phys. Chem.* **6** (2005) 526.

[10] C.P. Andrieux, J.M. Savéant, *J. Electroanal. Chem.* **28** (1970) 339.

[11] Z. Rongfeng, D.H. Evans, *J. Electroanal. Chem.* **385** (1995) 201.

[12] P. Zuman, S. Tang, *Col. Czech. Chem. Commun.* **28** (1963) 829.

[13] J. Jacq, *J. Electroanal. Chem.* **29** (1971) 149.

[14] C. Batchelor-McAuley, Q. Li, S.M. Dapin, R.G. Compton, *J. Phys. Chem.* B **114** (2010) 4094.

[15] E. Gileadi, *J. Electroanal. Chem.* **532** (2002) 181.

[16] J.O.M. Bockris, D. Drazie, A.R. Despic, *Electrochim. Acta* **4** (1961) 325.

[17] A.J. Bard, L.R. Faulkner, *Electrochemical Methods — Fundamental and Applications*, 2nd edn, Wiley, New York, 2007.

[18] D.H. Evans, *Chem. Rev.* **108** (2008) 2113.

5

Voltammetry at Microelectrodes

The introduction of microelectrodes — electrodes with at least one dimension of the order of microns, or less, in size — transformed the scope of voltammetric studies in the 1980s. Amongst various attributes, the possibility emerged of exploring significantly faster kinetic processes than was previously possible. The basic insight into why this is possible resides in the contrasting modes of diffusion to microelectrodes and to planar macroelectrodes and will be developed in the next two sections.

5.1 The Cottrell Equation for a Spherical or Hemispherical Electrode

In Chapter 3 we developed the current–time response resulting at a planar macroelectrode when it is subjected to a potential step from a value where no current flowed (no electrolysis) to one relating to a transport-limited current corresponding to a potential so large that the concentration at the electrode surface was reduced to zero. It was seen that the current varied as the reciprocal of the square root of time, t:

$$I \propto \frac{1}{\sqrt{t}} \tag{5.1}$$

such that at long times the current approached zero. It is illuminating to pursue the same exercise for a spherical or hemispherical electrode such as a mercury-drop electrode.

It is required to solve Fick's second law of diffusion written in spherical coordinates

$$\frac{\partial c}{\partial t} = D\left(\frac{\partial^2 c}{\partial r^2} + \frac{2}{r}\frac{\partial c}{\partial r}\right) \tag{5.2}$$

subject to the boundary conditions

$$
\begin{aligned}
t < 0, \quad & r \geq r_e, \quad & c = c^* \\
t \geq 0, \quad & r = r_e, \quad & c = 0 \\
t > 0, \quad & r \to \infty, \quad & c \to c^*,
\end{aligned}
$$

where r_e is the radius of the (hemi)sphere.

The change of variable

$$u = rc \tag{5.3}$$

converts Eq. (5.2) into

$$\frac{\partial u}{\partial t} = D\frac{\partial^2 u}{\partial r^2}, \tag{5.4}$$

which is of exactly the same form as the equation describing Fick's law in one dimension, and as such is readily solved to give

$$c = c^*\left\{1 - \frac{r_e}{r}erfc\left(\frac{r - r_e}{\sqrt{4Dt}}\right)\right\}, \tag{5.5}$$

where

$$erfc(x) = 1 - erf(x). \tag{5.6}$$

The function $erf(x)$ was introduced in Chapter 3. Figure 5.1 shows the evolution of c as a function of time, t, and radial distance, r, for two different values of D, and for two different values of r_e.

Equation (5.5) allows us to deduce the flux of reactant at the electrode surface:

$$j/\text{mol cm}^{-2}\,\text{s}^{-1} = D\left.\frac{\partial c}{\partial r}\right|_{r=r_e} \tag{5.7}$$

$$j/\text{mol cm}^{-2}\,\text{s}^{-1} = Dc^*\left\{\frac{1}{\sqrt{D\pi t}} + \frac{1}{r_e}\right\}. \tag{5.8}$$

Figure 5.2 shows plots of current density,

$$i = I/A = nFj,$$

against time for the case of $n = 1$ and for the four cases shown in Fig. 5.1.

It is insightful to consider the short-time and long-time limits of Eq. (5.8)

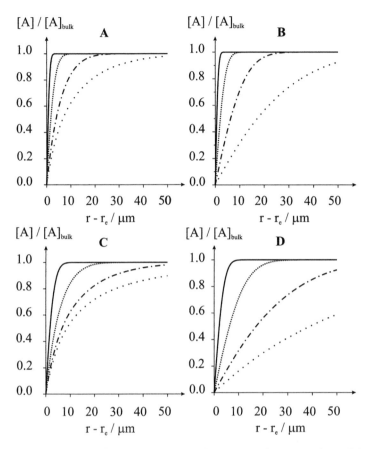

Fig. 5.1 Concentration profiles at varying times after a potential step at a spherical electrode into the diffusion controlled region for the reduction of *A* to *B*. Throughout $[C^*]_0 = 1\,\text{mM}$. For (A), $D = 5 \times 10^{-6}\,\text{cm}^2\text{s}^{-1}$ and $r_e = 10\,\mu\text{m}$. For (B), $D = 5 \times 10^{-6}\,\text{cm}^2\text{s}^{-1}$ and $r_e = 100\,\mu\text{m}$. For (C), $D = 5 \times 10^{-5}\,\text{cm}^2\text{s}^{-1}$ and $r_e = 10\,\mu\text{m}$. Finally for (D), $D = 5 \times 10^{-5}\,\text{cm}^2\text{s}^{-1}$ and $r_e = 100\,\mu\text{m}$. In all cases, the four curves correspond to (solid line) 0.001, (small-dotted line) 0.01, (dot-dashed line) 0.1 and (large-dotted line) 1 second.

Case (i): Short-time limit. Here,

$$j = \frac{c^*\sqrt{D}}{\sqrt{\pi t}},\tag{5.9}$$

corresponding exactly to the result obtained in Chapter 3 for the case of linear, one-dimensional diffusion. This will hold for

$$\sqrt{D\pi t} \ll r_e.\tag{5.10}$$

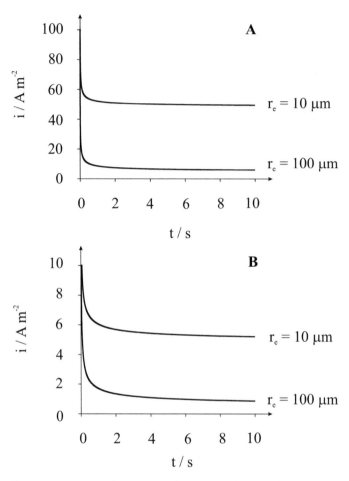

Fig. 5.2 Chronoamperometry for a potential step at a spherical electrode into diffusion-controlled region for the reduction of A to B. Parameters: $[C^*]_0 = 1\,\text{mM}$; (A) $D = 5 \times 10^{-5}\,\text{cm}^2\text{s}^{-1}$ and (B) $D = 5 \times 10^{-6}\,\text{cm}^2\text{s}^{-1}$.

Under these conditions, as is implicit in Fig. 5.3, the diffusion layer is small compared to the radius of the spherical electrode. Accordingly, the diffusion is approximately linear as illustrated in Fig. 5.3.

Case (ii): Long-time limit. Here

$$j = D\frac{c*}{r_e} \tag{5.11}$$

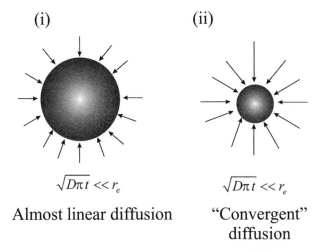

(i)

(ii)

$$\sqrt{D\pi t} \ll r_e \qquad\qquad \sqrt{D\pi t} \ll r_e$$

Almost linear diffusion   "Convergent"
                                   diffusion

Fig. 5.3 Diffusion to a sphere. Cases (i) and (ii) correspond to $\sqrt{D\pi t} \ll r_e$ or $\sqrt{D\pi t} \gg r_e$ for short times (larger radii) or long times (small radii).

or

$$I = 4\pi r_e DFc^* \quad \text{(sphere)} \tag{5.12}$$

$$I = 2\pi r_e DFc^* \quad \text{(hemisphere)}. \tag{5.13}$$

In this limit, a steady-state current (flux) is established! At first sight, this is paradoxical since it might be expected that the current decreases towards zero as the electrolysis proceeds, as happens for linear diffusion (Cottrell equation). The point is that as the electrolysis continues the diffusion layer thickness, as shown in Fig. 5.1, expands and its 'surface area' ($\propto 4\pi(r_e + \delta)^2$, where δ is the diffusion layer thickness) increases because of the spherical expansion. Accordingly, it meets more and more electroactive species as time proceeds and it is this material which 'feeds' the constant diffusion gradient at the electrode surface once steady-state is attained. Of course, this argument assumes that bulk solution is infinite in extent and of composition c^*. It is this latter boundary condition which generates the predicted behaviour; if the electrode is in a vessel of finite volume then a no flux ($\frac{\partial c}{\partial x} = 0$) boundary condition the walls of the vessel would ensure that concentration ultimately falls to zero after a near infinite period of electrolysis!

Returning to Eq. (5.8) it is clear that the steady-state current will dominate more relative to the time-dependent terms for electrodes of small r_e. That is to say, for a 'microelectrode' of spherical or hemispherical geometry, a steady-state current is rapidly established, the flux (mol cm^{-2} s^{-1}) being greater the smaller the electrode is. These are two fundamental characteristics of microelectrodes.

5.2 Potential Step Transients at Microdisc Electrodes

So far, we have considered the current–time transients resulting from potential steps, first at large planar 'macro' electrodes and second at spherical electrodes. We next consider the very important case of a microdisc electrode, as shown in Fig. 5.4.

This is a much harder problem since the flux is no longer constant over the electrode surface. The electrode is 'not uniformly accessible'.

The form of Fick's second law appropriate to cylindrical coordinates (Fig. 5.4) is

$$\frac{\partial c}{\partial t} = D\frac{\partial^2 c}{\partial r^2} + \frac{D}{r}\frac{\partial c}{\partial r} + D\frac{\partial^2 c}{\partial z^2},\tag{5.14}$$

where z is the perpendicular distance from the electrode surface and r is a radial coordinate from the centre of the disc. To solve the potential step problem of interest, we apply the following boundary conditions.

$$
\begin{array}{llll}
t < 0 & \text{all } r, & z & c = c^* \\
t \geq 0 & z = 0, & r < r_e & c = 0 \\
t \geq 0 & z \to \infty \text{ all } r & & c = c^* \\
t \geq 0 & r \to \infty \text{ all } z & & c = c^*,
\end{array}
$$

where r_e is the radius of the electrode. Analytical treatments of this problem require approximation and the resulting equations have been reported.[1] As with the spherical electrode, a transient response is observed leading at long times to a steady-state current

$$I = 4nFc^*Dr_e\tag{5.15}$$

for a n-electron electrode process. The flux at any radius, r, on the disc electrode is given by

$$j = \frac{2}{\pi}\frac{c^*D}{\sqrt{r_e^2 - r^2}}.\tag{5.16}$$

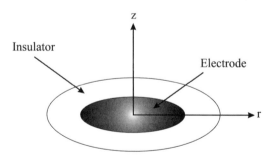

Fig. 5.4 A microdisc electrode and the cylindrical coordinates (r and Z) used to describe it.

In practice, of course, the flux cannot be infinite at the disc edges as predicted by this expression since the kinetics of the electrode reaction must be finite; nevertheless, at steady state the current is dominated by that flowing at the edge as reflected in the fact that the steady-state current scales with the perimeter (circumference) of the disc $(2\pi r_e)$.

The full current–time transient has been calculated as

$$I = 4nFc^*Dr_e f(\tau), \tag{5.17}$$

where the dimensionless time, $\tau = 4Dt/r_e^2$. For short times, when $\tau < 1$

$$f(\tau) = \left(\frac{\pi}{4\tau}\right)^{1/2} + \frac{\pi}{4} + 0.094\tau^{1/2} + \cdots, \tag{5.18}$$

whilst at long times, $\tau > 1$,

$$f(\tau) = 1 + 0.71835\tau^{-1/2} + 0.005626\tau^{-3/2} - 0.00646\tau^{-5/2} + \cdots. \tag{5.19}$$

It follows that at very short times the response is simply that predicted by the Cottrell equation for planar diffusion:

$$I = 4nFc * Dr_e \left(\frac{\pi}{4\tau}\right)^{1/2}$$

$$I = \frac{nFA\sqrt{D}c^*}{\sqrt{\pi t}}, \tag{5.20}$$

where $A = \pi r_e^2$. It follows that at these short times

$$4Dt \ll r_e^2,$$

the diffusion to the electrode surface is effectively planar (linear). This reflects the fact that under these conditions the diffusion layer contributes a very thin zone over the surface of the electrode; in other words the diffusion layer thickness is thin compared with the electrode radius. At longer times, the decay of current is less rapid then the Cottrellian, $1/\sqrt{t}$, and this is because radial diffusion contributes to the current. This implies that the diffusion layer has spread beyond the edge of the disc. At long times the shape of the diffusion layer approaches hemispherical, as shown in Fig. 5.5.

5.3 Microelectrodes Have Large Current Densities and Fast Response Times

In the previous section we have considered the current transients resulting from potential steps at spherical and microdisc electrodes. In both cases, at very short

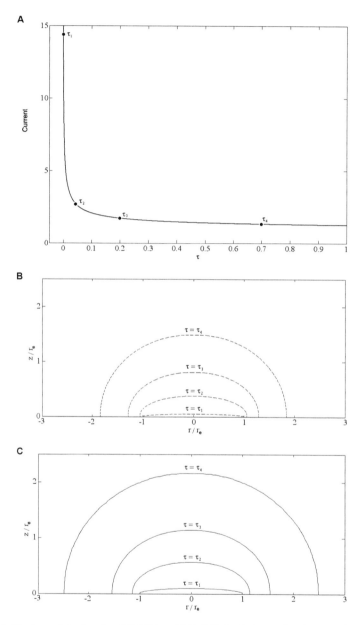

Fig. 5.5 Contour plots showing (B) 50% and (C) 90% depletion of the electroactive species at a disc electrode following a potential step at four times $\tau = 0.001, 0.04, 0.2$ and 0.7 on the current transient (A). Note that $\tau = tD/r_e$, where r_e is the disc radius and the x-axis in (B) and (C) is r/r_e and the y-axis z/r_e.

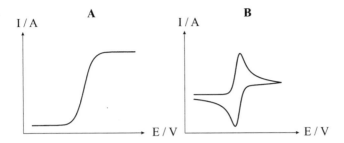

Fig. 5.6 Voltammetry at (A) micro- and (B) macro-electrodes; slow scan rates (see text).

times a Cottrellian ($I \propto 1/\sqrt{t}$) response was seen but at longer times the current tended to a steady-state value. In each case, the magnitude of the flux under steady-state conditions is merely proportional to the size of the electrode, *viz* the sphere or disc radius. It follows that the smaller the electrode the greater the current density and the faster material diffuses to (and from) the electrode surface. Moreover, since the transient response is typically a 'balance' between a Cottrellian term and a steady-state term, the timescale to reach steady-state is shorter the smaller the electrode is.

The fact that steady-state currents can be recorded at a microdisc electrode has a very significant implication for voltammetry using such electrodes. We saw in the previous chapter that cyclic voltammetry at large planar electrodes, where potential steps lead to a Cottrellian decay to zero current, is characterised by peak-shaped responses, the fall off resulting from depletion of material near the electrode surface such that the current is controlled by diffusion from bulk solution. In contrast, microelectrodes show voltammograms in which the current does not decay to zero but rather to the steady-state current predicted by our potential step analysis. Accordingly, if the voltage applied to the working electrode is scanned sufficiently slowly then no current peak is displayed, as shown schematically in Fig. 5.6.

We return to a more detailed consideration of microelectrode voltammetry later in this chapter.

5.4 Applications of Potential Step Chronoamperometry Using Microdisc Electrodes

The experiments described in Section 5.2 have very considerable experimental power. Specifically, they can measure

- unknown diffusion coefficients, D,
- unknown concentrations, c^*, and
- unknown numbers of electrons, n, transferred.

Moreover, and most significantly, they can measure D and either c^* or n provided the third parameter is known. How is this possible?

The current transient at a microdisc electrode of radius r_c is given by

$$I = 4nFc^*Dr_ef(\tau), \tag{5.21}$$

where short- and long-time approximates for $f(\tau)$ were given in Eqs. (5.18) and (5.19) above. A convenient single expression for $f(\tau)$ has been obtained empirically by Shoup and Szabo[1] and shown to correctly predict the current over the entire time domain with a maximum error of less than 0.6%, well inside typical experimental accuracy.

The Shoup and Szabo expression is

$$f(\tau) = 0.7854 + 0.8863\tau^{-1/2} + 0.2146 \exp\left(-0.7823\tau^{-1/2}\right). \tag{5.22}$$

It follows from Eqs. (5.21) and (5.22) that at short times the current scales with $D^{1/2}$, whereas at long times the current scales with D. In contrast, the current is directly proportional to the product nc^* at all times. It follows that provided the experimental data recorded is of high quality, it is possible to best-fit the measured transient to give simultaneously optimum values of (D) and (nc^*). Then, if n or c^* are known, the third parameter may be deduced. It is illuminating to give some examples, which illustrate the power of potential step chromoamperometry.

First, consider the voltammetry of a saturated solution of oxygen in a room temperature ionic liquid (RTIL). The latter are liquids composed entirely of ions; Table 5.1 shows the typical cations and anions which together from such liquids.

Table 5.1. Typical cations and anions that form room temperature ionic liquids.

| [C$_n$mim] | [N$_{abcd}$] | [C$_4$pyr] | [Py$_{1e}$] |

| [BF$_4$] | [PF$_6$] | [N(Tf)$_2$] | [FAP] |

Typically, neither the solubility nor the diffusion coefficient of oxygen will be known in the medium of interest. On the other hand, the reduction can be assigned to the one electron process

$$O_2 + e^- \longrightarrow O_2^{\bullet-}, \qquad (5.23)$$

in which the superoxide anion is generated. Accordingly, this is a problem in which n (=1) is known but c^* and D are not. However, this can be determined from analysis of suitable microdisc potential step 'chronoamperograms'. Figure 5.7 shows typical data obtained using a gold microdisc electrode for a (saturated) solution of oxygen in the room temperature ionic liquid [C$_2$mim][NTf$_2$].

Also shown are points corresponding to the best-fit simulation; excellent agreement is apparent. The fit allows the deduction that the concentration of oxygen in a saturated sample of the ionic liquid is 3.9 mM further permitting the deduction of the Henry's Law constant, K_H, via the following equation:

$$c^* = K_H P_{O_2},$$

where P_{O_2} is the partial pressure of oxygen. A value of $K_H = 3.9$ mM atm^{-1} was accordingly inferred. The diffusion coefficient value for O_2 in the same medium was determined as 8.3×10^{-6} cm^{2}s^{-1}.

A second example relates to the electro-reduction of the thioether shown in Fig. 5.8 in tetrahydrofuran at $-74°$C.[3]

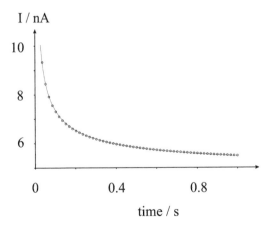

Fig. 5.7 Comparison of experiment (line) and simulated (circles) results of a chronoamperometric measurement for the reduction of oxygen at a gold electrode in [C$_2$mim][NTf$_2$]. Figure reproduced from Ref. [2] with permission.

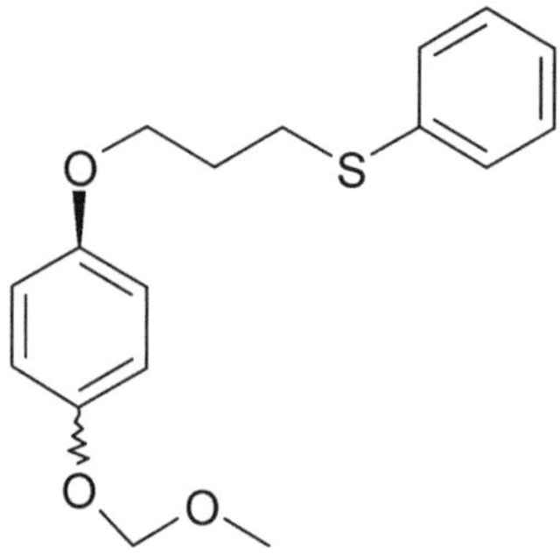

Fig. 5.8 Phenyl thioether: [(3-{[trans-4-(methoxymethyoxy)cyclohexyl]oxy}propyl)thio]
benzene, Ph-S-R.

In this case solutions of known concentrations can be studied but the diffusion
coefficient is unknown. Moreover, the nature of the electrode reaction is uncertain
and indeed was the purpose of the study. Several possibilities could be envisaged:

$$n = 2 \quad R - S - Ph + 2e^- \longrightarrow R^- + Ph - S^-$$
$$n = 2 \quad R - S - Ph + 2e^- \longrightarrow RS^- + Ph^-$$
$$n = 1 \quad R - S - Ph + e^- \longrightarrow RS^\bullet + Ph^-$$
$$n = 1 \quad R - S - Ph + e^- \longrightarrow RS^- + Ph^\bullet.$$

Potential step experiments using solutions of known concentrations and a
5 μm (radius) platinum microdisc electrode showed data such as is illustrated
in Fig. 5.9.

Simulation (also shown) revealed a diffusion coefficient of 2.3×10^{-6} cm^2 s^{-1}
and, most importantly, that $n = 2\,(\pm 0.2)$ so providing important mechanistic
insight which, along with other data, showed that the reaction pathway was

$$n = 2 \quad R - S - Ph + 2e^- \longrightarrow R^- + PhS^-.$$

The generic scope for using potential step transients for the determination of the
number of electrons transferred in voltammetric events is clear.

5.5 Double Potential Step Microdisc Chronoamperometry Exploring the Diffusion Coefficient of Electrogenerated Species

The previous section showed how potential step chronoamperometry is a powerful method for the determination of unknown diffusion coefficients especially

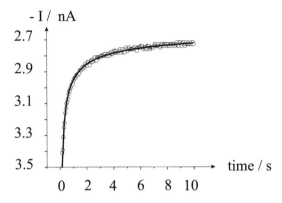

Fig. 5.9 Experimental (solid line) and fitted theoretical (circles) chronoamperometric curve for the two-electron reduction of 3 mM RSPh in THF (0.1 M TBAP) at a 5 μm platinum microelectrode and temperature of $-74°C$ ($\pm2°C$). Potential was stepped to -3.7 V vs. (Fc/Fc$^+$PF$_6^-$) in the plateau region of the reductive wave. Fitting by Origin^{TM} software. Diffusion coefficient, $D = 2.3$ (±0.25) $\times 10^{-6}$ cm^2 s^{-1} and the number of electrons transferred per molecules, $n = 2$ in THF at $-74°C$. Reprinted from Ref. [3] with permission from Elsevier.

if the species of interest is present at unknown concentrations, or if the number of electrons transferred in the electrochemical reactions is ambiguous. It is apparent that if the technique is extended to embrace two steps in potential then more information is available. Specifically, considering the electrode reaction

$$A \pm ne \longrightarrow B$$

in a solution containing only A, it is evident that whilst a step from a potential corresponding to zero current flow to one at which the electrolysis of A is transport controlled will give information about A, while a second step to potential where B is reconverted to A will give information about the diffusion coefficient of B. Figure 5.10 illustrates the potential–time dependence during the experiment, where it is assumed that:

- E_1 corresponds to zero current,
- E_2 causes the transport limited conversion of A to B and
- E_3 induces the diffusion controlled reconversion of B to A.

Mathematically, in order to predict the current flowing, the following equations have to be solved representing diffusion to a microdisc electrode and so are written

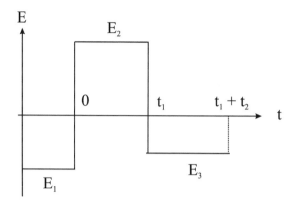

Fig. 5.10 Scheme of the potential–time dependence in double potential step chronoamperometry.

in the cylindrical coordinates, r and z:

$$\frac{\partial [A]}{\partial t} = D_A \left(\frac{\partial^2 [A]}{\partial r^2} + \frac{1}{r} \frac{\partial [A]}{\partial r} + D \frac{\partial^2 [A]}{\partial z^2} \right) \tag{5.24}$$

$$\frac{\partial [B]}{\partial t} = D_B \left(\frac{\partial^2 [B]}{\partial r^2} + \frac{1}{r} \frac{\partial [B]}{\partial r} + D \frac{\partial^2 [B]}{\partial z^2} \right), \tag{5.25}$$

where D_A and D_B are the diffusion coefficients of A and B, respectively. The relevant boundary conditions are

$$t < 0 \quad \text{all } r, z \quad [A] = [A]_{bulk}, \quad [B] = 0$$

$$t_1 > t \geq 0 \quad 0 < r < r_e \quad z = 0 \quad [A] = 0, \quad D_A \frac{\partial [A]}{\partial z} = -D_B \frac{\partial [B]}{\partial z}$$

$$t \geq t_1 \quad 0 < r < r_e \quad z = 0 \quad [B] = 0, \quad D_A \frac{\partial [A]}{\partial z} = -D_B \frac{\partial [B]}{\partial z},$$

where t_1 is defined in Fig. 5.10. Zero flux conditions are imposed at the surface of the insulating materials surrounding the disc and at the axis of symmetry

$$r > r_e \quad z = 0 \quad \frac{\partial [A]}{\partial z} = \frac{\partial [B]}{\partial z} = 0$$

$$r = 0 \quad z \geq 0 \quad \frac{\partial [A]}{\partial r} = \frac{\partial [B]}{\partial r} = 0,$$

and the bulk concentration is maintained at all times sufficiently distant from the electrode

$$\text{All } t, \quad r \to \infty, \quad z \to \infty, \quad [A] \to [A]_{bulk}, \quad [B] \to 0.$$

The solution to the problem requires numerical simulation and has been reported by Klymenko and colleagues[4] who provide tables of data allowing analysis of experimental transients without recourse to further simulation. The method provides a sensitive way of simultaneously measuring D_A and D_B, as is evident from the following examples.

First consider the electrode process

$$Fe(CN)_6^{4-}(aq) - e^- \rightarrow Fe(CN)_6^{3-}(aq). \tag{5.26}$$

Figure 5.11 shows a comparison of the fit between the observed and the simulated potential step transients for this system studied in 0.1 M aqueous KCl at a 100 μm diameter platinum disc electrode.

Excellent agreement is seen, which has permitted[4] the inference that the diffusion coefficient of $Fe(CN)_6^{4-}$ was 6.58 ($\pm$0.37) $\times$ 10^{-6} cm^2 s^{-1} and that of $Fe(CN)_6^{3-}$ was 7.51 ($\pm$0.99) $\times$ 10^{-6} cm^2 s^{-1} at 25°C. In this case both species are stable so it is possible to measure both diffusion coefficient directly via other techniques. Excellent agreement was seen, so validating the double step methods especially in terms of accuracy.

A second example concerns the following electrode processes in non-aqueous solution

$$TMPD - e \rightleftarrows TMPD^{\bullet+}$$

$$TMPD^{\bullet+} - e \rightleftarrows TMPD^{2+},$$

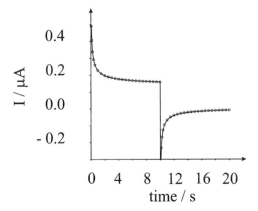

Fig. 5.11 Comparison of the fit between observed and simulated double potential step transients for the $Fe(CN)_6^{4-}$ system in aqueous 0.1 M KCl at a 100 μm diameter platinum disk. Reprinted from Ref. [4] with permission from Elsevier.

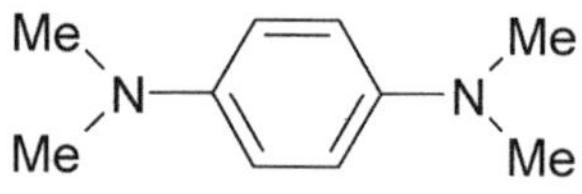

Fig. 5.12 The structure of TMPD.

where *TMPD* is *N,N,N',N'*-tetramethylphenylenediamine, the structure of which is shown in Fig. 5.12.

The molecule was selected as a probe for making a comparative study of diffusion rates in various ionic liquids and in the conventional aprotic solvent, acetonitrile, since it is established that it shows two one-electron oxidations in these media; the two oxidations occur at potentials more than *ca* 0.5 volts apart so can be studied without any problem of overlapping voltammetric waves (see Fig. 5.13); Table 5.2 shows the solvents studied and their viscosities.

Moreover, by study of the salt $TMPD^{\bullet+}BF_4^-$ and using double potential step chronoamperometry it was possible to measure the diffusion coefficients of all three species $TMPD$, $TMPD^{\bullet+}$ and $TMPD^{2+}$. In an initial set of experiments, steps were made first from potentials at which $TMPD^{\bullet+}$ was neither oxidised nor reduced to more positive potentials corresponding to the transport limited oxidation of $TMPD^{\bullet+}$ to $TMPD^{2+}$ and second back to the starting potential so giving the diffusion coefficients of $TMPD^{\bullet+}$ and $TMPD^{2+}$. A second set of double step experiments explored the inter-conversion of $TMPD$ and $TMPD^{\bullet+}$ so giving information about the diffusion coefficients of these two species. Figure 5.13 shows typical data; the excellent agreement between simulation and experiment is again evident.

The diffusion coefficients for all three species — $TMPD$, $TMPD^{\bullet+}$ and $TMPD^{2+}$ — were found[5] to accurately follow an Arrhenius type temperature dependency:

$$D = D_\infty \exp\left(\frac{-E_a}{RT}\right),\tag{5.26}$$

as discussed in Chapter 3. In any particular solvent E_a, the activation energy for diffusion was found to be closely similar for all three species and to that measured for the solvent viscosity,

$$\eta = \eta_\infty \exp\left(\frac{E_a}{RT}\right).\tag{5.27}$$

The relevant data are shown in Table 5.3.

These observations suggest

$$D \propto \frac{1}{\eta},\tag{5.28}$$

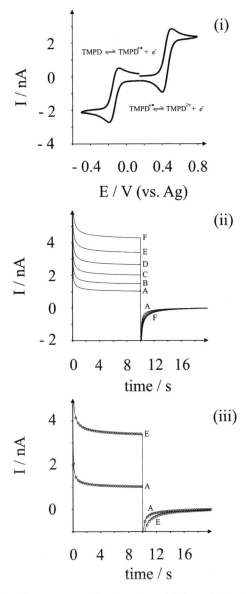

Fig. 5.13 (i) Cyclic voltammogram showing the oxidation of 20 mM $TMPD^+BF_4^-$ in [C_2mim][NTf$_2$] at a Au disk electrode (nominal radius 5 μm), scan rate of 0.2 Vs^{-1}. (ii) Double potential step chronoamperomograms measured on the same system across the redox couple $TMPD^{+\bullet}/TMPD^{2+}$ at temperatures of (A) 298K, (B) 308 K, (C) 318 K, (D) 328 K, (E) 338 K and (F) 348 K. (iii) Comparison of these experimental transients (solid line) with best fit theoretical transients (circles) for the temperatures (A) 298 K and (E) 338 K. Reprinted from Ref. [5] with permission from Wiley.

Table 5.2. Solvents used in the study of TMPD diffusion.[6]

Solvent	Viscosity at 25°C (mPa s)
MeCN	0.345
$[C_2mim][NTf_2]$	30.8
$[C_4pyr][NTf_2]$	72.5
$[C_{10}mim][NTf_2]$	127.5
$[P_{14,666}][NTf_2]$	399.9

consistent with the Stokes–Einstein relationship

$$D = \frac{k_B T}{6\pi\eta r}, \tag{5.29}$$

where k_B is the Boltzmann constant and r the molecular radius assuming the molecule is spherical. The Stokes–Einstein equation is derived by assuming that Stokes law for the viscous drag on a *macro*scopic sphere holds for molecules. This is an obvious approximation.

The data for the three species was also used to test the 'equal diffusion coefficient' approximation commonly adapted in the modelling of voltammetric data. This supposes that for the electrode reaction

$$A + e \rightarrow B,$$

then to a good approximation,

$$D_A \sim D_B,$$

unless there are obvious major structural differences between A and B. For conventional aprotic solvents such as acetonitrile, dimethylformamide or propylene carbonate this is usually an excellent approximation. For the case of acetonitrile solvent the diffusion coefficient of $TMPD^{\bullet+}$ was found to be within 12% of that of $TMPD$, and similar data has been reported for other radical cations and anions in comparison with their neutral parent molecules. In contrast, in the room temperature ionic liquid the diffusion coefficient of the charged cation was approximately one half that of the neutral parent, $TMPD$. The diffusion coefficient of the dication was approximately one third that of $TMPD$ in the solvents studied.

Table 5.3. Arrhenius parameters for the variation with temperature (298–348 K) in neat solvent of viscosity (see Eq. (5.27)) and diffusion coefficient (see Eq. (5.26)) of species *TMPD*, its radical cation, *TMPD*$^{•+}$, and dication *TMPD*$^{2+}$, in four room temperature ionic liquids. Diffusion parameters in regular font were derived from single step potential chronoamperograms, whilst those in italics were determined via the analysis of double potential step data.

| | Viscosity | | Diffusion | | | | | |
| | | | TMPD | | TMPD$^{•+}$ | | TMPD^{2+} | |
	η_∞ [mPa S]	$E_{a,\eta}$ [kJ mol^{-1}]	D_∞ [m^{2}s^{-1}]	$E_{a,D}$ [kJ mol^{-1}]	D_∞ [m^{2}s^{-1}]	$E_{a,D}$ [kJ mol^{-1}]	D_∞ [m^{2}s^{-1}]	$E_{a,D}$ [kJ mol^{-1}]
[C$_2$mim][NTf$_2$]	3.36×10^{-4}	28.5	4.54×10^{-6}	28.3	1.51×10^{-6}	27.0	8.71×10^{-7}	26.9
[C$_4$pyr][NTf$_2$]	4.61×10^{-4}	29.6	6.70×10^{-6}	30.5	2.19×10^{-6}	9.2	*1.86×10^{-6}*	30.0
[C$_{10}$mim][NTf$_2$]	1.89×10^{-4}	33.1	8.61×10^{-6}	32.7	5.86×10^{-6}	33.2	[a]	[a]
[P$_{14,666}$][NTf$_2$]	4.52×10^{-5}	39.5	*1.62×10^{-4}*	41.8	*1.24×10^{-6}*	43.2	[a]	[a]

[a] Determination prevented due to interfering homogeneous processes.

5.6 Cyclic and Linear Sweep Voltammetry Using Microdisc Electrodes

Description of the cyclic (and linear sweep) voltammetry experiment carried out using a microdisc electrode for the simple electrode reaction

$$A - e \rightleftarrows B$$

requires the solution of the following diffusion equations

$$\frac{\partial [A]}{\partial t} = D_A \left(\frac{\partial^2 [A]}{\partial r^2} + \frac{1}{r}\frac{\partial [A]}{\partial r} + D\frac{\partial^2 [A]}{\partial z^2} \right) \tag{5.30}$$

$$\frac{\partial [B]}{\partial t} = D_B \left(\frac{\partial^2 [B]}{\partial r^2} + \frac{1}{r}\frac{\partial [B]}{\partial r} + D\frac{\partial^2 [B]}{\partial z^2} \right), \tag{5.31}$$

which are coupled by the following boundary conditions

$$0 < t \le t_{switch} \quad E = E_1 + \upsilon t$$

$$t_{switch} < t \quad E = E_1 + \upsilon t_{switch} - \upsilon(t - t_{switch})$$
$$= E_1 + 2\upsilon t_{switch} - \upsilon t.$$

$$\text{All } t, \quad z = 0, \quad 0 < r < r_e \quad D_A\frac{\partial [A]}{\partial z} = -D_B\frac{\partial [B]}{\partial z}$$
$$= k_a[A]_0 - k_c[B]_0$$

$$t \le 0 \quad \text{all } r, \quad \text{all } z \quad [A] = [A]_{bulk}; \quad [B] = 0.$$

$$\text{All } t, \; r, \quad z \to \infty \quad [A] = [A]_{bulk}; \quad [B] = 0,$$

where in the most general cases

$$k_a = k^0 \exp\left(\frac{\beta F}{RT} \left\{ E - E_f^o(A/B) \right\} \right) \tag{5.32}$$

$$k_c = k^0 \exp\left(\frac{-\alpha F}{RT} \left\{ E - E_f^o(A/B) \right\} \right). \tag{5.33}$$

The first three boundary conditions describe the triangular potential sweep imposed on the working electrode which starts at E_1. The potential then sweeps in a positive direction at a known scan rate, υ, until a time, t_{switch}, at which point the sweep reverses direction. The fourth and fifth boundary condition require the loss of A at the interface to be compensated by the formation of B and second that the rate of the interfacial reaction is governed by Butler–Volmer kinetics with the rate constants in the usual form given above by Eqs. (5.32) and (5.33). The two remaining boundary conditions relate the initial concentrations and those pertaining to bulk solution.

Figure 5.14(A–F) shows the voltammograms for the electrochemically irreversible limit with $k_0 = 10^{-5}$ cm s^{-1} for different sized microdisc electrodes in the range 0.1 μm to 1000 μm with $D_A = D_B = 10^{-5}$ cm^2 s^{-1} and $E_f^0(A/B) = 0$ V.

The currents in each case have been normalised to the steady-state limiting current to allow the different curves to be visible on the same graph. In all cases, a scan rate of 1 Vs^{-1} was utilised. The larger electrodes show a peaked response not dissimilar to that seen for voltammetry at a planar electrode. In contrast, for the smaller electrodes, near steady-state behaviours are seen, and the characteristic 'peak' is lost. Figure 5.15 shows that the average current density over the disc increases significantly as the disc radius decreases.

Figures 5.16(A–F) and 5.17(A–F) show analogous voltammetry simulated for the fast electrode kinetic limit of $k_0 = 1$ cm s^{-1}. Again the transition for peak shaped cyclic voltammograms for small electrode is apparent as is the greatly increased current density on the smaller electrodes.

The cyclic voltammetry peak shaped response is characteristic of linear diffusion and when it occurs the diffusion layer is almost flat in shape whereas the steady-state limit is characteristic of convergent diffusion and an approximately hemispherical shaped diffusion layer, see Fig. 5.18 for the diffusion layers associated with different types of voltammogram.

It follows that if the time for reaching steady-state $\left(\sim r_e^2/D\right)$ is short compared to the time for sweeping the voltammogram $(\sim RT/Fv)$ then steady-state-like behaviour will be seen if:

$$\frac{r_e^2}{D} \ll \frac{RT}{Fv}. \tag{5.34}$$

It is therefore possible to define a parameter

$$p = \sqrt{Fr_e^2 v/RTD}, \tag{5.35}$$

so that $p \ll 1$ corresponds to steady-state behaviour while $p \gg 1$ gives the 'cyclic voltammetry' limit. For the electrochemically reversible limit the peak current can be empirically expressed as

$$I_p = 4Fr_e[A]_{bulk}[0.34e^{-0.66p} + 0.66 - 0.13e^{-11/p} + 0.351p]. \tag{5.36}$$

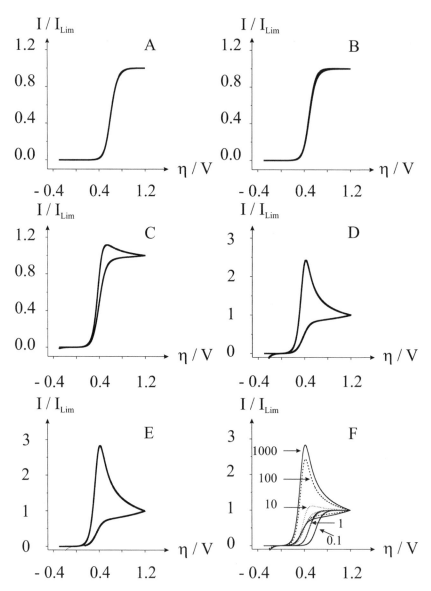

Fig. 5.14 Normalised current versus overpotential for the irreversible oxidation of A to B on a microdisc electrode. $E_f^0 = 0\,V$; $\alpha = 0.5$; $k_0 = 10^{-5}\,cm\,s^{-1}$; $\upsilon = 1\,V\,s^{-1}$; $[A]_0 = 1\,mM$; $D = 10^{-5}\,cm^2\,s^{-1}$. Plots A to E show the responses of electrodes of size 0.1, 1.0, 10, 100 and 1000 μm; Plot F shows the five plots superimposed. $\eta = E - E_f^0$.

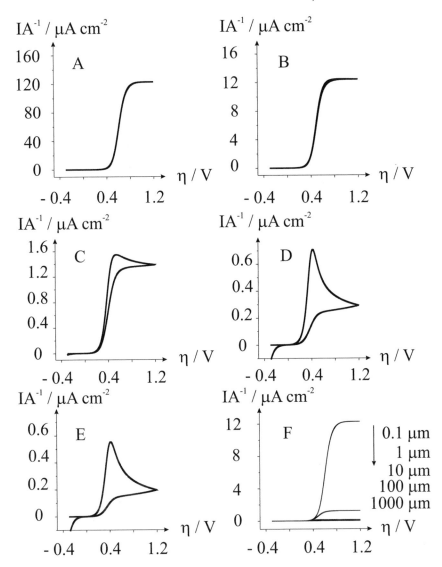

Fig. 5.15 Current density (I/A) versus overpotential for the irreversible oxidation of A to B on a microdisc electrode. $E_f^0 = 0\,\text{V}; \alpha = 0.5; k_0 = 10^{-5}\,\text{cm s}^{-1}; \upsilon = 1\,\text{V s}^{-1}; [A]_0 = 1\,\text{mM};$ $D = 10^{-5}\,\text{cm}^2\,\text{s}^{-1}$. Plots A to E show the responses of electrodes of size 0.1, 1.0, 10, 100 and 1000 μm; Plot F shows the five plots superimposed. $\eta = E - E_f^0$.

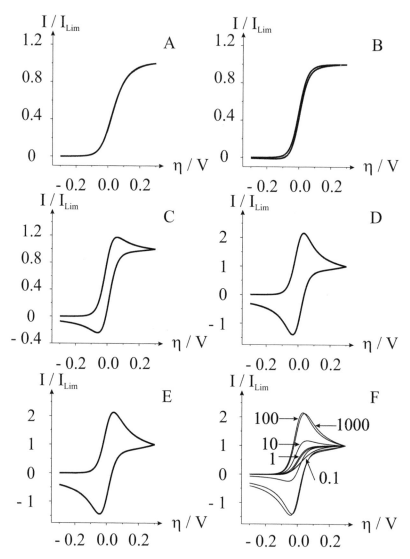

Fig. 5.16 Normalised current versus overpotential for the reversible oxidation of A to B on a microdisc electrode. $E_f^0 = 0\,\text{V}$; $\alpha = 0.5$; $k_0 = 1\,\text{cm s}^{-1}$; $\upsilon = 1\,\text{V s}^{-1}$; $[A]_0 = 1\,\text{mM}$; $D = 10^{-5}\,\text{cm}^2\,\text{s}^{-1}$. Plots A to E show the responses of electrodes of size 0.1, 1.0, 10, 100 and 1000 μm; Plot F shows the five plots superimposed. $\eta = E - E_f^0$.

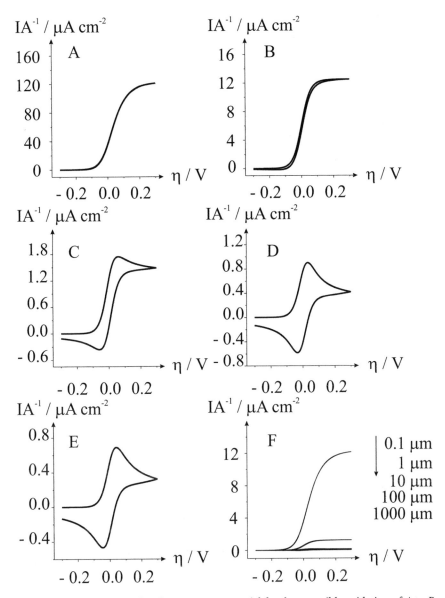

Fig. 5.17 Current density (I/A) versus overpotential for the reversible oxidation of A to B on a microdisc electrode. $E_f^0 = 0\,\mathrm{V}; \alpha = 0.5; k_0 = 1\,\mathrm{cm\,s^{-1}}; \upsilon = 1\,\mathrm{V\,s^{-1}}; [A]_0 = 1\,\mathrm{mM};$ $D = 10^{-5}\,\mathrm{cm^2\,s^{-1}}$. Plots A to E show the responses of electrodes of size 0.1, 1.0, 10, 100 and 1000 μm; Plot F shows the five plots superimposed. $\eta = E - E_f^0$.

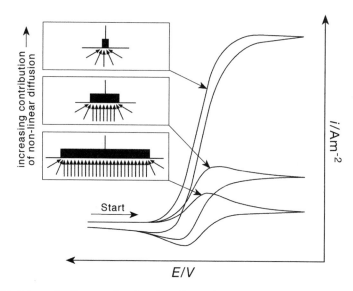

Fig. 5.18 Schematic diagram showing the relationship between the size of the electrode (with respect to the diffusion layer thickness) and the contribution of convergent diffusion to the observed voltammetry.

The limits are ready seen to be

$$p \to 0, \quad I_p \to 4Fr_e[A]_{bulk} \tag{5.37}$$

$$p \to \infty, \quad I_p \to 1.4Fr_e[A]_{bulk}p. \tag{5.38}$$

The current in the $p \to 0$ limit is the expected state current, whilst the expression for I_p as $p \to \infty$ corresponds to the expected peak current at a large, planar electrode. Figure 5.19(A) shows this transition graphically. Also shown (B) is the corresponding behaviour in the irreversible limit. Table 5.4 summarises the behaviour in the two limits.

We have noted above that the transition from steady-state to 'cyclic voltammetric' type response at microdisc electrodes depends on the rate of diffusion relative to the voltage sweep rate. This point can be emphasised with reference to the electro-reduction of O_2.

$$O_2 + e \to O_2^{\bullet -}$$

in the RTIL $[N_{6222}][NTf_2]$ where the two species have markedly different diffusion coefficients:

$$D(O_2) = 1.48 \times 10^{-10} m^2 s^{-1}$$

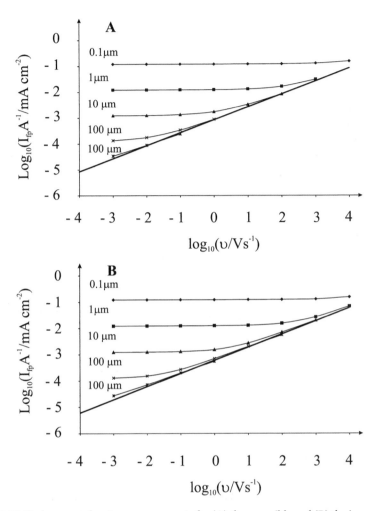

Fig. 5.19 Peak current density versus scan rate for (A) the reversible and (B) the irreversible oxidation of A to B. Parameters for (A): $E^0 = 0\,V$; $\alpha = \beta = 0.5$; $k = 1\,cm\,s^{-1}$; $[A]_0 = 1\,mM$; $D_A = D_B = 10^{-5}\,cm^2\,s^{-1}$. Parameters same for (B) as (A) except $k = 10^{-5}\,cm\,s^{-1}$. The solid line shows the expected macro planar electrode behaviour.

$$D(O_2^{\bullet -}) = 4.66 \times 10^{-12}\,m^2\,s^{-1}.$$

This leads to the curious microdisc voltammetry shown in Fig. 5.20 where for a scan rate of 1 Vs^{-1} both steady-state and transient behaviours can be seen.

On the forward scan, corresponding to the reduction of the faster moving O_2, a near steady-state current voltage curve is seen. On the reverse scan, corresponding

Table 5.4. Cyclic voltammetry characteristics associated with planar and radial diffusion.

Property	Dominant form of mass transport	
	Planar diffusion	Radial diffusion
δ versus r_e	$\delta \ll r_e$	$\delta \gg r_e$
Type of response	clear peak $\rightarrow I_p$	steady-state $\rightarrow I_{lim}$
Scan rate dependence?	Yes	No
Current dependence	$I_p \propto v^{0.5}; I_p \propto D^{0.5}; I_p \propto A$	$I_{lim} \propto r_e; I_{lim} \propto D$

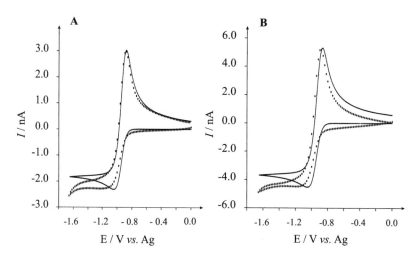

Fig. 5.20 Comparison of experimental and simulated voltammograms ($1\,\mathrm{Vs^{-1}}$ scan rate) for the reduction of oxygen in $[N_{6222}][N(Tf)_2]$ in (A) 50% O_2 and 50% N_2 and (B) 90% O_2 and 10% N_2 gas mixtures. The open circles represent experimental data and the solid line represents simulated data. Reprinted with permission from Ref. [6]. Copyright (2003) American Chemical Society.

to the re-oxidation of the much more slower moving $O_2^{\bullet-}$ a peak response is evident! Figure 5.21 shows the concentration profiles for the two species at different points in the voltammogram. It is clear that steady-state diffusion profiles of O_2 are established on the forward scan but that this is not the case on the reverse scan. Note that the slowness of the diffusion of $O_2^{\bullet-}$ leads to a large concentration build up of this species near the electrode; the concentration reaches several times that of the bulk concentration of O_2.

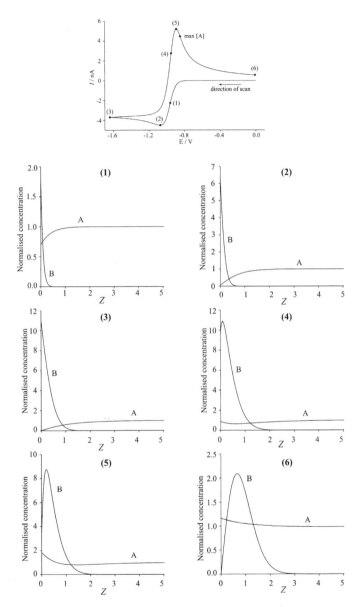

Fig. 5.21 Simulated cyclic voltammogram for the scan rate of $1\,\mathrm{Vs^{-1}}$ with the specific points marked. The concentration distributions corresponding to the points are also shown. The point marked as 'max $[A]$' corresponds to the maximum surface concentration of oxygen, O_2. Note that A = oxygen and B = superoxide. Also $Z = z/r_e$ and the concentration is normalised to bulk A. Reprinted with permission from Ref. [6]. Copyright (2003) American Chemical Society.

5.7 Steady-State Voltammetry at the Microdisc Electrode

Microdisc electrode voltammograms run at sufficiently low voltage scan rates so that authentic steady-state behaviour can be analysed to give information about the electrode kinetics of the electrode process, provided the microdisc is sufficiently small that electrochemically irreversible or quasi-reversible behaviour is seen. This requires the mass transport coefficient to be large compared with k^0:

$$m_T \sim \frac{D}{r_e} \gg k^0. \tag{5.39}$$

Microdiscs of a few microns in size are routinely available in most laboratories so that $r_e \sim 10^{-4}$ cm and given a typical diffusion coefficient of $D \sim 10^{-5}$ cm^2 s^{-1} it is clear that quite fast electrochemical rate constants can be measured in this way, although clearly not the fastest (see Chapter 2).

The shape of the voltammetric wave when the diffusion coefficients of the reduced and oxidised species are equal, and when only one species is present in solution, has been summarised by Aoki[7] building on the perceptive work of Oldham and Zoski[8] by means of the approximate equation

$$\frac{I}{I_{\lim}}(1 + \exp(\pm\varsigma)) = \frac{\lambda}{\left\{\lambda + \frac{2\lambda+12}{\lambda+3\pi}\right\}}, \tag{5.40}$$

where

$$\lambda = \frac{k^0 r_e}{D}\left\{e^{-\alpha\xi} + e^{\beta\xi}\right\} \tag{5.41}$$

and

$$\varsigma = \frac{F}{RT}(E - E_f^0). \tag{5.42}$$

The term I_{ss} is the steady-state limiting current, α and β are the transfer coefficients. Note that as k^0 becomes large, $\lambda \to \infty$ and the voltammetric wave shape is

$$\frac{I}{I_{\lim}} = \frac{1}{1 + \exp(\pm\varsigma)}, \tag{5.43}$$

where the + corresponds to a reduction and − to an oxidation. Equation (5.43) applies to any electrochemically reversible voltammogram measured under diffusionally steady-state conditions.

A much used method for extracting kinetic data was introduced by Mirkin and Bard.[9] They presented a simple way of determining values for k^0, α and E_f^0 from the experimental parameters $\Delta E_{1/4}$ and $\Delta E_{3/4}$, extracted directly from quasi-reversible steady-state voltammograms as shown in Fig. 5.22.

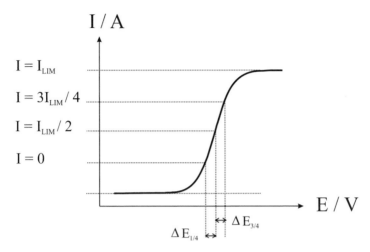

Fig. 5.22 Diagram explaining how the parameters, $\Delta E_{1/4}$ and $\Delta E_{3/4}$ are extracted from a steady-state voltamogram.

There are two sets of tables produced by Mirkin and Bard, one applicable to uniformly accessible electrodes such as the rotating disc electrode (see Chapter 8), and another that is applicable to non-uniformly accessible microdisc electrodes. The required kinetic data can be read directly from the tables by inputing values for $\Delta E_{1/4}$ and $\Delta E_{3/4}$. For values of these parameters that fall between the ones given in the tables, there is an interpolation method given in the paper.

The greatest advantage of this method is its ease of use, requiring only *one* voltammogram to output all the kinetic data required without any complicated graphical or numerical analysis (but accuracy can be improved by taking several scans and averaging the results). As the voltammogram is recorded under steady-state conditions, the quartile potentials can be determined with a high degree of accuracy, without the errors that can be introduced using transient voltammetry by charging currents and ohmic potential drop. The accuracy is also improved as the results are independent of concentration of electroactive species and electrode area. However, there is an upper limit of around 0.05 cm s^{-1} to the k^0 values that can be measured using this method, enforced by diffusion limited kinetics at *ca* micron-sized microdisc electrodes.

5.8 Microelectrodes versus Macroelectrodes

One definition of a microelectrode is an electrode which has its characteristic dimension on the scale of micrometers (see Fig. 5.23).

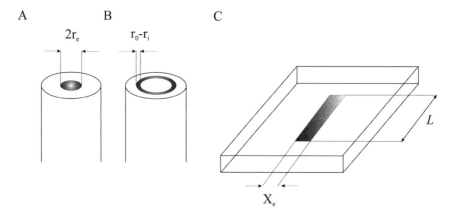

Fig. 5.23 Commonly used microelectrode geometries: (A) microdisc, (B) microring, and (C) microband. In each case the characteristic dimension(s) are indicated.

Smaller electrodes may also be included in this category, or may be called 'nanoelectrodes' or nanodes. A macroelectrode is any electrode larger than a microelectrode, usually having a characteristic dimension on the scale of millimetres or centimetres.

Microelectrodes differ in their behaviour from macroelectrodes in several key ways. These key differences favour the use of microelectrodes in the measurement of fast electrochemical processes.

- Non planar diffusion. As noted earlier in this chapter microelectrodes enjoy faster rates of mass transport than macroelectrodes, facilitating the measurement of faster kinetics both heterogeneous (k^0, see previous section) and homogeneous (see Chapter 7).
- Reduced capacitance. The capacitance of the double layer is directly proportional to the area of the electrode. Thus, decreasing the area of the electrode immediately reduces the capacitance, C, of the system. Note that the time constant of the electrochemical cell is given by the product $R_s C$, where R_s is the solution resistance. Accordingly, when the potential applied to the electrode is changed, the currents that flow simply due to the movement of the ions comprising the supporting electrolyte only decay away on the timescale of $R_s C$ so that for any electrode there is an upper limit on the voltage scan rate that can be usefully employed, corresponding to the point at which these charging currents obscure the Faradic process of interest. Note that, since by definition of capacitance as

$$C = q/E,$$

where q is charge and E is voltage, then

$$I_{cc} = Cv,$$

if C is a constant. Accordingly, the charging current I_{cc}, scales directly with the voltage scan rate v. Since in the limit of planar diffusion the Faradaic current, I_F, scales as

$$I_F \propto v^{1/2},$$

it follows that

$$\frac{I_{cc}}{I_F} \propto v^{1/2}$$

and at some sufficiently large voltage sweep rate the capacitative currents will swamp the Faradaic currents of interest.

- Reduced ohmic drop (see Chapter 2). The ohmic drop is proportional to the total current passed. The smaller electrode size decreases this current. This has the advantage (Chapter 2) that under certain circumstances a two-electrode arrangement may be used in place of the more common three-electrode systems, on the basis that the potential of a combined reference/counter electrode is unlikely to shift significantly if the currents passed are of the order of nanoampere or less.

The use of microelectrodes, utilising all of the above advantages, for state-of-the-art fast scan cyclic voltammetry is described in the next section. Before that, however, we point that one further significant advantage of microelectrodes over macroelectrodes is the fact that they permit voltammetric studies in high resistance organic solvents such as toluene, benzene and heptane. Figure 5.24 shows illustrative data for the one-electron oxidation of ferrocene in toluene and in benzene.

Tetra-n-butylammonium perchlorate was used as the supporting electrolyte in each case. Although the observation of voltammetric waves in these circumstances is impressive, the distorting power of ohmic drop is almost always ever present. Last, we note that microelectrode voltammetry has even been attempted in the gas phase![11] Figure 5.25 shows 'voltammograms' in which the 'steps' are attributed to the following processes:

$$Fe^+(g) + e \rightleftarrows Fe(s)$$

$$Cu^+(g) + e \rightleftarrows Cu(s).$$

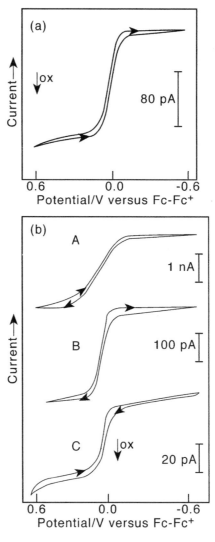

Fig. 5.24 Voltammetry in high resistance aromatic hydrocarbon solvents. (a) Oxidation of 5×10^{-5} M ferrocene at 25°C in toluene (0.1 M Hex$_4$NClO$_4$) at a 5 micron radius platinum electrode using a scan rate of 20 mVs^{-1}. (b) Oxidation of 1×10^{-5} mM, A; 1×10^{-4} mM, B; 1.6×10^{-5} mM ferrocene at 22°C in benzene (0.1 M Hex$_4$NClO$_4$) at a 5 micron radius platinum electrode using a scan rate of 50 mVs^{-1}. Reprinted from Ref. [10] with permission from Elsevier.

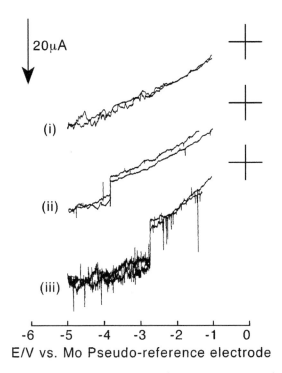

Fig. 5.25 Cyclic voltammograms between $-5\,\text{V}$ and $-1\,\text{V}$ versus Mo pseudo-reference electrode: (i) with deionised water, (ii) with 0.05 M iron sulphate solution in deionised water, (iii) 0.05 M copper sulphate nebulised into a flame. There is no smoothing of the data . The scan was started at $-1\,\text{V}$ at a rate of $0.6\,\text{V\,ss}^{-1}$. Reprinted from Ref. [11] with permission from Elsevier.

5.9 Ultrafast Cyclic Voltammetry: Megavolts per Second Scan Rates

As a result of pioneering work in Paris,[12,13] instrumentation has been designed in the form of a three-electrode potentiostat involving feedback compensation to facilitating the study of nanosecond timescales by allowing the recording of ohmic drop-free voltammograms at scan rates of a few megavolts per second. This range of scan rates corresponds to the development of diffusion layers whose dimensions are only a few nanometers thick.

Figure 5.26 shows data attained at scan rates of up to and in excess of $2 \times 10^6\ \text{Vs}^{-1}$ for the reduction of anthracene in acetonitirile containing *ca.* 1.0 M

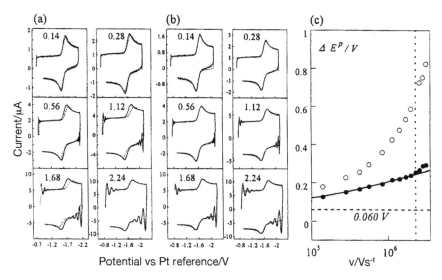

Fig. 5.26 Cyclic voltammetry of anthracene (14.3 mM) reduction in acetonitrile (0.9 M NEt$_4$BF$_4$) at a gold disk electrode (25 microns) at different scan rates at 20°C: (a) 100% compensated (solid curves) versus un-compensated (dashed curves) voltammograms; (b) 100% compensated (solid curves) versus simulated (dashed curves) voltammograms. (c) Variations of the peak to peak potential as a function of scan rate for uncompensated (open circles) or 100% compensated (solid circles) modes. Solid curves are the theoretical predictions. Reprinted from Ref. [12] with permission from Elsevier.

tetraethylammonium tetrafluroborate as supporting electrolyte:

As can be seen cyclic voltammetric peak shaped response are seen as the very thin diffusion layers ensure planar diffusion. Modelling of the peak-to-peak potential separation using theory as given in Chapter 4 enabled the inference of the very fast rates constant (k^0) of 5.1 cm s^{-1} and a transfer coefficient (α) of 0.45.

5.10 Ultrasmall Electrodes: Working at the Nanoscale

One message of the previous section has been that as far as electrode dimensions are concerned — 'the smaller the better'! We saw that reduction of size conferred enhanced mass transport rates, reduced *IR* drop and also diminished double layer charging effects. Whilst micron-sized electrodes are commercially available there is

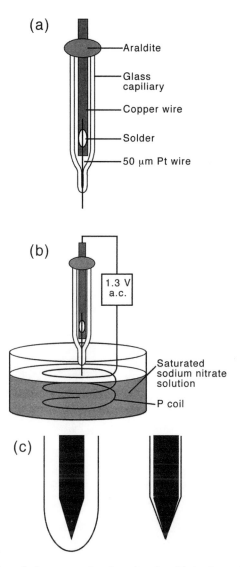

Fig. 5.27 Schematic of the stages involves in the fabrication of platinum ultra-microelectrodes. (a) A platinum wire electrode is constructed. (b) The platinum is etched to a fine point by placing the wire in a saturated sodium nitrate solution and applying an alternating voltage of 1.2 V between the wire and a platinum coil in which the platinum wire is located centrally. (c) The electrode is insulated by the electrodeposition of a paint (i), which retracts from the tip end during curing to expose a nanoscopic platinum (ii). Reprinted from Ref. [14] with permission from Elsevier.

significant research into the manufacture and use of electrodes with characteristic dimensions in the sub-micrometer to nanometer range. These electrodes have been used to study rapid electron transfer kinetics, amongst other applications. A decrease in electrode dimension also allows measurements to be made in micro-environments such as single cells and in living tissue where they cause minimal physical damage.

One popular and effective method for the production of 'nanodes' is due to Macpherson and Unwin.[14] The electrodes are prepared by etching a fine platinum wire and then coating the etched wire with electrophoretically deposited paint which retracts from the tip end during curing to expose a nanoscale platinum electrode. The size of the electrochemically active area is controlled by the number of paint coatings applied during the insulation procedure. Figure 5.27 summarises the manufacture process which produces electrodes with effective radii in the range 1 μm to 10 nm.

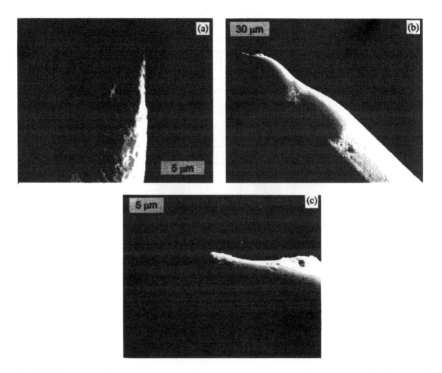

Fig. 5.28 Scanning electron micrographs of an etched wire (a) and an anodically coated platinum wire at low resolution (b) and high resolution (c). Reprinted from Ref. [14] with permission from Elsevier.

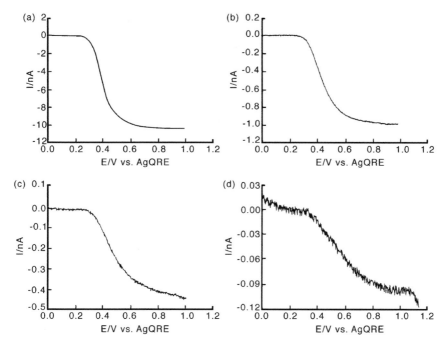

Fig. 5.29 Steady-state voltammograms for the oxidation of 0.2 mM ferrocyanide in 0.2 M potassium chloride using electrodes of radii (a) 1.2 microns, (b) 120 nm, (c) 49 nm and (d) 13 nm. The reference electrode was a silver quasi-reference electrode (Ag QRE) with a scan rate of 20 mVs^{-1}. Reprinted from Ref. [14] with permission from Elsevier.

Figure 5.28 Shows images of these electrodes and Fig. 5.29 illustrates voltammograms for the oxidation of ferricyanide, $Fe(CN)_6^{4-}$ at such electrodes. Other procedures for making gold and carbon nanodes have been reported.[15,16] We note that an alternative approach to voltammetry at nanoscale involves the use of arrays of nano-particles partitioned sufficiently apart as to confer appreciate diffusional isolation;[17] arrays of electrodes are considered in the next chapter.

References

[1] D. Shoup, A. Szabo, *J. Electroanal. Chem.* **140** (1982) 237.

[2] M.C. Buzzeo, DPhil. Thesis, University of Oxford, UK, 2005.

[3] C.A. Paddon, F.L. Bhatti, T.J. Donohoe, R.G. Compton, *J. Electroanal. Chem.* **589** (2006) 187.

[4] O.V. Klymenko, R.G. Evans, C. Hardacre, I.B. Svir, R.G. Compton, *J. Electroanal Chem.* **571** (2004) 211.

[5] R.G. Evans, O.V. Klymenko, P.D. Price, S.G. Davis, C. Hardacre, R.G. Compton, *Chem. Phys. Chem.* **6** (2005) 526.

[6] M.C. Buzzeo, O.V. Klymenko, J.D. Wadhawan, C. Hardacre, K.R. Seddon, R.G. Compton, *J. Phys. Chem. A.* **107** (2003) 8872.

[7] K. Aoki, *Electroanalysis* **5** (1993) 627.

[8] K.B. Oldham, C.G. Zoski Oldham, *J. Electroanal. Chem.* **313** (1991) 17.

[9] M.V. Mirkin, A.J. Bard, *Anal. Chem.* **84** (1992) 2293.

[10] A.M. Bond, T.F. Mann, *Electrochimica Acta* **32** (1987) 863.

[11] D.J. Caruana, S. P. McCormack, *Electrochem. Commun.* **2** (2000) 816.

[12] C. Amatore, E. Maisonhaute, G. Simonneau, *Electrochem. Commun.* **2** (2000) 81.

[13] C. Amatore, E. Maisonhaute, G. Simonneau, *J. Electroanal. Chem.* **486** (2000) 141.

[14] C.J. Slevin, N.J. Gray, J.V. Macpherson, M.A. Webb, P.R. Unwin, *Electrochem. Commun.* **1** (1999) 282.

[15] D-H. Woo, H. Kang, S. M. Park, *Anal. Chem.* **75** (2003) 6732.

[16] C. Wang, Y. Chen, F. Wang, X. Hu, *Electrochemica Acta* **50** (2005) 5588.

[17] C.M. Welch, R.G. Compton, *Anal. Bioanal. Chem.* **384** (2006) 601.

6

Voltammetry at Heterogeneous Surfaces

The purpose of this chapter is to further develop the conceptual insights from the previous two chapters in the context of electrodes which can be described as spatially heterogeneous. That is to say, the electrochemical activity varies over the surface of the electrode. Specific examples embrace

- Partially blocked electrodes
- Microelectrode arrays
- Electrodes made of composite materials
- Porous electrodes
- Some modified electrodes.

The simulation of the voltammetric response at such surfaces is challenging and interesting because of the surface variation and because of the often random distribution of the zones of different electrode activity. We start by considering the case of partially blocked electrodes.

6.1 Partially Blocked Electrodes

Figure 6.1 illustrates the concept of a partially blocked electrode. It shows a macro-electrode covered with inert particles of a material different to that of the underlying electrode surface and which, when the electrode is immersed in solution, 'block' the diffusional path of the electroactive species to the electrode surface. At first

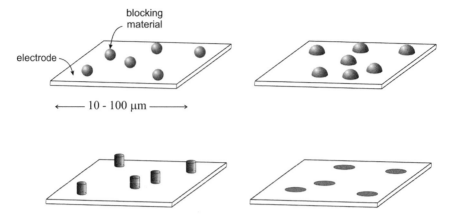

Fig. 6.1 Schematic representation of partially blocked electrodes shown for the case of spherical particles (top left), hemispheres (top right), cylinders (bottom left) and flat discs (bottom right). Reprinted from Ref. [1] with kind permission of Springer Science and Business Media.

sight this appears to be a straightforward issue of the relative geometrical areas of the 'blocked' and exposed parts of the electrode surface. However, we will see below that this conclusion is only correct if both zones of the electrodes — blocked and exposed — are of macro size. If they are of micron-sized dimensions, then the voltammetric response is much more difficult to predict. We will use the ideas developed in this section later in the chapter to describe the voltammetric response of microelectrode arrays and of electrode surfaces modified with electro-catalysts and further demonstrate that voltammetry can be used for particle sizing.

We first focus on partially blocked electrodes assuming the blocks to be the flat discs shown in Fig. 6.1. In terms of practical voltammetry, we might adopt the view that we are posing the question 'how free from dirt must my electrode be in order to see the 'ideal' voltammetric response predicted in Chapters 3 and 4?' The answer, surprisingly, is that even a substantially contaminated electrode can give a near ideal signal! This is both a strength and a weakness of voltammetry.

Figure 6.2 shows an array of electrochemically inactive flat circular blocks (discs) supported on a flat macro-electrode.

The discs are arranged in a cubic distribution. We pose the question that, for the single electrode process

$$A(aq) + e \underset{k_a}{\overset{k_c}{\rightleftarrows}} B(aq),$$

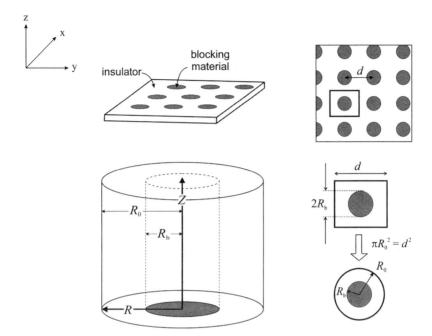

Fig. 6.2 The diffusion domain problem for electrochemically inert blocks in a cubically spaced distribution. Reprinted from Ref. [1] with kind permission of Springer Science and Business Media.

what is the effect of the number and size of the blocks on the observed cyclic voltammetry? We assume that the electrode kinetics follows the Butler–Volmer model (Chapter 2)

$$k_c = k^0 \exp\left(\frac{-\alpha F}{RT}\eta\right) \tag{6.1}$$

$$k_a = k^0 \exp\left(\frac{\beta F}{RT}\eta\right) \tag{6.2}$$

where α and β are transfer coefficients such that $\alpha + \beta = 1$, k^0 is the standard electrochemical rate constant and η is the overpotential where

$$\eta = E - E_f^0(A/B), \tag{6.3}$$

E is the applied working potential and $E_f^0(A/B)$ is the formal potential of the A/B redox couple.

The sought cyclic voltammetric response can be found from

$$\frac{\partial [A]}{\partial t} = D_A \nabla^2 [A] \tag{6.4}$$

and

$$\frac{\partial [B]}{\partial t} = D_B \nabla^2 [B], \tag{6.5}$$

where

$$\nabla^2 \equiv \frac{\partial^2}{\partial x^2} + \frac{\partial^2}{\partial y^2} + \frac{\partial^2}{\partial z^2},$$

subject to the boundary conditions (6.1), (6.2) and (6.3) with E varying between the desired sweep range limits of E_1 and E_2 as in Chapter 4. Unlike the problems solved in Chapter 4 the above problem is posed in three Cartesian directions, x, y and z, as defined in Fig. 6.2. This makes the simulation computationally challenging, and anticipating the need to subsequently address the problem of randomly distributed blocks where the computational challenge would be significantly yet greater, we adopt the diffusion domain approximation.

The diffusion domain approximation is shown in Fig. 6.2 for cubic arrays of blocks. The electrode surface is first split into square cells each of equal area. The diffusion domain approximation then substitutes each unit cell by a circular domain of identical area to the square unit cell. Accordingly, if the centre-to-centre distance between adjacent blocks is p, then the radius R_0 of the circular domain is given by

$$p^2 = \pi R_0^2, \tag{6.6}$$

so that

$$R_0 = 0.564p. \tag{6.7}$$

Similarly, for a hexagonal array of blocks (see Fig. 6.3), where the unit cells are rectangular, if $x = p$ then $y = \sqrt{3/2}\, p^2$, so that

$$\pi R_0^2 = \sqrt{3/2}\, p^2 \tag{6.8}$$

and hence

$$R_0 = 0.5258p. \tag{6.9}$$

With this approximation, the diffusion Eqs. (6.4) and (6.5) can be solved within the cylindrical domain illustrated in Fig. 6.2, adopting the coordinates Z as the direction normal to the electrode surface and R as the radial coordinate. R_0 is the radius of the domain and R_b the radius of the block. Accordingly, the three

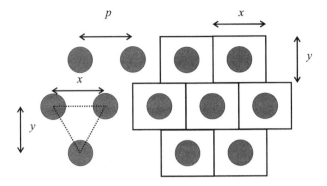

Fig. 6.3 Hexagonal block packing arrangement.

dimensional problem of Eqs. (6.4) and (6.5) is reduced by the diffusion domain approximation to the tractable two dimensional problem of solving the equations

$$\frac{\partial [A]}{\partial t} = D_A \frac{\partial^2 [A]}{\partial R^2} + \frac{D_A}{R} \frac{\partial [A]}{\partial R} + D_A \frac{\partial^2 [A]}{\partial Z^2} \tag{6.10}$$

$$\frac{\partial [B]}{\partial t} = D_B \frac{\partial^2 [B]}{\partial R^2} + \frac{D_B}{R} \frac{\partial [B]}{\partial R} + D_B \frac{\partial^2 [B]}{\partial Z^2} \tag{6.11}$$

subject to the same boundary conditions as above but with additional boundary condition of 'no flux' being applied to the walls of the cylinder:

$$R = R_0, \text{ all } Z : D_A \frac{\partial [A]}{\partial R} = D_B \frac{\partial [B]}{\partial R} = 0,$$

and another specifying the bulk solution concentrations,

$$0 < R < R_b, \ Z \to \infty \quad [A] \to [A]_{bulk}; \ [B] \to 0.$$

In this manner, each cylindrical domain is diffusionally independent of its neighbours and so the voltammetric response of the whole macroelectrode is simply that of a single diffusion domain scaled up by the total number of blocks on the surface, assuming these to be significantly numerous that 'edge effects' are relatively insignificant.

The insightful results of the diffusion domain calculations can be exemplified by posing a simple question: 'If I take a macroelectrode of typical dimensions, say 4 mm × 4 mm, what is the effect on a typical voltammogram if one half of the total surface area is covered by inert blocks in the form of flat discs?' To answer this question we assume that a representative voltammogram relates to the following parameters:

$$k^0 = 10^{-2} \text{cm s}^{-1}; \ \alpha = 0.5,$$
$$D_A = D_B = 10^{-5} \text{cm}^2 \text{ s}^{-1}$$

and assume a voltage scan rate of 0.1 Vs^{-1}. We note that the 50% blockage stipulated requires a surface coverage,

$$\Theta = 0.5 = \frac{\pi R_b^2}{\pi R_0^2} = \left(\frac{R_b}{R_0}\right)^2, \tag{6.12}$$

but that we can realise the coverage by either having a small number of large blocks or a large number of small blocks. Accordingly, we consider the results for three different cases:

(a) the electrode has a single macroblock of size $R_0 = 0.18$ cm, noting that $\pi \times 0.18^2 = 0.5 \times 0.4 \times 0.4$

(b) the electrode is covered with 50 μm radius blocks with centre-to-centre separations (p) of 200 μm, and

(c) the electrode is half covered with a cubic array of 1 μm radius blocks.

The three different cyclic voltammetric responses are shown in Fig. 6.4 along with the response of the unblocked electrode and the contrasts are startling. For the case (a) of a single macroblock ($R_0 = 0.18$ cm) the currents flowing are very close to one half of those seen at the unblocked electrode. In contrast — case (c) — the effect of blocking one half ($\Theta = 0.5$) at the electroactive surface with 1 μm blocks is tiny: the peak current is almost that seen at the unblocked electrode! The case (b), where the block size is 50 μm, is intermediate between these two limits of effectively unblocked and geometrically blocked behaviour. The differing behaviour can be understood if the concentration profiles of A, corresponding to the peak current in the voltammograms, are considered. These are shown in Fig. 6.5.

In the case of the single block, there is a linear diffusion to the active part of the electrode surface ($R_b < R < R_0$), whilst the concentration of A over the blocking disc remains effectively at its bulk value. That is because on the timescale of the experiment there is insufficient time for radial diffusion to remove material over a distance of 0.18 cm for the block and bring it to the active part of the electrode for electrolysis. For case (b), where there are 50 μm radius blocks, it can be seen that the timescale of the voltammetric sweep is such that the linear diffusion to the

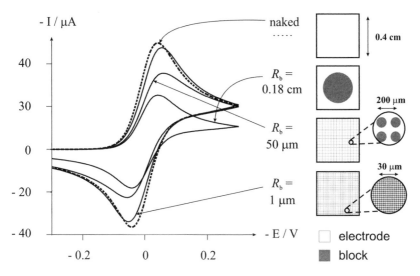

Fig. 6.4 The expected voltammetric responses for different sized blocks each corresponding to a total coverage of $\theta = 0.5$. Reprinted from Ref. [1] with kind permission of Springer Science and Business Media.

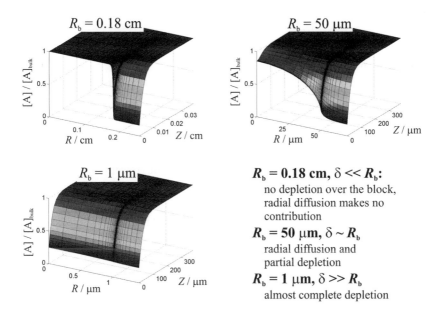

Fig. 6.5 Concentration profiles of A within different diffusion domains. Reprinted from Ref. [1] with kind permission of Springer Science and Business Media.

active part of the electrode is significantly augmented by radial diffusion and that there is sufficient time for significant but not complete depletion of A in the zone of the blocks. Finally, in case (c) where the blocks are 1 μm in radius the timescale of the experiment is ample to allow very efficient radial diffusion and so there is near complete electrolysis of the material over the entire electrode surface — both blocks and active zones.

Can we relate the timescale of the voltammetric experiment to the distances diffused? To do this, we return to the discussion in Chapter 3 and recall that the approximate distance δ diffused by a species with a diffusion coefficient D in a time t is

$$\delta = \sqrt{\pi D t}.$$

In considering the cyclic voltammetry problem posed above, we estimate the relevant value of t by considering the appropriate 'potential width' of the voltammogram, ΔE, so that

$$\delta = \sqrt{\pi D \frac{\Delta E}{v}},$$

where v is the voltage sweep rate and for a typical nearly electrochemically reversible voltammogram a value of ΔE of the order of 0.10 V is appropriate, as illustrated in Fig. 6.6. Accordingly, we can estimate

$$\delta \approx 55 \, \mu m,$$

using the scan rate adopted in the calculation of Fig. 6.4 (0.1 Vs^{-1}).

We can now rationalise the concentration profiles shown in Fig. 6.5 and consequently the voltammograms in Fig. 6.4. First, notice that 55 μm roughly corresponds to the thickness of the diffusion layer (in the Z direction) over the active part of the electrode. Second, 55 μm corresponds also to the appropriate distance diffused in the radial direction, so for case (a) the block radius is very large compared to the distance of radial depletion and hence the concentration of A in the region of the blocking disc is largely undepleted. In contrast, for case (c) where the block radius is 1 μm, this is a distance small compared to the scale of radial diffusion and so the concentration of A over the blocking disc is depleted almost as effectively as if the block were replaced by an active electrode. It follows that the voltammogram seen for case (c) will be almost the same as for the unblocked 4 × 4 mm electrode even though it is 50 % covered with inert material! Conversely, the voltammogram for case (a) will be scaled down in terms of current approximately by the fraction $(1 - \theta)$ corresponding to the fraction of active electrode. Case (b), where the block size is comparable to the diffusion distance, is clearly intermediate in behaviour.

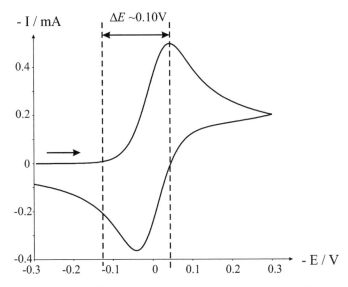

Fig. 6.6 The potential 'width' of a voltammogram. Reprinted from Ref. [1] with kind permission of Springer Science and Business Media.

At this juncture, it is worth pausing to ask about the accuracy and success of the diffusion domain approximation. Brookes *et al.*[2] and Davies *et al.*[3] have reported experiments with lithographically constructed partially blocked electrodes formed with both cubic and hexagonal arrays of circular blocking discs of various radii and coverages. Experimental voltammetry showed excellent agreement with the diffusion domain theory, so validating that as an approach used at various places in this chapter.

To further illustrate the power of the diffusion domain approximation, we next address the issue of randomly blocked electrodes. Figure 6.7 shows how the approximation is applied in this case, again assuming the blocks to be flat discs that are voltammetrically inactive. Here, the surface is initially decomposed into a set of Voronoi cells, as shown in Fig. 6.7.

The decomposition can be understood as follows. Each of the N discs on the surface is treated as a point located at the disc centre. The nearest neighbours of every point are then identified and the distances between each set of neighbours are bisected. Polygonic Voronoi cells are then generated by linking together all the 'midway' points surrounding any particular disc centre. Each disc then occupies its own Voronoi cell of area A_n, where n is a label identifying each of the N disc points

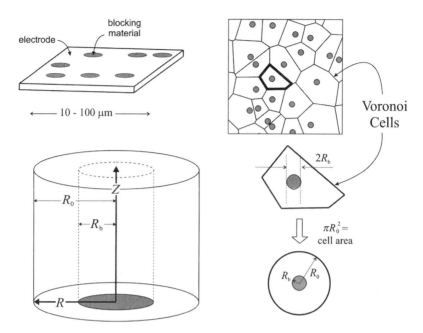

Fig. 6.7 Illustration of blocking material randomly distributed over an electrode surface, and the construction of Voronoi cells. Reprinted from Ref. [1] with kind permission of Springer Science and Business Media.

$(n = 1, 2, \ldots N - 1, N)$ so that the total area of the electrode, A_{elec} is given by

$$A_{elec} = \sum_{n=1}^{N} A_n.$$

Next, as with the cubic and hexagonal arrays, it is evident that a Voronoi cell is an awkward shape to simulate because of the intrinsic three-dimensional nature of the simulation, further complicated in the case of random distribution by the irregular shape of the cell. Accordingly, we again adopt the diffusion approximation and replace the Voronoi cells by circular domains of equal area, as shown in Fig. 6.7. As before, this results in a cylindrical diffusion domain in which to conduct the simulation of the diffusional response, now made tractable by the reduction of the problem to a two-dimensional simulation.

The simulation of the random distribution requires solving the transport Eqs. (6.4) and (6.5) as before, but now for a range of R_0 values. In the case of randomly distributed blocks, the nearest neighbours show a Poisson distribution:

$$P(R_o) = \frac{2\pi R_0 N}{A_{elec}} \exp\left(\frac{-\pi R_o^2 N}{A_{elec}}\right), \tag{6.13}$$

where the probability of finding domains of radius R_o is given by $P(R_0)$ and where the probability of finding domains of radius between R_0 and $R_0 + dR_0$ is given by $P(R_0)dR_0$. The cyclic voltammograms simulated for different values of R_0 are weighted accordingly and summed to produce the final voltammogram reflecting the entire surface; Fig. 6.8 illustrates this procedure.

Figure 6.9(A) shows an example of a lithographically constructed randomly blocked array in which a gold electrode has been 'blocked' with circular insulating disc, the coordinates of the centre of which have been assigned via random number generation. Figures 6.9(B) and (C) show examples for two different coverages of the excellent fit between theory and experiment in using the diffusion domain approximation in the very challenging area of randomly distributed blocks.

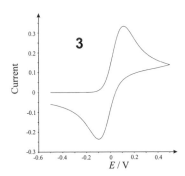

1. Simulate CVs for the range of domains present on the electrode surface.

2. Use a function to determine the number of each domain present.

3. Scale each CV and add together.

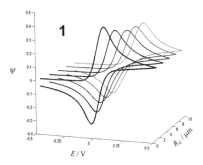

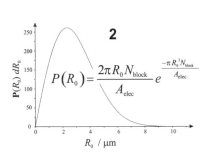

Fig. 6.8 Illustration of the steps involved in the simulation of cyclic voltammograms for a random ensemble of diffusion domains. Reprinted from Ref. [1] with kind permission of Springer Science and Business Media.

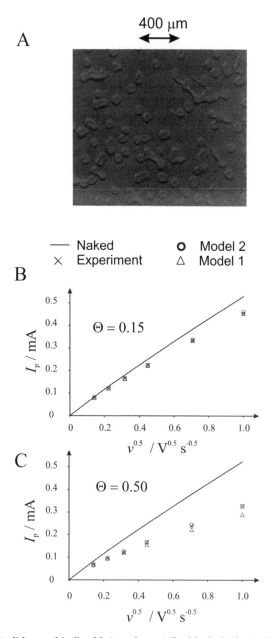

Fig. 6.9 (A): A lithographically fabricated partially blocked electrode with random arrays of discs on the surface. (B) and (C): The measured peak current as a function of the square root of the voltage scan rate using the experimental system of N,N,N',N'-tetramethylphenylenediamine in acetonitrile to verify the diffusion domain theory. Reprinted from Ref. [1] with kind permission of Springer Science and Business Media. Model 1 is that simulated for a regular array whilst model 2 is for a random array.

Why is the diffusion domain approximation so successful? Consider the walls of the Voronoi cells in Fig. 6.7. Because the walls are equidistant from the surrounding blocks, the concentrations of *A* and *B* on both sides of each wall will be similar. Thus, little or zero net flux of *A* or *B* will pass through the cell walls. Accordingly, each cell can be considered to be diffusionally independent — but not diffusionally isolated, since the zero flux boundary conditions allows for significant changes of concentrations from the bulk values at the cell wall — and the voltammogram of the whole surface can be simulated by simply summing the voltammograms of each unit cell on the electrode surface.

Last, we consider the effect of block radius and block coverage on the cyclic voltammogram observed at partially blocked electrodes and identify four limiting cases which are illustrated in Fig. 6.10.

Case 1

In this limit, the blocks and the electrochemically active parts of the surface are both of 'macro' dimension. Consequently, the active part of the electrode experiences linear (planar) diffusion, whilst the concentration of the electroactive species remains effectively that of the bulk solution in the zone of the blocking discs.

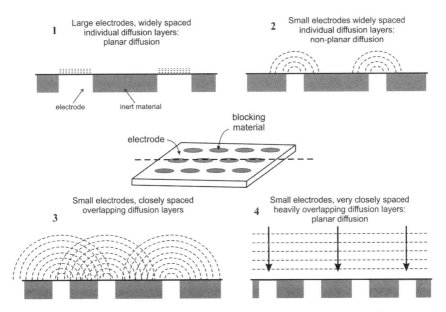

Fig. 6.10 Voltammetric behaviour for a heterogeneous electrode with active and inert parts illustrating Cases 1, 2, 3 and 4 (see text).

This limit is illustrated by the case of the single 0.18 cm radius block in Fig. 6.4. The voltammogram seen in this limit is exactly the same as seen at an unblocked electrode, except that the current is reduced by a factor of $(1 - \theta)$, where θ is the fraction of the surface covered with blocks.

Case 2

In this limit, the electroactive zones are of 'micro' dimensions but the zones of the inert blocks remain of macro size. The electrode therefore behaves as a collection of isolated microelectrodes, each of which experiences both radial and normal diffusion.

Case 3

Here, the electroactive zones of the electrode, as in case 2, behave with microelectrode characteristics but now the non-active (insulating and blocking) parts of the surface are sufficiently small, so that the diffusion fields of the two electrodes begin to overlap. This case is illustrated in Fig. 6.4 by the case of the 50 μm radius blocks.

Case 4

This represents the extreme limit of case 3, where the spacing between the electroactive zones is now so small that a very strong overlap of diffusional fields occurs, with the effect that the heterogeneous surface behaves almost like an unblocked electrode. The case is exemplified by the 1 μm radius blocks in Fig. 6.4 and the entire electrode effectively experiences planar diffusion. This limit has been explored using approximate analytical theory[4] with the conclusion that the electrode response is that of an unblocked electrode of the same size but that the Butler–Volmer rate constant is apparently reduced from k^0 to $(1 - \theta)k^0$. Simulations based on the diffusion domain approximation fully confirm this insight.[1]

Two significant insights result from generic studies of partially blocked electrodes. First, case (3) is of significance in studies of graphite electrodes, as discussed later in this chapter. In particular, whereas case 1 and case 4 show the characteristics of planar diffusion and hence voltammograms that can be simulated using the one dimensional theory of Chapter 4, and whereas case 2 shows the convergent diffusion of an array of isolated microelectrode, case 3 is significant since it is associated with cyclic voltammograms which *cannot* be simulated using planar diffusion, nor do the voltammograms resemble typical microelectrode voltammograms. Figure 6.11 shows an attempted best fit of a simulated case 3 voltammogram using a one-dimensional (planar) diffusion model.

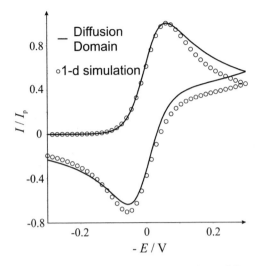

Fig. 6.11 Attempt to model a case 3 voltammogram using planar diffusion. Reprinted from Ref. [1] with kind permission of Springer Science and Business Media.

The best fit has been constructed so that the forward peak matches the case 3 voltammogram. Under these conditions, it can be seen that, first, the one-dimensional simulation underestimates the current flowing in the diffusional tail; second, the planar simulation overestimates the size of the reverse peak in comparison with the case 3 situation. Both of these distinguishing features are consistent with the electroactive zones of the electrode retaining vestigial micro-electrode characteristics — the pure microelectrode limit would be expected to give a steady plateau current in the diffusion controlled extreme and would have no reverse peak, so that the observed case 3 behaviour is part way to those expectations.

The second insight concerns the question we posed earlier as to the extent to which a blocked, contaminated or physically defective electrode might deviate from ideal electrode behaviour or not. Figure 6.12 shows SEM images of an electrode partially blocked with elevated cylindrical blokes with the latter forming a hexagonal array.

Each cylinder block has a radius of 50 μm and a height of 5.5 μm. The centre-to-centre distance, p, is 123 μm and the electrode has 550 blocks in an area of 7.3 $\times$ 10^{-2} cm^2. Current–voltage scans were measured at this electrode (Fig. 6.13(a)) using the well documented redox couple

$$Ru(NH_3)_6^{3+}(aq) + e \rightleftarrows Ru(NH_3)_6^{2+}(aq).$$

(a)

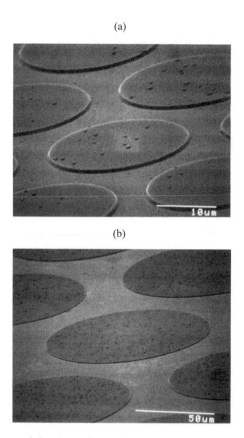

(b)

Fig. 6.12 SEM images of the electrode used in the study of electrodes partially blocked with elevated cylindrical blocks. Reprinted from Ref. [1] with kind permission of Springer Science and Business Media.

It was found that the peak currents varied with the square root of the scan rate as shown in Fig. 6.13(b).[5] Notice that the expected results were in near perfect agreement with those simulated using the diffusion domain approach throughout the range of scan rates used — 0.01 to 3 Vs^{-1}! More significantly, although the voltammograms were measured over this wide range of scan rates, which were changed by a factor of 300, a reasonable linear correlation between the peak current and the square root of the voltage scan rate is evident, despite the fact that ca 59% of the electrode surface is covered with inert cylindrical blocks some 5 μm in height! Of course, the slope of the peak current versus square root of scan rates plot is much less than that of the unblocked electrode (see Fig. 6.13), but the message is that to an unwary experimenter the voltammetric data in isolation might have implied

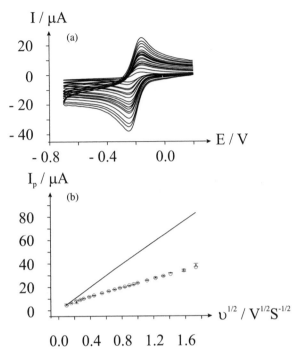

Fig. 6.13 (a) Experimental cyclic voltammograms for BG at varying scan rates in 1.08 mM $Ru(NH_3)_6^{3+}$/0.1 KCl solution. (b) Comparison between experimental and theoretical results for BG. The parameters used for the simulation are as follows: $[A]_{bulk} = 1.08$, $D = 6.2 \times 10^{-6}$ cm s^{-1}, $k_0 = $ cm s^{-1}. Shown in (b), the solid line represents the response of a planar electrode, squares (with error bars) are the experimental data and circles are the simulation data. Reprinted from Ref. [5] with permission from Elsevier.

a clean, unblocked electrode and a diffusion coefficient for $Ru(NH_3)_6^{3+}$ *ca.* four times less than reality (since the peak current scales with $D^{1/2}$). The implications for experimental practice are obvious.

6.2 Microelectrode Arrays

As explained in Chapter 5, the use of single microelectrodes has revolutionised voltammetry giving access to hitherto inaccessible short timescales. Nevertheless, for many practical purposes, most notably analytical measurements, it has proved convenient and advantageous to employ arrays containing hundreds or even thousands of microdisc electrode wires in parallel. Such arrays may be produced lithographically, in which case regular arrays, typically with cubic or hexagonal distribution of microdiscs, may be readily realised with disc radii of the micron or larger scale (Fig. 6.14).

On the other hand, an ingenious and relatively cheap and simple approach to the fabrication of microelectrode arrays is the adoption of 'random assemblies of microdiscs' or RAM^{TM} electrodes by Fletcher and Horne[6] in which the term 'assembly', rather than 'arrays', emphasises that the microdiscs are not regularly spaced. In a typical RAM *ca.* 3000 randomly dispersed carbon microfibres of a few

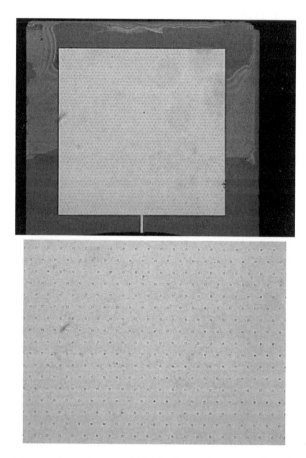

Fig. 6.14 A gold microelectrode array 'chip' is shown (top picture) with magnification (bottom picture) of the array where the hexagonal arrangement of the microelectrodes is clearly observed. The individual gold ultra-microelectrodes are 2.5 (radius) in a hexagonal arrangement and are separated from their nearest neighbours by 55 microns.

microns diameter are randomly embedded in epoxy resin. The microdiscs are the sectioned ends of the carbon fibres and between 20% and 40% of them are electrically connected. Regardless of the method of construction, microelectrode arrays provide an attractive alternative to macroelectrodes because they can provide — as explained below — a voltammetric response of similar magnitude to their macro counterparts, but with a considerably less background capacitive current. At the other extreme, in the case of regular arrays of electrodes produced lithographically, it is possible by careful design and section of experimental timescales to ensure that the microelectrodes behave independently of each other in the diffusional sense.

Figure 6.15(a) shows a regular cubic array of microelectrodes and the use of the diffusion domain approximation in rendering tractable the description of the mass transport to the array of electrodes.

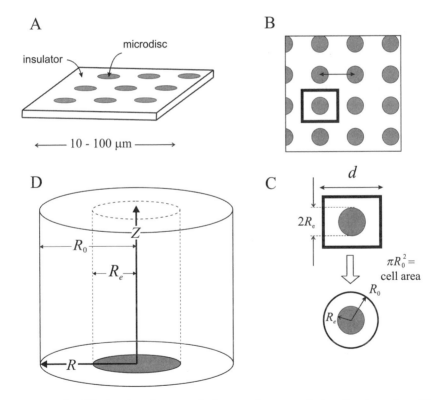

Fig. 6.15 The diffusion domain approach for a regular array of microdisc electrodes with cubic packing geometry. Reprinted from Ref. [1] with kind permission of Springer Science and Business Media.

On the basis of the ideas in Chapter 5 together with those of the previous section, it can be appreciated that in a linear sweep voltammetry experiment, as the redox reaction is driven by the applied potential, a depletion (diffusion) layer grows in the region of the electrode surface. In the case of an array of microdisc electrodes separated by insulating material, individual diffusion layers will develop, and continue to grow throughout the experiments. It transpires[7] that the voltammetry of the array is highly dependent on the size of the individual diffusion layers, δ, versus the size of the discs themselves, and on the size of the diffusion layers versus the centre-to-centre separation of the discs, d, (see Fig. 6.15). Based on these two factors Fig. 6.16 shows the four categories to which an array can be assigned and Table 6.1 summarises the associated linear sweep and cyclic voltammetric characteristics associated with each category, which we next consider in turn, noting the complexity of the cases of partially blocked electrodes discussed in the previous section.

Category 1

In this limit, the microelectrodes are so far apart that their diffusion layers are fully independent and the timescale of the voltammetry is so short (fast voltage scan rate) that there is planar diffusion to the microelectrodes: this requires $\delta \ll R_e$, where R_e is the radius of the microelectrode (Fig. 6.15) and δ is the size of the diffusion layer.

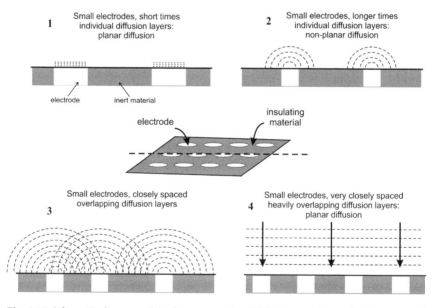

Fig. 6.16 Schematic diagram of the four categories of diffusion profile to which an array of microelectrodes may belong.

Table 6.1. Linear sweep and cyclic voltammetry characteristics associated with the 4 categories in Fig. 6.16, where δ is the size of the diffusion zone, R_e is the microdisc radius, d is the centre-to-centre separation, I_p is the peak current, I_{lim} is the limiting current and ν is the scan rate.

Property	Category 1	2	3	4
δ versus R_e	$\delta < R_e$	$\delta > R_e$	$\delta > R_e$	$\delta > R_e$
δ versus d	$\delta < d$	$\delta < d$	$\delta > d$	$\delta >> d$
Type of response	Clear peak$\rightarrow I_p$	Steady-state$^a \rightarrow I_{lim}$	Slight peak-to-clear peak$\rightarrow I_p$	Clear peak$\rightarrow I_p$
Scan rate dependence?	Yes	No	Yes	Yes
Current dependence	$I_p \propto \nu^{0.5}$	$I_{lim} \propto R_b$	–	$I_p \propto \nu^{0.5}$

^{a}for certain scan rates only.

Consequently, peaks are seen in the voltammetry and the array response is simply that of an isolated macroelectrode multiplied by the total number of electrodes in the array.

Category 2

In this case, the electrodes are again widely spaced ($\delta \ll d$) to ensure independence but now the timescale of the voltammetry is such that $\delta > R_e$, so that steady-state behaviour is developed at each microelectrode. Again, the array response is that of the individual microelectrode scaled by the total number of microdiscs in the array.

Category 3

In this limit, the microelectrodes are closer together so that the diffusional fields partially overlap. The result is that a peak develops in the voltammetric response of the array which cannot be related to the voltammetric properties of isolated microelectrodes. The behaviour of category 3 is easiest appreciated as intermediate between categories 2 and 4 to which we next turn.

Category 4

In this case, $\delta \gg d$, so that there is strong overlap of the diffusional fields leading to the entire array behaving as a single macroelectrode of the same total area as the microdisc plus insulator array. Physically, this arises since electrode material can diffuse from the zone of the insulator to the microdisc surface on the timescale of the experiment. Accordingly, peak-shaped responses are seen, with a peak current scaling with the square root of the voltage scan rates as expected for a single macroelectrode.

The insights above provide two 'take-home' messages. First they echo a conclusion of the last section, that it is possible to have a macroelectrode response, providing the voltammetric timescale is correctly chosen, from an electrode surface of which the greater part is insulating rather than made of the electrode material of interest! This insight is leading current research to the design of electrode surfaces modified with relatively small amounts of electro-catalysts but which nevertheless act as if they were a macroelectrode made of the same material.

Second, in the characterisation of the microelectrode arrays it has been common practice to make steady state measurements to 'count' the number of active electrodes. This is often necessary since lithographic procedures are commonly imperfect so that not all of the microelectrodes in any array are actually actively wired up. The problem is more serious for RAM^{TM} electrodes. Typically, it is assumed that category 2 applies, so that the number of electrodes is inferred by dividing the total array current by that expected for a single isolated microelectrode

$$I = 4nFDR_e[A]_{bulk}. \tag{6.14}$$

In practice, many arrays are constructed so that category 3 applies; typically, this is evident by a value for the total number of microdiscs in the array that depends on the voltage scan rate used for the determination. Application of the correct category 3 theory allows the rigorous determination of the number of discs for both regular[8] and random[9] arrays.

6.3 Voltammetry at Highly Ordered Pyrolytic Graphite Electrodes

Edge plane (EPPG) and basal plane pyrolytic graphite (BPPG) electrodes are fabricated from highly ordered pyrolytic graphite (HOPG). As shown in Fig. 6.17, HOPG consists of layers of graphite with an interlayer spacing of 3.35 Å.

If the graphite crystal is cut — as shown — perpendicular to these sheets, an 'edge plane' electrode is formed, whereas if the crystal is cut parallel to the sheets then a 'basal plane' electrode is precluded. In the latter case, the surface should

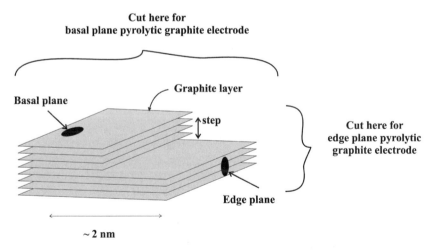

Fig. 6.17 Schematic representation of a highly ordered pyrolytic graphite electrode of a four layer step edge. Reproduced from Ref. [10] with permission of The Royal Society of Chemistry.

not be thought of as perfect; rather, as illustrated in Fig. 6.17, there will be surface defects in the form of steps exposing the edges of the graphite layer. Such steps are typically 1–4 sheets of graphite in height. In carefully prepared samples of basal plane HOPG, the distance between the edge steps can be as much as 1–10 μm. Due to the nature of the chemical bonding, the two planes — edge and basal — exhibit completely different electrochemical properties. For voltammetry, the edge plane usually exhibits considerably faster electrode kinetics in comparison with the basal plane. This means that an electrode consisting entirely of edge plane, an 'edge plane pyrolytic graphite electrode', will show a nearly reversible voltammogram whilst an electrode consisting mainly of basal plane will show irreversible behaviour but crucially depending on the number of edge plane step defects on the surface. For this reason, edge plane electrodes are often preferable to basal plane or other for electroanalytical purposes.[11] Figure 6.18 illustrates the benefits in respect of the reduction of chlorine, Cl_2, in aqueous nitric acid.

The figure shows the voltammograms for

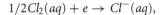

$$1/2\,Cl_2(aq) + e \rightarrow Cl^-(aq),$$

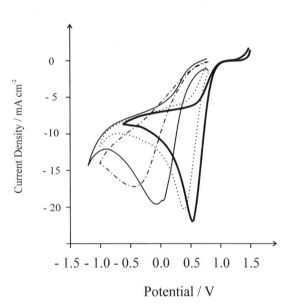

Fig. 6.18 Cyclic voltammograms for reduction of chlorine in 0.1 M nitric acid solution using edge-plane pyrolytic graphite (thick line), which is compared with a glassy carbon (dotted), basal plane pyrolytic graphite (thin line) and boron-doped diamond (dotted and dashed line) electrodes. All scans recorded at 100 mV s^{-1} versus Saturated Calomel Electrode. Reprinted from Ref. [11] with kind permission of Springer Science and Business Media.

~ 75 Å

~ 50 Å

Fig. 6.19 The structure of glassy carbon.

recorded at EPPG, BPPG, glassy carbon and boron-doped diamond electrodes. Note that glassy carbon (Fig. 6.19) is composed of a structure of interwoven ribbons of graphite structure.

Glassy carbon is a much harder form of graphite and it typically prepared from polymeric resin via heat treatment (1000–3000°C), usually under pressure, forming an extremely conjugated sp^2 carbon structure. Boron-doped diamond is another carbon based electrode in which *ca* one carbon atom in one thousand in a diamond structure has been replaced by boron giving a material with a metallic conductivity. An SEM image of commercial BDD is shown in Fig. 6.20.

The voltammetric data in Fig. 6.18 shows that the Cl_2/Cl^- couple has significantly faster electrode kinetics on the EPPG electrode as compared to the other three materials: the reduction peak occurs at much less negative potentials.

Figure 6.21 shows a typical cyclic voltammogram for the oxidation of 1 mM ferrocyanide in 1 M aqueous potassium chloride recorded at a bppg HOPG electrode and at a EPPG electrode. Overlaid as a dashed curve is the simulated

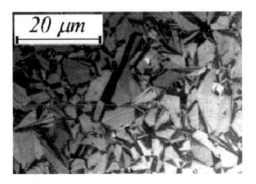

20 μm

Fig. 6.20 SEM image of a piece of commercial BDD.

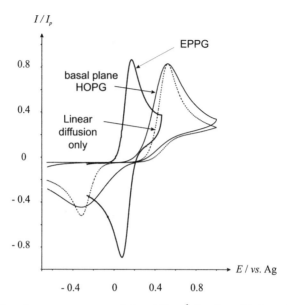

Fig. 6.21 Cyclic voltammograms recorded at 1 V s^{-1} for the oxidation of 1 mM ferro-cyanide in 1 M KCl solution at a basal plane HOPG electrode and an edge plane pyrolytic graphite electrode. The dashed voltammogram is the simulated fit using linear diffusion only. Reproduced from Ref. [10] with permission of The Royal Society of Chemistry.

'linear diffusion only' voltamogram that best fits the experimental bppg HOPG voltammogram.

This simulation considers only the spatial dimension normal to the electrode (as in Chapter 4) and so assumes all parts of the electrode are equally and uniformly active. The figure illustrates the two main features of BPPG voltammetry. First, there is a significant increase in the peak-to-peak separation as compared to the EPPG voltammogram consistent with the substantially slower electrode kinetics. Second, the fit to the 'linear diffusion only' theory is poor. Noticeably, there is an enhancement of the diffusional tail of the experimental voltammogram over theory: the currents after the peak fall off less sharply than predicted. Also the size of the reverse scan current peak is over estimated by the theory and the experimental peak is noticeably less than predicted by linear diffusion. This behaviour is entirely reminiscent of the case 3 behaviour discussed in Sec. 6.1, that is of a partially blocked electrode in which zones where electrolysis could occur were of microelectrode size and character and were separated by distances such that there was some, but not strong, overlap of their diffusional fields.

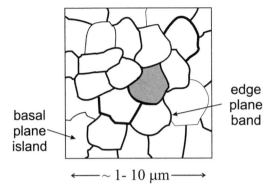

Fig. 6.22 Schematic representation of a basal plane surface. Reprinted from Ref. [12] with permission from Elsevier.

The correct voltammetric shape of the BPPG voltammogram in Fig. 6.20 was correctly and quantitatively simulated by considering the basal plane surface is schematically depicted in Fig. 6.22.

In this, the edge plane steps are inferred to be primarily the sites of electrolytic action, whereas the basal plane 'islands' are relatively electrolytically inert. The surface thus behaves as a random array of microband electrodes which can be shown to produce exactly the voltammetric characteristic displayed by BPPG HOPG surface. The implication is that the BPPG voltammogram is caused by electrolysis taking place at only a small fraction of the total electrode surface, namely at the edge-plane-like steps.[10]

6.4 Electrochemically Heterogeneous Electrodes

In the previous sections, we have considered partially blocked electrode and also microelectrode arrays. We next generalise these ideas by introducing the concept of an 'electrochemically heterogeneous' electrode as shown schematically in Fig. 6.23.

It comprises a macroelectrode with two or more types of spatial zones, either, both or all of which being electroactive but characterised by different electrode kinetics of the Butler–Volmer type: $\{k_1^0, \alpha_1\}, \{k_2^0, \alpha_2\}, \ldots$. The simulation of such electrodes can be carried out using the diffusion domain approximation.[3,13]

As illustration of the behaviour of electrochemically heterogeneous electrode, we consider the case of a EPPG electrode surface modified with an adsorbed layer of anthraquinone and on which has been immobilised a partial layer of gold particles. We consider the response of this doubly modified electrode — illustrated

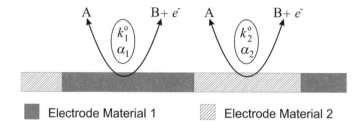

Electrode Material 1   Electrode Material 2

Fig. 6.23 An electrochemical reaction occurring on the same electrode surface with different Butler–Volmer characteristics at different spatial locations. Reprinted from Ref. [1] with kind permission of Springer Science and Business Media.

schematically in Fig. 6.24 — towards the oxidation of ferrocyanide, $Fe(CN)_6^{4-}$, in aqueous solution.

The oxidative voltammetry at a pure gold electrode is shown in Fig. 6.25 along with that seen at a pure anthraquinone modified EPPG electrode.

The closer peak-to-peak separation in the former case indicates that the electrode kinetics are faster on gold than on the modified carbon. Standard electrochemical rate constants (25°C) of 0.013 cm s^{-1} and 0.0011 cm s^{-1} were measured in each case. It is expected that the composite electrode (Fig. 6.24) will show a behaviour intermediate in character between that of pure gold and pure modified carbon limits, as shown schematically in Fig. 6.26.

Figure 6.27 shows the results of diffusion domain based simulations showing how the calculated peak-to-peak separation varies with voltage scan rate and coverage, Θ, of gold, where $\Theta = 0$ corresponds to the pure EPPG limit and $\Theta = 1$ to the pure gold extreme. Experiments in which the voltammetric response of ferrocyanide was used to calculate the gold coverage gave results in excellent agreement with independent data obtained via SEM imaging of the modified surfaces.[14]

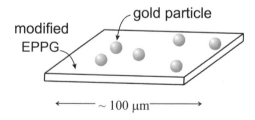

Fig. 6.24 Schematic diagram of the blocking gold particles on a modified edge plane pyrolytic graphite electrode. Reprinted from Ref. [12] with permission from Elsevier.

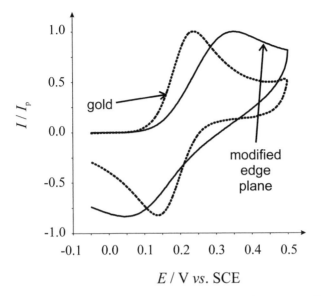

Fig. 6.25 Cyclic voltammetric response of aqueous ferrocyanide at pure gold and pure anthraquinone-modified edge plane pyrolytic graphite electrode. Reprinted from Ref. [1] with kind permission of Springer Science and Business Media.

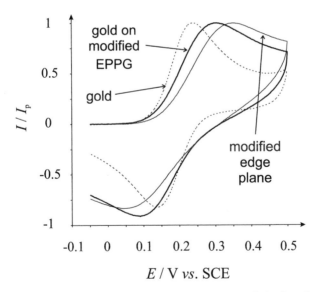

Fig. 6.26 Typical voltammetric response of gold on the modified edge plane pyrolytic graphite electrode with that of the bare gold and anthraquinone-modified edge plane electrode. Reprinted from Ref. [1] with kind permission of Springer Science and Business Media.

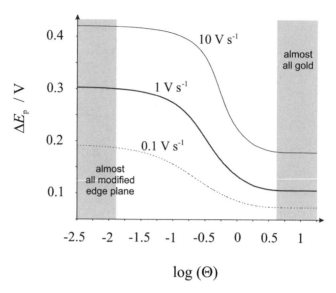

Fig. 6.27 Simulated working curves showing the relationship between ΔE_p and $\log \theta$ for cyclic voltammograms recorded at the gold particle anthraquinone-modified edge plane pyrolytic graphite electrode. Reprinted from Ref. [1] with kind permission of Springer Science and Business Media.

6.5 Electrodes Covered with Porous Films

First, it is instructive to briefly consider the voltammetric response of a planar macroelectrode covered with an insulating surface coating containing non-interconnected cylindrical pores, such that electroactive species have diffusional access to the electrode only via the pores. Suppose that the pore geometry is such that its radius is r_p and its depth Z_p. The voltammetric behaviour is controlled by the size of the diffusion layer, δ, relative to Z_p and r_p.[14] In particular, if $\delta < Z_p$ then diffusion to the electrode surface is necessarily linear (planar), since the concentration changes are entirely confined to the pore walls which preclude radial diffusion, and this is the case regardless of the size of r_p. Accordingly, even for microelectrode-sized values of r_p, conventional one-dimensional cyclic voltammetry is observed, provided δ is constrained to within the pore. Figure 6.28(b) shows that for a deep pore linear diffusion prevails, whereas for a relatively shallow pore there can be a contribution from radial diffusion.

Once $\delta > Z_p$ then the response becomes sensitive to r_p and to the average distance between adjacent pores on the electrode surface.

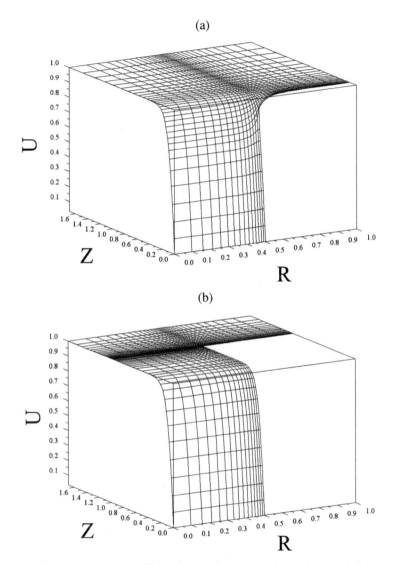

Fig. 6.28 Concentration profiles illustrating the influence of increasing pore height on the mass transport properties of a single diffusion domain. For a deep pore ($\delta < Z$) linear diffusion is seen (b), whereas for $\delta < Z$ radial diffusion also contributes, (a). Reprinted from Ref. [14] with permission from Elsevier. U is a dimensionless concentration, Z and R are cylindrical coordinates.

Semi-infinite diffusion

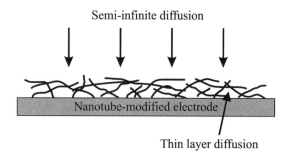

Thin layer diffusion

Fig. 6.29 Schematic of the two types of diffusion that contribute to current at a carbon nanotube modified electrode. Reprinted from Ref. [15] with permission from Elsevier.

Second, a different type of porous coating on an electrode surface can be formed by immobilising a layer of nanoparticles or nanotubes onto a planar electrode. Alternatively, polymer films can be cased onto the electrode, sometimes with a not dissimilar effect. Figure 6.29 shows a nanotube modified electrode in schematic form. The surface can be thought of as a porous layer in which pockets of the electroactive species are trapped in between multiple layers of nanotubes.

It has been proposed that the electrolysis of electroactive species trapped within these pockets can be usefully described using a model of a thin layer cell of high electrode area reflecting the large surface area of the nanotubes in the porous layer; the electrode is envisaged as in contact with a finite, 'thin layer' of solution.[15] In contrast, electrolysis at the unmodified electrode is described using the semi-infinite diffusion model of Chapter 4.

Figure 6.30 shows a comparison of linear sweep voltammetry using both the semi-infinite and thin layer diffusion models under the assumption of Butler–Volmer kinetics coupled to Fickian diffusion with $k^0 = 10^{-4}$ cm s^{-1} and a value of 10^{-5} cm^2 s^{-1} for the diffusion coefficient of the electroactive species. The surface area of the porous material (area of the thin layer electrode) was assumed to be 30 times that of the geometric area of the underlying supporting electrode. Figure 6.29 shows that the thin layer model produces a smaller peak-to-peak separation than the semi-infinite model. This is true in the electrochemically reversible limit as well, where for the thin layer situation with a tiny electrode thickness the forward and back peaks would occur with the same potential whereas there would be a separation of *ca.* 57 mV at 298 K (see Sec. 4.5) in the case of semi-infinite diffusion. However for quasi- and irreversible systems the difference between the two models can be more significant in terms of the magnitude of the different models as seen in Fig. 6.30.

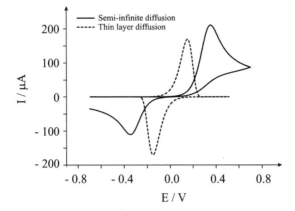

Fig. 6.30 Comparison of linear sweep voltammetry using semi-infinite and thin layer planar diffusion models. For both models, $k^0 = 10^{-4}$ cm s^{-1}; $D = 10^{-5}$ cm^2 s^{-1}; $v = 0.1$ Vs^{-1}; $c = 10^{-6}$ mol cm^{-3}. Semi-infinite diffusion electrode area, $A = 1$ cm^2; thin layer area, $A = 30$ cm^2; thickness, $l = 1$ μm. Reprinted from Ref. [15] with permission from Elsevier.

An implication of the above is the following caveat. If an electrode is modified with a porous layer then the mass transport characteristics can be changed. If the contribution to the voltammetry from material within the porous layer dominates then it is possible for the naïve to be misled with thinking that the reduction in peak-to-peak voltage separation might reflect an electro-catalytic effect of the material forming the porous coating rate than constituting a mass transport phenomena, as implicit in Fig. 6.30 where the same electrochemical rate constant is used for both current-voltage curves. These effects are thought to have implications for understanding 'electro-catalysis' (or otherwise) in the following

- Carbon nanotube electrodes[15,16]
- C$_{60}$ modified electrodes,[17] and
- Some polymer modified electrodes.[18]

6.6 Voltammetric Particle Sizing

It has been a recurring theme throughout this chapter and the two preceding chapters, that the size of the diffusion layer of an electrode can be controlled by means of the voltage scan rate. Implicit in this is the notion that voltammetric measurements might be able to provide spatial information. This section, and the one following, will illustrate how this can be realised.

We start by considering the voltammetric response of a microdisc electrode on which an inert sphere is positioned on the electrode centre as shown in Fig. 6.31.

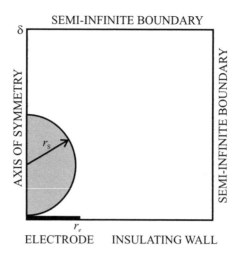

Fig. 6.31 Schematic diagram of a sphere on an electrode. Reprinted from Ref. [19] with permission from Wiley.

We suppose that the sphere radius is r_s and the microelectrode radius r_e. Clearly, the effect of the presence of the sphere will be to reduce the diffusional current to the electrode. However, the relative extent to which the current is reduced from that of the naked electrode is a function of voltage scan rate. Consider a very fast voltage scan rate such that the microelectrode diffusion layer size δ is small compared to r_e and r_s. Under these conditions, the diffusion layer is a thin 'skin' on the electrode surface and peaks are seen in the voltammogram. Since the sphere touches the electrode only at a point, the influence of the sphere is relatively minimal. As the voltage sweep rate is reduced, however, the diffusion layer 'fattens' and sees increasingly more of the sphere and so the peak current is relatively more reduced compared to the current on the naked electrode unblocked by the sphere. Ultimately, as the scan rate is slowed so that $\delta > r_e$, a steady-state current is established which is significantly less than that seen for the naked electrode. Figure 6.32 shows the peak current as a function of scan rate for an electrode with $r_e = 59\,\mu$m and a sphere of radius $125\,\mu$m.

Experimental and simulated data are shown.[19] The expected convergence at high scan rates of the data for the naked electrode and the electrode plus sphere can be seen along with the establishment of significantly different steady-state (scan rate independent) data at low voltage scan rates. Interestingly, and significantly, comparison of simulation and experiments showed the method to be a highly sensitive method of sizing the sphere.

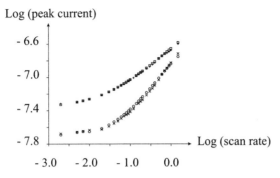

Fig. 6.32 Comparison between experimental and theoretical results. The parameters in common used for the true microdisc simulation and the simulation of the electrode with a sphere positioned in the centre are as follows: $[A]_{bulk} = 3 \times 10^{-6}\,\text{mol cm}^{-3}$, $r_e = 59\,\mu\text{m}$, $\alpha = 0.5$, $D = 0.63 \times 10^{-5}\,\text{cm}^2\,\text{s}^{-1}$, $k_o = 0.05\,\text{cm s}^{-1}$, $E_{start} = -0.2\,\text{V}$, $E_{stop} = 0.5\,\text{V}$, $E_0 = 0.19\,\text{V}$. Additionally, for the simulation of the sphere positioned in the centre of the electrode, a sphere radius of 125 m was used. Crosses are the naked electrode experimental data, circles are the electrode and sphere experimental data, squares the naked electrode simulation, and crosses the electrode and sphere simulation. Reprinted from Ref. [19] with permission from Wiley.

We next build on the discussion above, and that in the previous sections, to show how the average spherical diameter of *ca* micron-sized particles can be measured voltammetrically by means of an experiment in which a macroelectrode was modified with known masses, m_{block} of monodisperse and approximately spherical particles of alumina of diameter 1 μm using the procedure shown in Fig. 6.33.

In this way, a partially blocked electrode was formed in which the blocks were random alumina spheres. Since the latter are non-conducting and electrochemically inert, they simply acted to modify the diffusional characteristics of the macroelectrode. The oxidation of aqueous ferrocyanide, $Fe(CN)_6^{4-}$, was studied with those modified (partially blocked) electrodes and with the corresponding unblocked electrodes. Figure 6.34 shows representative voltammograms. It can be seen that, as the mass of alumina on the electrode surface increases, the peak currents are somewhat diminished and the peak-to-peak separation of the voltammogram is increased. This is consistent with the formation of a partially blocked electrode (see Section 6.1).

Data such as that in Fig. 6.34 were analysed on the basis of the diffusion domain theory in Sec. 6.1 assuming the blocks to be monodisperse inert spheres of radius r_b. For a given radius and mass of material used to modify the electrode, the number

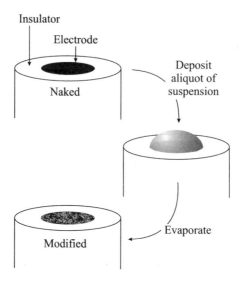

Fig. 6.33 Procedure for modifying an edge plane pyrolytic graphite electrode with alumina particles for voltammetric particle sizing. Reprinted from Ref. [1] with kind permission of Springer Science and Business Media.

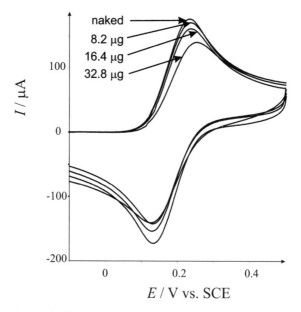

Fig. 6.34 The observed effect on the voltammetric response of increasing the mass of alumina on the voltammetric response. Reprinted from Ref. [1] with kind permission of Springer Science and Business Media.

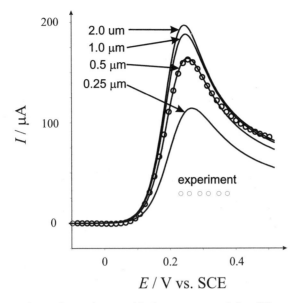

Fig. 6.35 Comparison of experiment with theory generated for different values of R_b. Reprinted from Ref. [1] with kind permission of Springer Science and Business Media.

of spheres is given by

$$n_{block} = \frac{3m_{block}}{4\rho\pi r_b^3},\qquad(6.15)$$

where ρ is the density of alumina. In this way, the experimental voltammetry could be compared with that simulated for different values of n_{block} and r_b. Given that m_{block} is known, these are not two independent parameters but rather are linked via Eq. (6.15). Figure 6.35 shows comparison of experimental and simulation for different values of r_b.

An excellent fit is seen for $2r_b = 1\mu$m corresponding to the particle size.

The ability of voltammetric methods to explore shape and size as well as redox chemistry should be evident from the above. This possibility is developed further in the next section.

6.7 Scanning Electrochemical Microscopy (SECM)

A scanning electrochemical microscope is an instrument in which a moveable microelectrode in an electrolyte solution is scanned in close proximity to a solid surface to characterise the topography and/or the redox activity of the solid–liquid

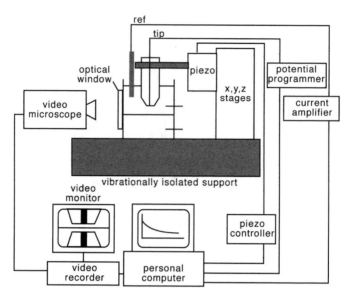

Fig. 6.36 Block schematic of the typical instrumentation for SECM with an amperometric microelectrode tip. The tip position may be controlled with various micropositioners, as outlined in the text (piezoelectric elements in the schematic). The tip potential is controlled with a potential programmer with respect to a reference electrode, and the current is measured with a simple amplifier device. The tip position may be viewed using a video microscope. Typically, phase 1 is liquid and phase 2 solid. Reprinted from Ref. [22] with permission from Elsevier.

interface. The origins of the technique lie in the work of Engstrom and co-workers[20,21] who were the first to show that a microelectrode could be employed as a local probe to map the concentration profiles of species near a larger, active electrode. The technique has since been applied to a variety of other interfaces, including the liquid–liquid and the liquid–air interfaces[22,23] and has subsequently developed as a technique producing commercial instruments. Figure 6.36 shows a typical instrument. In the simplest usage, the effect of the interface on a diffusion-controlled current at the microelectrode due to a solution-phase electroactive species is monitored.

The ability for mapping surface topography can be appreciate by considering Figs. 6.37(a) and (b) which shows a microelectrode in solution and near an insulating surface — clearly the latter case will result in reduced currents so that movement of the electrode over the surface can in principle resolve the topography with a resolution of an order of the electrode radius.

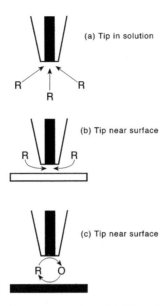

Fig. 6.37 Feedback mode of SECM operation. (a) The microelectrode tip is far from the substrate. The steady-state tip current is produced by diffusion of R to the tip and electrode reaction $R \rightarrow O + ne^-$. (b) With the microelectrode approaching an insulating substrate, hindered diffusion of R to the tip leads to decreased current (negative feedback). (c) With the microelectrode near a conductive substrate, positive feedback of R from the reverse reaction at the substrates leads to increase in current. Reprinted from Ref. [22] with permission from Elsevier.

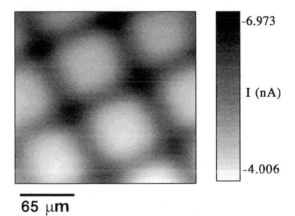

Fig. 6.38 SECM image of copper minigrid made with an amperometric Pt tip ($a = 5$ mm) in ruthenium hexamine solution at about 10 mm distance. High current (darker image) is the result of positive feedback. Reprinted from Ref. [24] with kind permission of Springer Science and Business Media.

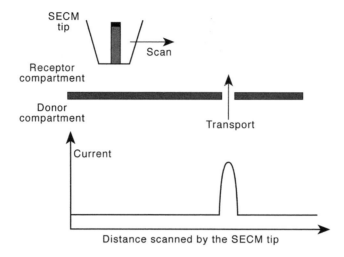

Fig. 6.39 Examining permeability with SECM. Transport of species by convection, diffusion or migration (promoted by pressure, concentration or electric field gradients) can be detected as an increase in the transport-limited current versus tip position can then be converted to a corresponding permeability map of the target interface. Reprinted from Ref. [22] with permission from Elsevier.

Figure 6.37(c) shows that if the electrode is brought close to a conducting surface, 'positive feedback' from the surface reversing the microelectrode reaction leads to amplification of the current. Figure 6.38 shows a typical image, that of a minigrid imaged using the $Ru(NH_3)_6^{3+/2+}$ couple in aqueous solution.

A 5 micron radius electrode positioned ca 10 microns above the grid was used to obtain the image. The method can also be used to probe fluxes of redox active material. For example, Fig. 6.39 shows the principle of the method adapted for the detection of pores in membranes or surfaces via SECM.[20]

References

[1] T.J. Davies, C.E. Banks, R.G. Compton, *J. Solid State Electrochem.* **9** (2005) 797.
[2] B.A. Brookes, T.J. Davies, A.C. Fisher, R.G. Evans, S.J. Wilkins, K. Yunus, J.D. Wadhawan, R.G. Compton, *J. Phys. Chem. B* **107** (2003) 1616.
[3] T.J. Davies, B.A. Brookes, A.C. Fisher, K. Yunus, S.J. Wilkins, P.R. Greene, J.D. Wadhawan, R.G. Compton, *J. Phys. Chem. B* **107** (2003) 6431.
[4] C. Amatore, J.M. Savéant, D. Tenner, *J. Electroanal. Chem.* **147** (1983) 39.
[5] F.G. Chevallier, N. Fietkau, J. Del Campo, R. Mas, F. X. Muñoz, L. Jiang, T.G.J. Jones, R.G. Compton, *J. Electroanal. Chem.* **596** (2006) 25.
[6] S. Fletcher, M. Horne, *Electrochem. Commun.* **1** (1999) 502.

[7] T.J. Davies, S. Ward-Jones, C.E. Banks, J. del Campo, R. Mas, F.X. Muñoz, R.G. Compton, *J. Electroanal. Chem.* **585** (2005) 51.

[8] O. Ordeig, C.E. Banks, T.J. Davies, J. del Campo, R. Mas, F.X. Muñoz, R.G. Compton, *Analyst* **131** (2006) 440.

[9] O. Ordeig, C.E. Banks, T.J. Davies, J. del Campo, R. Mas, F. X. Muñoz, R.G. Compton, *J. Electroanal. Chem.* **592** (2006) 126.

[10] C.E. Banks, T.J. Davies, G.G. Wildgoose, R.G. Compton, *Chem. Comm.* **7** (2005) 829.

[11] C.E. Banks, R.G. Compton, *Anal. Sci.* **21** (2005) 1263.

[12] T.J. Davies, R.R. Moore, C.E. Banks, R.G. Compton, *J. Electroanal. Chem.* **574** (2004) 123.

[13] F.G. Chevallier, T.J. Davies, O.V. Klymenko, L. Jiang, T.G.J. Jones, R.G. Compton, *J. Electroanal. Chem.* **577** (2005) 211.

[14] F.G. Chevallier, L. Jiang, T.G.J. Jones, R.G. Compton, *J. Electroanal. Chem.* **587** (2006) 254.

[15] I. Streeter, G.G. Wildgoose, L. Shao, R.G. Compton, *Sens. Act. B* **113** (2008) 462.

[16] G.P. Keely, M.E.G. Lyons, *Int. J. Electrochem. Sci.* **6** (2009) 794.

[17] L. Xiao, G.G Wildgoose, R.G. Compton, *Sens. Act. B* **138** (2009) 524.

[18] M.C. Henstridge, E.J.F. Dickinson, M. Aslanoglu, C. Batchelor-McAuley, R.G. Compton, *Sens. Act. B* **145** (2010) 417.

[19] N. Fietkau, F.G. Chevallier, L. Jiang, T.G.J. Jones, R.G. Compton, *Chem. Phys. Chem.* **7** (2006) 2162.

[20] R.C. Engstrom, M. Weber, D. J. Wunder, R. Burgess, S. Winquist, *Anal. Chem.* **58** (1986) 844.

[21] R.C. Engstrom, T. Meaney, R. Tople, R. M. Wightman, *Anal. Chem.* **59** (1987) 2005.

[22] A. L. Barker, M. Gonsalves, J. V. Macpherson, C. J. Slevin, P. R. Unwin, *Anal. Chimica Acta* **385** (1999) 223.

[23] M.V. Mirkin, B.R. Horrocks, *Anal. Chimica Acta* **406** (2000) 119.

[24] G. Nagy, L. Nagy, *Fres. J. Anal. Chem.* **366** (2000) 735.

7

Cyclic Voltammetry: Coupled Homogeneous Kinetics and Adsorption

Previous chapters have introduced the voltammetric study of simple electrode processes of the form

$$A(aq) + e^- \rightleftarrows B(aq),$$

where both A and B are chemically stable on the timescale of the experiment. In this chapter, we start by considering the influence of the possible chemical instability of B on the observed voltammetry and show how cyclic voltammetry can be used to study the reaction kinetics and mechanism of electro-generated species. We conclude the chapter by saying a little about voltammetry of adsorbed species.

7.1 Homogeneous Coupled Reactions: Notation and Examples

Electrode reaction mechanisms observed in this chapter are described using the Testa and Reinmuth notation.[1] In this convention, the letter E is used to denote a heterogeneous electron transfer step, and the letter C to indicate a homogeneous step.

Example 1: E reaction

This is the trivial case of heterogeneous one-electron oxidation (or reduction) to form a stable radical cation (or anion). An example is the oxidation of ferrocene (dicyclopentadienyl-iron):[2]

$$FeCp_2 - e^- \longrightarrow FeCp_2^{\bullet+}.$$

Example 2: EC reaction

Often, an electron transfer yields a product that is unstable. In these cases, a homogeneous chemical reaction will follow electron transfer. In an EC reaction, a chemical step (if the chemical step is irreversible the label EC_{irr} is sometimes used) follows an electron transfer, yielding an electrochemically inactive product. An example is the oxidation of 1,4-aminophenol in acidic aqueous solution:[3]

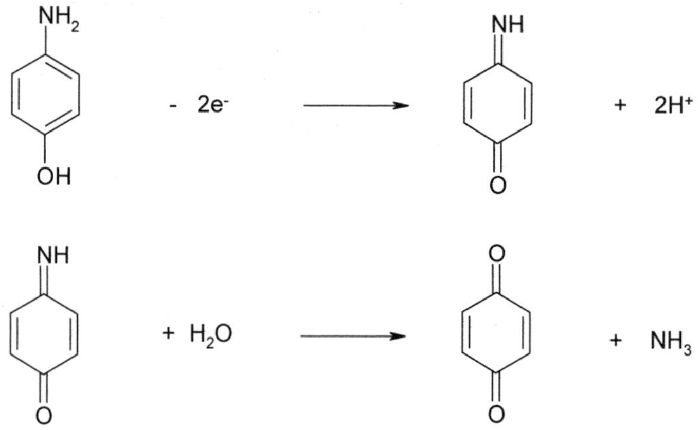

Example 3: EC₂ reaction

Example 3: EC_2 reaction

In this case a dimerisation follows the initial oxidation or reduction

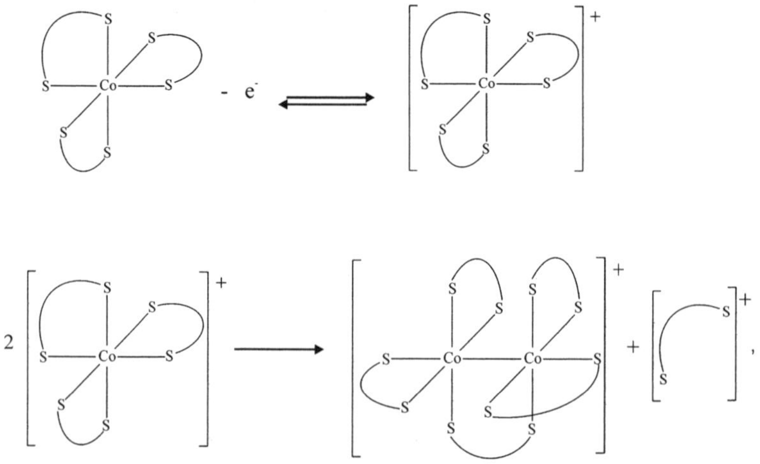

where the ligand S__S is S_2CNMe_2.[4]

Example 4: ECE reaction

This is where the chemical step yields a product which is itself electro-chemically active in the experimental potential range, e.g. the reduction of 1, 2-bromonitrobenzene in aprotic solvents:[5]

7.2 Modifying Fick's Second Law to Allow for Chemical Reaction

In Chapter 4 we saw that the cyclic voltammetry for the simple redox process

$$A(aq) + e^- \rightleftarrows B(aq)$$

could be predicted, when both A and B are stable, by solution of the equations defining Fick's second law with the application of suitable boundary conditions. In one dimension (x), such as is relevant for cyclic voltammetry at a planar macro-electrode, the equations are

$$\frac{\partial [A]}{\partial t} = D_A \frac{\partial^2 [A]}{\partial x^2}$$

and

$$\frac{\partial [B]}{\partial t} = D_B \frac{\partial^2 [B]}{\partial x^2}.$$

In the case that the species are chemically reactive, it is necessary to modify the equation accordingly. For example, if in homogeneous solution, say B, undergoes nth order kinetics with a homogeneous rate constant, k_n, then

$$\frac{\partial [B]}{\partial t} = D_A \frac{\partial^2 [A]}{\partial x^2} - k_n [B]^n,$$

where the last term reflects the reactivity. Recall that k_1 has units of s^{-1} and k_2 of dm^3 mol^{-1} s^{-1}.

7.3 Cyclic Voltammetry and the EC Reaction

The EC mechanism can be summarised by the following general scheme

$$A + e \rightleftarrows B$$

$$B \overset{k_1}{\rightleftarrows} C,$$

where k_1 is the first-order or pseudo-first order rate constant for the reaction of B to form C, and the equilibrium constant

$$K = \frac{[C]}{[B]}.$$

The mass transport equations for planar diffusion to a macroelectrode relating to the above scheme are

$$\frac{\partial [A]}{\partial t} = D_A \frac{\partial^2 [A]}{\partial x^2}$$

$$\frac{\partial [B]}{\partial t} = D_B \frac{\partial^2 [B]}{\partial x^2} - k_1 [B] + \frac{k_1}{K} [C]$$

$$\frac{\partial [C]}{\partial t} = D_C \frac{\partial^2 [C]}{\partial x^2} + k_1 [B] - \frac{k_1}{K} [C].$$

These can be solved, in the most general case by assuming Butler–Volmer kinetics for the electrode reaction

$$DA \left. \frac{\partial[A]}{\partial x} \right|_{x=0} = k_c[A]_{x=0} - k_a[B]_{x=0}$$

$$DA \left. \frac{\partial[A]}{\partial x} \right|_{x=0} = -D_B \left. \frac{\partial[B]}{\partial x} \right|_{x=0},$$

where

$$k_c = k^0 \exp\left(-\frac{\alpha F}{RT} \left[(E - E_f^0(A/B)) \right] \right)$$

and

$$k_a = k^0 \exp\left(\frac{\beta F}{RT} \left[(E - E_f^0(A/B)) \right] \right).$$

A no flux boundary condition for C is assumed at the electrode surface:

$$x = 0 \quad D_C \frac{\partial[C]}{\partial x} = 0$$

and in bulk solution

$$x \to \infty \quad [A] \to [A]_{bulk} \quad [B] = [C] = 0.$$

Then, providing the limits of the potential sweep are set sufficiently wide such that the starting or switching potential does not influence the voltammogram, the resulting cyclic voltammogram can be shown to depend on just three parameters, taking α as fixed (typically 0.5 and assuming $D_A = D_B = D$). These are

$$\Lambda = k^0 \sqrt{\frac{RT}{D \upsilon F}},$$

$$K_1 = \frac{k_1}{\upsilon} \left(\frac{RT}{F} \right)$$

and

$$K = \frac{[C]_{eqm}}{[B]_{eqm}},$$

as defined above where the subscript 'eqm' implies equilibrium concentrations. In the limit of an EC_{irr} process, the voltammetry becomes independent of K as in effect $K \to \infty$ and the diffusion problem is reduced to a two species (A and B) only problem:

$$\frac{\partial[A]}{\partial t} = D_A \frac{\partial^2[A]}{\partial x^2}$$

$$\frac{\partial[B]}{\partial t} = D_B \frac{\partial^2[B]}{\partial x^2} - k_1[B].$$

The parameter Λ controls the electrochemical reversibility of the voltammetry much as in Chapter 4. The dimensionless parameter K_1 effectively describes the ratio of the first order rate control for the decomposition of B to the voltage scan rate ($K_1 \propto k_1/v$). Figure 7.1 shows voltammograms calculated for the reversible limit ($\Lambda = 100$; fast electrode kinetics) but for different values of K_1 assuming $E_f^0(A/B) = 0\,\mathrm{V}$.

It can be seen that, as the value of K_1 changes, the current peak on the reverse scan is either present or is lost. At slow scan rates, corresponding to a large value of K_1, no reverse peak is seen since the B produced during the forward scan is lost before the reverse scan can be completed. As the value of K_1 decreases (corresponding to the sequence A to F in Fig. 7.1) the back peak appears once the time taken to scan the voltammogram is comparable with the lifetime of B.

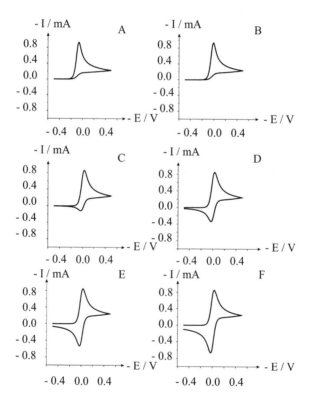

Fig. 7.1 Cyclic voltammograms at varying values of K for an EC_{irr} process with $\Lambda = 100$, $E^0 = 0\,\mathrm{V}$; $\alpha = 0.5$, $k^0 = 1.973\,\mathrm{cm\ s^{-1}}$; $[A]_0 = 1\,\mathrm{mM}$; $A = 1\,\mathrm{cm^2}$; $D_A = D_B = D = 10^{-5}\,\mathrm{cm^2\ s^{-1}}$. (A) $K_1 = 10^3$; (B) $K_1 = 10^1$; (C) $K_1 = 10^{-1}$; (D) $K_1 = 10^{-1.5}$; (E) $K_1 = 10^{-2}$; (F) $K_1 = 10^{-3}$.

For very fast scan rates (Fig. 7.1(F)), the voltammogram is unchanged from that expected if B were stable, since negligible loss of B occurs during the voltammetric experiment. Note that in Fig. 7.1 the value of $K_1 = 10^{-1}$ (Fig. 7.1(C)) can be realised, for example, by a rate constant $k_1 = 4\,\mathrm{s}^{-1}$ and a voltage sweep rate $v = 1\,\mathrm{Vs}^{-1}$. On the other hand if $k_1 = 400\,\mathrm{s}^{-1}$ then a voltage sweep rate of $100\,\mathrm{Vs}^{-1}$ is required to obtain the voltammetric trace shown in Fig. 7.1(C).

As well as the presence or absence of a back peak, one further feature of the voltammetry shown in Fig. 7.1 is that when the loss of B is fast the peak for the reduction of B occurs at potentials positive of $E_f^0(A/B)$! This apparent paradox is a consequence of the assumed electrochemical reversibility of the A/B couple and the effect of the loss of B via chemical reaction is to 'pull over' the redox equilibrium (in the sense of Le Chatelier's Principle) which thus occurs at a less negative potential (for a reduction):

$$A + e \rightleftharpoons B$$

$$B \xrightarrow{k_f} products.$$

The variation of the forward peak potential, E_p, in this 'fast homogeneous kinetics, no back peak' limit is

$$\frac{\partial E_p}{\partial \log_{10} v} = \frac{2.303RT}{2F},$$

so that the wave shifts towards positive potentials by ca 30 mV (at 25°C) for a ten-fold decrease in the scan rate. In the sense of Le Chatelier's Principle, the slower the scan rate the greater the extent of kinetic decay of B and so the more the A/B equilibrium is shifted.

Figure 7.2 shows the corresponding plots to Fig. 7.1 but for the limit of electro-chemically irreversibility ($\Lambda = 0.01$). The parameter K_1 again controls the presence or absence of a back peak and large values of this parameter, where $k_1 \gg \frac{vF}{RT}$, corresponds to the absence of a back peak.

Note that the low value of Λ ensures that there is a significant overpotential for the reduction of B to A so that the forward (reductive) peak occurs at a potential negative of $E_f^0(A/B)$. Similarly, if the back (oxidation) peak is seen, it occurs at potentials positive of the formal potential. Note that for the case of irreversible electrode kinetics, $\Lambda \ll 1$, there is no variation of the formal peak potential with voltage scan rates, in contrast to what was seen for reversible electrode kinetics (Fig. 7.1):

$$\frac{\partial E_p}{\partial \log_{10} v} = 0.$$

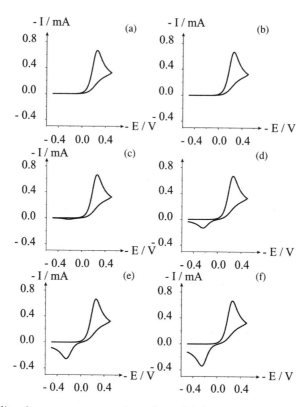

Fig. 7.2 Cyclic voltammograms at varying values of K for an EC_{irr} process with $\Lambda = 0.1$, $E^0 = 0\,V$; $\alpha = 0.5$, $k^0 = 1.973 \times 10^{-4}\,cm\,s^{-1}$; $[A]_0 = 1\,mM$; $A = 1\,cm^2$; $D_A = D_B = D = 10^{-5}\,cm^2\,s^{-1}$. (a) $K_1 = 10^3$; (b) $K_1 = 10^1$; (c) $K_1 = 10^{-1}$; (d) $K_1 = 10^{-1.5}$; (e) $K_1 = 10^{-2}$; (f) $K_1 = 10^{-3}$.

This is because no redox equilibrium is established at the electrode surface and so there is no influence of the follow up reaction on the electrochemically irreversible reduction of A to B. In terms of Le Chatelier's Principle, no equilibrium is established and so cannot shift in response to the perturbation of the chemical reaction of B.

7.4 How Do the Parameters K_1 and Λ Emerge?

To appreciate the origins of the two parameters controlling the voltammetry for the EC_{irr} process we re-visit the mathematical formulation of the problem.

In particular, it is required to solve the equation

$$\frac{\partial [A]}{\partial t} = D_A \frac{\partial^2 [A]}{\partial x^2}$$

and

$$\frac{\partial [B]}{\partial t} = D_B \frac{\partial^2 [B]}{\partial x^2} - k_1 [B],$$

subject to the boundary conditions

$$x \to \infty \quad [A] \to [A]_{bulk} \quad [B] \to 0$$

$$x = 0 \quad D_A \frac{\partial [A]}{\partial x} = -D_B \frac{\partial [B]}{\partial x}$$

$$D_A \frac{\partial [A]}{\partial x} = k^0 \exp \left(-\frac{\alpha F}{RT} \left[E - E_f^0(A/B) \right] \right) [A]_{x=0}$$

$$- k^0 \exp \left(-\frac{(1-\alpha)F}{RT} \left[E - E_f^0(A/B) \right] \right) [B]_{x=0},$$

where

$$E = E_1 + \upsilon t \quad (t < t_{switch})$$

$$E = E_1 + \upsilon t_{switch} - \upsilon (t - t_{switch})(t > t_{switch}).$$

The scan rate, υ, will be negative for a reduction ($A + e \to B$) and all the parameters used were introduced in Chapter 4.

We now introduce the dimensionless parameters listed below as well as make the assumption $D_A = D_B = D$:

$$\text{Dimensionless time, } \tau = t \frac{F\upsilon}{RT}$$

$$\text{Dimensionless distance, } \chi = x \sqrt{\frac{F\upsilon}{RTD}}$$

$$\text{Dimensionless homogeneous rate constant, } K_1 = \frac{k_1}{\upsilon} \frac{RT}{F}$$

$$\text{Dimensionless heterogeneous rate constant, } \Lambda = \frac{k^0}{D} \sqrt{\frac{RTD}{F\upsilon}}$$

$$\text{Dimensionless concentrations, } a = \frac{[A]}{[A]_{bulk}}, \ b = \frac{[B]}{[A]_{bulk}}$$

$$\text{Dimensionless potential, } \Theta = \frac{FE}{RT}.$$

With these substitutions the transport equations becomes

$$\frac{\partial a}{\partial \tau} = \frac{\partial^2 a}{\partial \chi^2}$$

and

$$\frac{\partial b}{\partial \tau} = \frac{\partial^2 b}{\partial \chi^2} - K_1 b.$$

The boundary conditions become

$$\chi \to \infty, \quad a \to 1, \quad b \to 0$$

$$\chi = 0, \quad \frac{\partial a}{\partial \chi} = -\frac{\partial b}{\partial \chi}$$

$$\frac{\partial a}{\partial \chi} = \Lambda\left(\exp\left(-\alpha\left[\Theta - \Theta_f^0(A/B)\right]\right)\right)a_{x=0}$$

$$- \exp\left((1-\alpha)\left[\Theta - \Theta_f^0(A/B)\right]\right)b_{x=0}$$

where

$$\Theta_f^0(A/B) = \frac{F}{RT}E^0(A/B),$$

$$\Theta = \Theta_1 + \tau \quad (0 < \tau < \tau_{switch}),$$

$$\Theta = \Theta_1 + 2\tau_{switch} - \tau(\tau_{switch} < \tau),$$

$$\Theta_1 = \frac{FE_1}{RT},$$

and

$$\tau_{switch} = t_{switch}\frac{Fv}{RT}.$$

Since the problem is entirely formulated in terms of a limited number of variable it follows that the dimensionless concentrations a and b are only functions of $\tau, K, \chi, \Theta - \Theta_f^0(A/B), \alpha, \Lambda$ where Θ itself depends on τ_{switch} and Θ_1. We assume that the latter two parameters are selected so as to define a potential window for the voltammogram which is specifically wide so that they do not influence the form or forms of the latter, in respect of peak potentials, currents, etc. (see Chapter 4).

We note that the electrode current is given by

$$I = FAD\left.\frac{\partial[A]}{\partial x}\right|_{x=0},$$

where A is the electrode area. The above equation converts the Fick's law flux into the current (see Chapter 2). It follows that

$$I = FAD[A]_{bulk}\sqrt{\frac{Fv}{RTD}}\frac{\partial a}{\partial \chi}. \tag{7.1}$$

It is therefore helpful to define the dimensionless current

$$\psi = \frac{I}{FA[A]_{bulk}\sqrt{D}\sqrt{\frac{Fv}{RT}}} = \left.\frac{\partial a}{\partial \chi}\right|_{x=0}.$$

It follows that, since ψ is the dimensionless concentration gradient evaluated at $\chi = 0$, the plot of ψ versus Θ, *viz* dimensionless current versus dimensionless potential, is a function only of Λ and K_1. Note that τ is linked to Θ via Eq. (7.1) so is not an independent variable. Mathematically,

$$\psi = \psi(\Theta, \Lambda, K).$$

It follows from this that in the absence of any homogeneous kinetics ($K_1 = 0$), the dimensionless current–voltage case (ψ versus Θ) depends only on Λ.

7.5 Cyclic Voltammetry and the EC$_2$ Reaction

If the chemical reaction of B is irreversible then the EC$_2$ mechanism is defined by the scheme

$$A + e \rightleftarrows B$$

$$B + B \xrightarrow{k_2} products.$$

The diffusion equations, assuming planar diffusion to a macroelectrode are

$$\frac{\partial [A]}{\partial t} = D_A \frac{\partial^2 [A]}{\partial x^2}$$

and

$$\frac{\partial [B]}{\partial t} = D_B \frac{\partial^2 [B]}{\partial x^2} - k_2[B]^2.$$

The resulting cyclic voltammetry, for a fixed value of α, assuming $D_A = D_B = D$ and the operation of Butler–Volmer kinetics, is again a function of Λ and a kinetic parameter defined slightly differently from the previous example of an EC process:

$$K_2 = \frac{k_2}{v}\left(\frac{RT}{F}\right)[A]_{bulk}.$$

Note that K_2 is identical to the definition of K except for the introduction of the bulk concentration of A; in the case of K_2, the rate constant k_2 is second order (units: dm^3 mol^{-1} s^{-1}) so that the parameter K_2 is dimensionless.

Figure 7.3 shows simulations of an EC$_2$ process, with an irreversible chemical step for a range of values of K_2 for the electrochemically reversible limit ($\Lambda = 100$). Note that as for the EC mechanism described earlier in this chapter the presence or absence of a back peak is controlled by the magnitude of the dimensionless kinetic rate constant, in the case of the EC$_2$ process, K_2.

Notice that for large K_2, corresponding to rapid loss of B relative to the time taken to scan the voltammogram, there is no back peak. In contrast, for tiny K_2 the voltammogram is essentially indentical to that seen for a simple E process. Again, in the limit of relatively fast kinetics, the voltammetric peak shifts to more positive

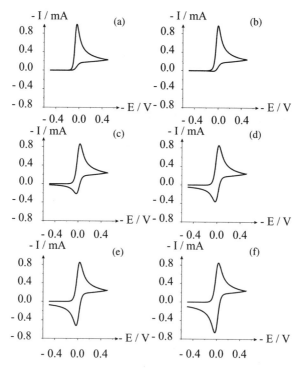

Fig. 7.3 Cyclic voltammograms at varying values of K_2 for an EC$_2$ process with $\Lambda = 100$, $E^0 = 0$ V; $\alpha = 0.5$, $k^0 = 1.973$ cm s^{-1}; $[A]_0 = 1$ mM; $A = 1$ cm^2; $D_A = D_B = D = 10^{-5}$ cm^2 s^{-1}. (a) $K_2 = 10^3$; (b) $K_2 = 10^1$; (c) $K_2 = 10^{-1}$; (d) $K_2 = 10^{-1.5}$; (e) $K_2 = 10^{-2}$; (f) $K_2 = 10^{-3}$.

potentials for large values of K_2. Quantitatively in this limit,

$$\frac{\partial E_p}{\partial \log_{10} \upsilon} = \frac{2.303RT}{3F},$$

provided $K_2 \gg 1$. Note that this implies E_p shifts by *ca* 20 mV (at 25°C) for a ten-fold change in the voltage scan rate, υ. This is a different value from that seen in the corresponding limit for an EC process.

Figure 7.4 illustrates the expected voltammetry for an EC_2 process in the electrochemically irreversible limit ($\Lambda = 0.01$). Again, K_2 controls the presence or absence of a back peak but now no shift in the formal peak potential is seen since

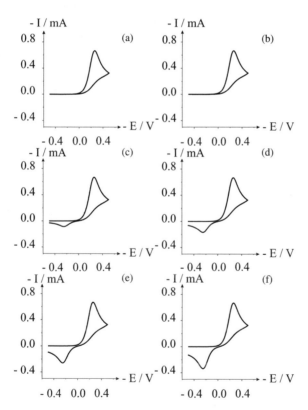

Fig. 7.4 Cyclic voltammograms at varying values of K_2 for an EC_2 process with $\Lambda = 0.01$, $E^0 = 0$ V; $\alpha = 0.5$, $k^0 = 1.973 \times 10^{-4}$ cm s^{-1}; $[A]_0 = 1$ mM; $A = 1$ cm^2; $D_A = D_B = D = 10^{-5}$ cm^2 s^{-1}. (a) $K_2 = 10^3$; (b) $K_2 = 10^1$; (c) $K_2 = 10^{-1}$; (d) $K_2 = 10^{-1.5}$; (e) $K_2 = 10^{-2}$; (f) $K_2 = 10^{-3}$.

the electrochemical irreversibility means that the follow up kinetics has no impact on the rate determining electron transfer kinetics.

A significant difference between EC and EC_2 processes occurs in respect of the voltammetry seen as a function of the concentration of A. In the case of the EC process, neither Λ nor K_1 depends on $[A]_{bulk}$ so that the electrochemical reversibility and the presence or absence of a back peak simply depends on the voltage scan rate as an experimental parameter, not $[A]_{bulk}$. On the other hand, in the case of an EC_2 process, since

$$K_2 \propto \frac{[A]_{bulk}}{v},$$

it follows that the observed voltammetry will be influenced *both* by the voltage scan rate and by the concentration of A. Figure 7.5 shows how in the electrochemically reversible limit, the voltammetry of a process with $k_2 \sim 400$ dm^3 mol^{-1} s^{-1} changes simply as the concentration of A varies from 10^{-3} to 10^{-2} M; an increased concentration promotes the follow-up kinetics and hence, the loss of the reverse peak, all other parameters remaining equal.

The scope for 'fingerprinting' second-order, or more strictly non-first order processes, coupled to the interfacial electron transfer, simply by altering the concentration of the electroactive species, is evident.

7.6 Examples of EC and EC$_2$ Processes

In this section we consider two examples to illustrate the results of the previous sections. The first concerns the one-electron reduction of the 2,6-diphenylpyrylium cation in acetonitrile solution. The radical formed in this process is known to rapidly dimerise as shown in Fig. 7.6.

The process has been studied using fast scan cyclic voltammetry (see Chapter 5) with voltage scan rates in the range 75,000–250,000 Vs^{-1}[16] as shown in Fig. 7.7.

It can be seen that only at very fast scan rates is a full back peak seen indicating the rapidity of the follow-up dimerisation. The experiments were conducted using microelectrodes of radius 5 microns; however, given the very fast scan rates considered, the diffusion layer thickness is so small that linear diffusion is approximated (see Chapter 5). Figure 7.8 shows a planar diffusion simulation assuming a rate constant of 2.5×10^7 dm^3 mol^{-1} s^{-1}. The similarity with Fig. 7.7 is evident.

It is educational to consider what happens if a second voltammetric cycle is performed immediately after the first. The corresponding simulations are shown in Fig. 7.9.

It can be seen that, except for the very fastest scan rates where there is almost a full back peak, there is a very significant diminution of the current peak in the

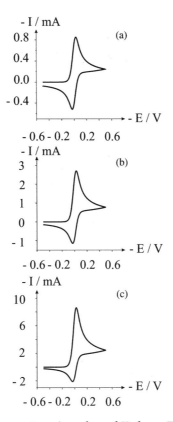

Fig. 7.5 Cyclic voltammograms at varying values of K_2 for an EC_2 process with $\Lambda = 100$, $E_f^0 = 0\,\mathrm{V}$; $\alpha = 0.5$, $k^0 = 1.973\,\mathrm{cm\,s^{-1}}$; $A = 1\,\mathrm{cm^2}$; $D_A = D_B = D = 10^{-5}\,\mathrm{cm^2\,s^{-1}}$; $k_2 = 389.3\,\mathrm{dm^3\,mol^{-1}}$. (a) $\log K_2 = -2$; $[A]_0 = 1\,\mathrm{mM}$; (b) $\log K_2 = -1.5$; $[A]_0 = 3.16\,\mathrm{mM}$; (c) $\log K_2 = -1$; $[A]_0 = 10\,\mathrm{mM}$.

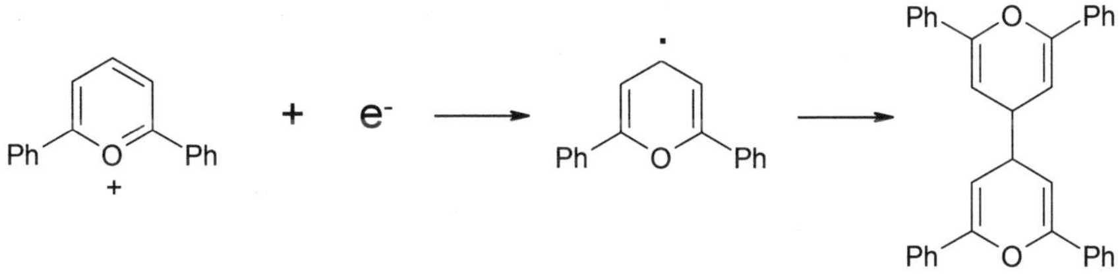

Fig. 7.6 The one-electron reduction of the 2,6-diphenylpyrylium cation, DPP^+.

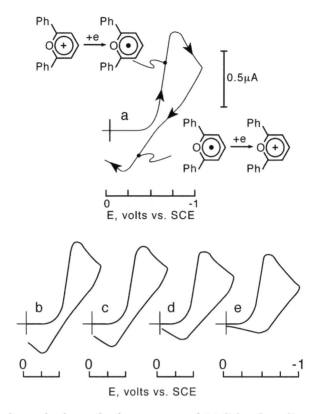

Fig. 7.7 Background subtracted voltammograms of 2,6-diphenyl-pyrylium perchlorate, 10 mM, in acetonitrile, 0.1 M NBu_4BF_4 at a platinum disk ultramicroelectrode (radius 10 microns) at 20°C. Scan rates: (a) 250, (b) 200, (c) 150, (d) 100 and (e) 75 kV s^{-1}. Reprinted from Ref. [6] with permission from Elsevier.

second scan as compared to the first scan. This is of course because, when the voltammetry is conducted on such a timescale as to allow significant dimerisation, there will be significant depletion of B from the interfacial region at the end of the first cycle. Equally, since A is not fully replenished by the oxidation of B back to A as the back (reverse) peak is absent, then there is also significant depletion of B near the electrode surface at the end of the first voltammogram and therefore at the start of the second. Consequently, the reduction peak for A is much smaller in the second cycle than in the first cycle.

Figure 7.10 shows the concentration profiles at various points in the two cycles for the scan rate of 1000 Vs^{-1}, where only a vestigial back peak is seen.

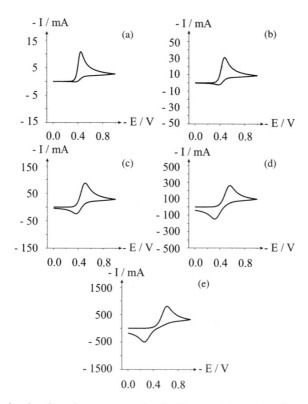

Fig. 7.8 Simulated cyclic voltammograms for the EC_2 reduction of 2,6-diphenyl pyrylium in MeCN. $E_f^0 = -0.435\,V$; $\alpha = 0.5$, $k^0 = 1\text{ cm s}^{-1}$; $[A]_0 = 1\text{ mM}$; $A = 1\text{ cm}^2$; $D_{DPP+} = D_{DPP} = 1.438 \times 10^{-5}\text{ cm}^2\text{ s}^{-1}$; k_f [2DPP$^-$ → (DPP)$_2$] $= 2.5 \times 10^7\text{ dm}^3\text{ mol}^{-1}\text{s}^{-1}$. Scan rates (Vs^{-1}) are: (a) 10^2; (b) 10^3 (c) 10^4; (d) 10^5; (e) 10^6.

The above illustrates the essential importance of refreshing the solution in the interfacial region between measurements. The practice, often encountered with novice students, of repeated potential cycling 'to stabilise the response' can be seen to be probably unhelpful to say the least!

As a second example, we consider the reduction of the esters diethylfumarate (DEF) and diethylmaleate (DEM) in dimethyformide solvent containing tetra-n-butyl ammonium tetrafluoroborate as supporting electrolyte. The structures are shown below

DEF =

DEM =

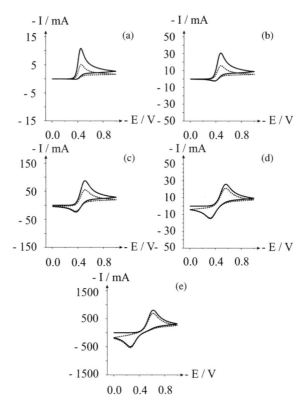

Fig. 7.9 Simulated cyclic voltammograms (2 cycles) for the reduction of 2,6-diphenyl-pyrylium in MeCN. $E_f^0 = -0.435\,V$; $\alpha = 0.5$, $k^0 = 1\,cm\,s^{-1}$; $[A]_0 = 1\,mM$; $A = 1\,cm^2$; $D_{DPP+} = D_{DPP}^{\cdot} = 1.438 \times 10^{-5}\,cm^2\,s^{-1}$; $D_{(DPP)2} = 9.485 \times 10^{-6}\,cm^2\,s^{-1}$; k_f $[2DPP^- \rightarrow (DPP)_2] = 2.5 \times 10^7\,dm^3\,mol^{-1}s^{-1}$; $K_{BC} = 10^{10}dm^3\,mol^{-1}$. Scan rates (Vs^{-1}) are: (a) $= 10^2$; (b) $= 10^3$ (c) 10^4; (d) $= 10^5$; (e) $= 10^6$. The solid line shows the first voltammetric cycle; the dashed line the second.

Voltammetric experiments conducted at platinum macroelectrode using scan rates up to ca 100 Vs^{-1} have been reported.[7] Figure 7.11 shows the voltammetry of DEF conducted at a voltage scan rates of 50 mVs^{-1}.

The electrochemically reversible one electron reduction of the substrate is evident:

$$DEF + e^- \rightleftarrows DEF^{\bullet-} \quad E_f^0 = -1.38\,V.$$

Clearly the radical anion, $DEF^{\bullet-}$ is stable on the voltammetric timescale studied. Figure 7.12 shows the corresponding voltammetry of DEM but measured at 5 Vs^{-1}. Here, the voltammetry is more complex.

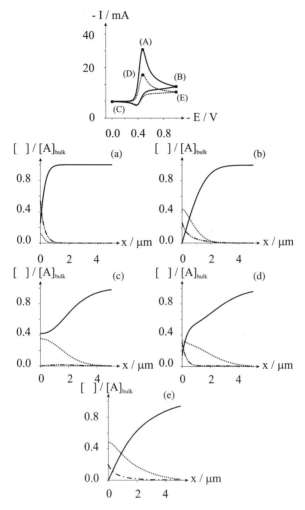

Fig. 7.10 Cyclic voltammograms (two voltammetric cycles) and concentration profiles for the reduction of 2,6-diphenyl-pyrylium in MeCN. $E_f^0 = -0.435$ V; $\alpha = 0.5$, $k^0 = 1$ cm s^{-1}; $[A]_0 = 1$mM; $A = 1$ cm^2; $D_{DPP+} = D_{DPP}^* = 1.438 \times 10^{-5}$ cm^2 s^{-1}; $D_{(DPP)2} = 9.485 \times 10^{-6}$ cm^2 s^{-1}; k_f [2DPP$^{\bullet-} \rightarrow$ (DPP)$_2$] $= 2.5 \times 10^7$ dm^3 mol^{-1}s^{-1}; Scan rates is 1000 Vs^{-1}. In the concentration profiles, the solid line shows the concentrations of DPP^+, the dashed line that of the dimer $(DPP)_2$ and the dash-dot line that of the reduced DPP. The five concentration profiles, (a)–(e), correspond to the positions shown in the voltammogram.

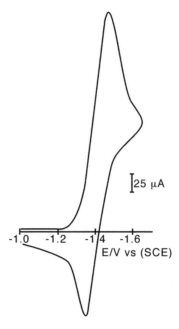

Fig. 7.11 Cyclic voltammogram of diethylfumarate (50 mVs^{-1}) in 0.1 M TBAF-DMF. Reproduced from Ref. [7] with permission of The Royal Society of Chemistry.

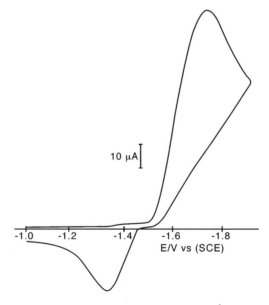

Fig. 7.12 Cyclic voltammogram of diethylmaleate (5 Vs^{-1}) in 0.1 M TBAF-DMF. Reproduced from Ref. [7] with permission of The Royal Society of Chemistry.

This can be understood with reference to the following proposed mechanism,

$$DEM + e \rightleftarrows DEM^{\bullet -} \quad E_f^0 = -1.58\,\text{V}$$

$$DEM^{\bullet -} \xrightarrow{k} DEF^{\bullet -}$$

$$DEF^{\bullet -} - e \rightleftarrows DEF \quad E_f^0 = -1.38\,\text{V},$$

where $k \sim 10\,\text{s}^{-1}$. Under the conditions shown in Fig. 7.12, the voltammetric timescale is still relatively slow compared to the rate of decomposition of $DEM^{\bullet -}$. Accordingly, on the forward scan the radical ion $DEM^{\bullet -}$ is formed but almost quantitatively dimerises to $DEF^{\bullet -}$ before the reverse peak is scanned. Accordingly, the reverse scan shows a peak corresponding to

$$DEF^{\bullet -} - e \rightleftarrows DEF,$$

as can be seen by comparison of Figs. 7.11 and 7.12. Figure 7.13 shows the effect of increasing the voltage scan rates. At low scan rates, Figs. (a) and (b), only the re-oxidation of $DEF^{\bullet -}$ is seen on the reverse scan.

However, at $10\,\text{Vs}^{-1}$ (c) two peaks are evident: one due to the oxidation of $DEF^{\bullet -}$ and the other due to $DEM^{\bullet -}$, some of which survives to be re-oxidised owing to the increased voltage scan rate. For faster scan rates still, Figs. (d) and (e) of 100 and $1000\,\text{Vs}^{-1}$, the voltammetric timescale is now so fast that the only process seen is

$$DEM + e \rightleftarrows DEM^{\bullet -}.$$

In effect, the kinetics of the isomerisation reaction of $DEM^{\bullet -}$ have been 'outrun.' It can be seen therefore that the voltammetry measured over a wide range of scan rates is qualitatively consistent with the scheme proposed above.

Again, it is interesting to consider what happens if a second voltammetric cycle is recorded immediately after the first. Figure 7.14 shows the second cycles (dashed line) superimposed on the first cycles (solid lines) of Fig. 7.13.

It can be seen that in the fast scan rate limit the voltammograms are effectively identical as is expected for a simple one-electron reduction with no homogeneous kinetics operating on the timescales considered. At scan rates of $10\,\text{Vs}^{-1}$ however, the exercise is informative in that two peaks are seen in the reductive sweep of the second cycle. This arises since if the voltammetric timescale allows the formation of $DEF^{\bullet -}$ from $DEM^{\bullet -}$, then the reverse sweep of the first cycle will form DEF near the electrode surface. It is this which is re-reduced to give the fast reductive peak of the forward scan of the second potential cycle. Of course, the second peak in the scan is due to the reduction of DEM which will have diffused from bulk solution to the electrode surface during the reverse scan of the first cycle. Figure 7.15 shows

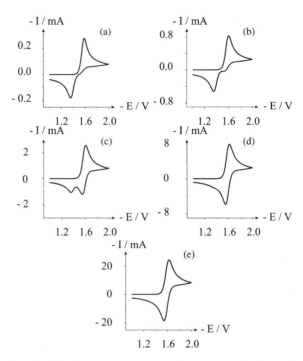

Fig. 7.13 Simulated cyclic voltammograms at varying scan rates for the reduction of DEM in DMF. $E^0_{f,DEM} = -1.58\,V$; $E^0_{f,DEF} = -1.38\,V$ $\alpha = 0.5$, $k^0 = 1\,cm\,s^{-1}$; $[DEM]_0 = 1\,mM$; $A = 1\,cm^2$; $D_{DEM} = D_{DEM^-\bullet} = D_{DEF} = D_{DEF^-\bullet} = 9.1 \times 10^{-6}\,cm^2\,s^{-1}$; $k_f\,[DEM^- \rightarrow DEF^-] = 10\,s^{-1}$; Scan rates: (a) $= 0.1\,Vs^{-1}$; (b) $= 1\,Vs^{-1}$; (c) $= 10\,Vs^{-1}$; (d) $= 100\,Vs^{-1}$; (e) $= 1000\,Vs^{-1}$.

the concentration profiles during the two scans at $1\,Vs^{-1}$. The observation of two peaks on the second cycles thus provides partial confirmation of the mechanism proposed above.

7.7 ECE Processes

Following the Reinmuth and Testa notation of Sec. 7.1 we can write the following general scheme for an ECE process:

$$A + e \rightleftarrows B$$

$$B \xrightarrow{k_1} C$$

$$C + e \rightleftarrows Z.$$

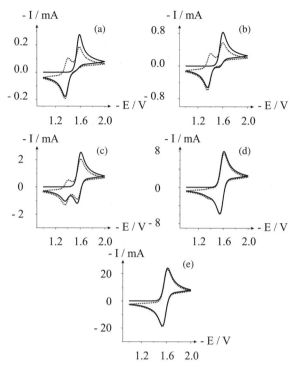

Fig. 7.14 Simulated cyclic voltammograms (2 cycles) at varying scan rates for the reduction of DEM in DMF. $E^0_{f,DEM} = -1.58$ V; $E^0_{f,DEF} = -1.38$ V $\alpha = 0.5$, $k^0 = 1$ cm s^{-1}; $[A]_0 = 1$ mM; $A = 1$ cm^2; $D_{DEM} = D_{DEM-\bullet} = D_{DEF} = D_{DEF-\bullet} = 9.1 \times 10^{-6}$ cm^2 s^{-1}; k_f [DEM$^{\bullet-} \rightarrow$ DEF$^{\bullet-}$] $= 10$ s^{-1}; Scan rates: (a) $= 0.1$ Vs^{-1}; (b) $= 1$ Vs^{-1}; (c) $= 10$ Vs^{-1}; (d) $= 100$ Vs^{-1}; (e) $= 1000$ Vs^{-1}. The solid line shows the first voltammetric cycle, whilst the dashed line depicts the second.

Even a relatively simple kinetic scheme as this is replete with voltammetric variations depending not only on the first order rate constant k_1 and the voltage scan rate employed but also on the thermodynamics (E^0_f) of the two redox couples involved, A/B and C/Z and their electrode kinetics; that is whether the couples are electrochemically reversible or not on the voltammetric timescale employed. Let us in this section suppose both couples are electrochemically reversible.

If both couples show fast electrode kinetics then the voltammetry is significantly influenced by the relative values of $E^0_f(A/B)$ and E^0_f (C/Z). If C is more easily reduced than A,

$$E^0_f(A/B) < E^0_f(C/Z)$$

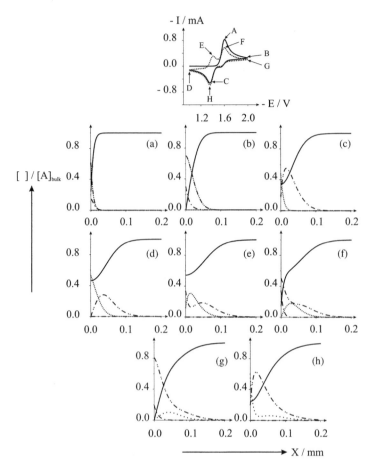

Fig. 7.15 Cyclic voltammogram and concentration profiles for the reduction of DEM in DMF. $E^0_{f,DEM} = -1.58$ V; $E^0_{f,DEF} = -1.38$ V $\alpha = 0.5$, $k^0 = 1$ cm s^{-1}; $[A]_0 = 1$ mM; $A = 1$ cm^2; $D_{DEM} = D_{DEM^{\bullet-}} = D_{DEF} = D_{DEF^{\bullet-}} = 9.1 \times 10^{-6}$ cm^2 s^{-1}; k_f [DEM$^{\bullet-} \rightarrow$ DEF$^{\bullet-}$] $= 10$ s^{-1}; Scan rate $= 1$ Vs^{-1}. In the voltammogram the solid and dashed lines correspond to the first and second cycles, respectively. The eight concentration profiles depict the concentrations of *DEM* (solid line), *DEM*$^{\bullet-}$ (dash-dot), *DEF*$^{\bullet-}$ (dash-dot-dot-dash) and *DEF* (dotted line).

and only one peak is seen in the voltammogram, the character of which is controlled by the magnitude of k_1 or more strictly the size of the dimensionless parameter

$$K_1 = \frac{k_1 RT}{vF}.$$

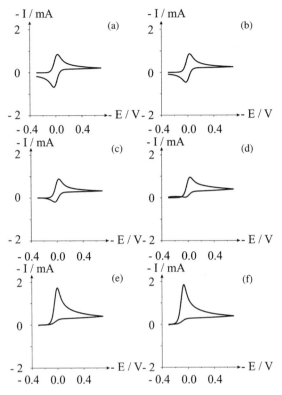

Fig. 7.16 Voltammograms simulated at varying values of k_1 for an ECE process with $\Lambda_1 = \Lambda_2 = 100$. $E_{f,1}^0 = 0\,V$; $E_{f,2}^0 = 0.4\,V$, $\alpha = 0.5$, $k^0 = 1.973\,cm\,s^{-1}$; $[A]_0 = 1\,mM$; $A = 1\,cm^2$; $D_A = D_B = D_C = D_D = 10^{-5}\,cm^2\,s^{-1}$; Scan rate $= 1\,Vs^{-1}$. Values of k_1 : (a) $= 0.0389\,s^{-1}$; (b) $= 0.389\,s^{-1}$; (c) $= 1.23\,s^{-1}$; (d) $= 3.89\,s^{-1}$; (e) $= 389$ s^{-1}; (f) $= 38930\,s^{-1}$; Corresponding values of K_1: (a) $= 10^{-3}$; (b) $= 10^{-2}$; (c) $= 10^{-1.5}$; (d) $= 10^{-1}$; (e) $= 10$; (f) $= 10^3$.

Figure 7.16 shows the effect of k_1 (and hence K_1) on the voltammetry when

$$E_f^0(A/B) = 0\,V \quad E_f^0(C/Z) = +0.4\,V$$

$$D_A = D_B = D_C = D_Z = 10^{-5}\,cm^2\,s^{-1}$$

for a 1 cm^2 electrode and $[A]_{bulk} = 1\,mM$. A voltage scan rate of 1 Vs^{-1} is used throughout the six voltammograms presented.

For low values of k_1, (a) and (b), a simple one-electrode reversible voltammogram is seen corresponding to the process

$$A + e \rightleftarrows B$$

since the kinetics are slow compared to the voltage scan rate. However, as they speed up the back peak is increasingly lost, (c), (d) and (e), and the forward peak due to the reduction of A increases in size and, in the limit of (d) and (e) becomes a two-electron process,

$$A + 2e \longrightarrow Z.$$

In experimental practice, of course, the homogeneous kinetics would be fixed but the value of K_1 would be altered by changing the voltage sweep rate to reduce the transformation from a one-electrode reduction with a full back peak at fast scan rates to a two-electron reduction with no back peak at relatively slow scan rates.

Figure 7.16(d) shows how complex voltammetry can be in intermediate cases. This voltammogram corresponds to a value of $K_1 = 0.1$. It is strange since it shows almost no peak on the reverse scan but the current on the reverse scan re-crosses that on the forward scan at potentials just positive of $E_f^0(A/B)$. This is no artefact. The cross-over occurs since there is still some small amount of B formed around the electrode during the outgoing scan and on the reverse scan at potentials negative of $E_f^0(A/B)$ which is reacting to form C which is then reduced to form Z. This is only visible at potentials positive of $E_f^0(A/B)$ since at these potentials the reduction of A to B is not possible whereas negative of this threshold the current of this process swamps the tiny current due to the reduction of C.

We next explore the case where C is less easily reduced than A:

$$E_f^0(A/B) > E_f^0(C/Z).$$

Figure 7.17 shows simulation for the case where

$$E_f^0(A/B) = 0 \,\text{V} \quad E_f^0(C/Z) = -0.4 \,\text{V},$$

with both redox couples assumed electrochemically reversible and that the chemical species share a common diffusion coefficient of 10^{-5} cm^2 s^{-1}. An electrode area of 1 cm^2 is presumed and a voltage scan rate of 1 Vs^{-1} has been used for all six voltammograms presented.

In the case of slow rate constants ($K_1 \ll 1$), only one voltammetric feature is evident corresponding to

$$A + e \rightleftarrows B,$$

as seen in Fig. 7.17(a). If the rate constant for the decomposition of B increases (shown in b, c and d), a second feature becomes evident at more negative potential than are required for the first case and this is due to

$$C + e \rightleftarrows Z.$$

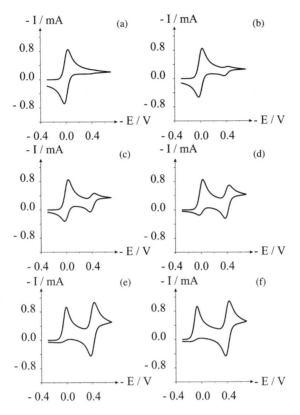

Fig. 7.17 Voltammograms simulated at varying values k_1 for an ECE process with $\Lambda_1 = \Lambda_2 = 100$. $E^0_{f,1} = 0\,\text{V}$; $E^0_{f,2} = 0.4\,\text{V}$, $\alpha = 0.5$, $k^0 = 1.973\,\text{cm s}^{-1}$; $[A]_0 = 1\,\text{mM}$; $A = 1\,\text{cm}^2$; $D_A = D_B = D_C = D_D = 10^{-5}\,\text{cm}^2\,\text{s}^{-1}$; Scan rate $= 1\,\text{Vs}^{-1}$. Values of k_1 : (a) $= 0.0389$ s^{-1}; (b) $= 0.389\,\text{s}^{-1}$; (c) $= 1.23\,\text{s}^{-1}$; (d) $= 3.89\,\text{s}^{-1}$; (e) $= 389\,\text{s}^{-1}$; (f) $= 38,930\,\text{s}^{-1}$; Corresponding values of K_1: (a) $= 10^{-3}$; (b) $= 10^{-2}$; (c) $= 10^{-1.5}$; (d) $= 10^{-1}$; (e) $= 10$; (f) $= 10^3$.

The size of the second wave increases with K_1 corresponding to an enhanced formation of C near the electrode surface. For large K_1 (e and f) the second wave has reached its maximum size and its peak current is comparable to the peak current seen for the reduction of A. Note however that since some A still reaches the electrode to undergo reduction at the potentials required for the reduction of C, it is the case that the current peak of the second wave is higher than that of the first. More significantly, however, the emergence of the second wave can be seen to be associated with the loss of the back peak of the first voltammetric feature. The latter is due to the oxidation of B to form A so, since C is formed directly from B, the loss of the back peak of the first feature necessarily correlates with the emergence of the C/Z peak.

The voltammograms in Fig. 7.17 correspond to different values of K_1 ranging from 10^{-3} to 10^3 and were obtained by changing the value of k_1. In experimental practice, however, the value of k_1 would be of course constant and the voltage scan rate changed in order to switch between at one limit, the one wave due to the A/B redox couple with a back peak due to the oxidation of B at fast scan rates over to the other limit of two voltammetric waves due to the A/B and C/Z couples and no back peaks on the first wave at slow scan rates.

This transition is shown in Fig. 7.18 for the reduction of 4-bromobenzophenone in acetonitrile containing 0.1 M tetrabutylammonium perchlorate as supporting electrolyte[8] for which the following ECE mechanism can be proposed:

$$\text{Br-}\bigcirc\!\!-\!\!\underset{O}{C}\!\!-\!\!\bigcirc \quad + \quad e \quad \rightleftharpoons \quad \left[\text{Br-}\bigcirc\!\!-\!\!\underset{O}{C}\!\!-\!\!\bigcirc \right]^{\bullet -}$$

4-BBP   **4-BBP$^{\bullet-}$**

$$\left[\text{Br-}\bigcirc\!\!-\!\!\underset{O}{C}\!\!-\!\!\bigcirc \right]^{\bullet -} \xrightarrow{\ k_1\ } \text{Br}^- \ + \ {}^{\bullet}\bigcirc\!\!-\!\!\underset{O}{C}\!\!-\!\!\bigcirc$$

4-BBP$^{\bullet-}$

$${}^{\bullet}\bigcirc\!\!-\!\!\underset{O}{C}\!\!-\!\!\bigcirc \ + \ \text{HS} \ \longrightarrow \ \bigcirc\!\!-\!\!\underset{O}{C}\!\!-\!\!\bigcirc$$

BP

$$\bigcirc\!\!-\!\!\underset{O}{C}\!\!-\!\!\bigcirc \quad + \quad e \quad \rightleftharpoons \quad \left[\bigcirc\!\!-\!\!\underset{O}{C}\!\!-\!\!\bigcirc \right]^{\bullet -}$$

BP$^{\bullet-}$

where $E_f^0(\text{4-BBP/4-BPP}^{\bullet-}) \simeq -1.21\,\text{V}$ and $E_f^0(\text{BP/BP}^{\bullet-}) \simeq -1.75\,\text{V}$. A value of $k_1 \sim 2400\ \text{s}^{-1}$ has been reported. Figure 7.18 shows the simulated voltammetry at voltage scan rates in the range 10 to 10^5 Vs^{-1}. At the faster sweep rates only the peak due to the 4-BBP/4-BPP$^{\bullet-}$ redox couple is evident since the kinetics of

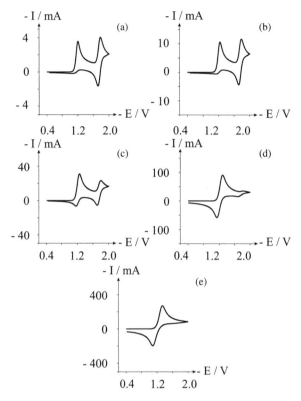

Fig. 7.18 Cyclic voltammograms at varying values scan rates for an ECE process. $E^0_{f,4-BBP} = -1.21$ V; $E^0_{f,BP} = -1.75$ V, $\alpha = 0.5$, $k^0 = 1\,\mathrm{cm\,s^{-1}}$; $[4 - BBP]_0 = 1$ mM; $A = 1\,\mathrm{cm^2}$; $D_{4-BBP} = D_{4-BBP^{\bullet-}} = 1.55 \times 10^{-5}\,\mathrm{cm^2\,s^{-1}}$; $D_{BP} = D_{BP^{\bullet-}} = 1.65 \times 10^{-5}\,\mathrm{cm^2\,s^{-1}}$. $k_f\,[4\text{-BBP}^{\bullet-} \rightarrow BP^{\bullet-}] = 2400\,\mathrm{s^{-1}}$; Scan rates: (a) $= 10\,\mathrm{Vs^{-1}}$; (b) $= 100\,\mathrm{Vs^{-1}}$; (c) $= 1000\,\mathrm{Vs^{-1}}$; (d) $= 10^4\,\mathrm{Vs^{-1}}$; (e) $= 10^5\,\mathrm{Vs^{-1}}$.

the debromination step are 'outrun'. Under these conditions the back peak due to the oxidation of 4-BPP$^{\bullet-}$ on the reverse scan is present. As the scan rate is decreased, however, the BP/BP$^{\bullet-}$ feature emerges at potentials near -1.75 V and this emergence correlates with the loss of the back peak in the 4-BBP/4-BPP$^{\bullet-}$ voltammetric wave. Again, it is instructive to see what happens when a second voltammetric cycle is investigated. Figure 7.19 shows the first and second scans for the 4-BBP system.

At the faster scan rates the 4-BPP/4-BPP$^{\bullet-}$ feature is almost unchanged on the second scan indicating again the stability of the radical anion 4-BPP$^{\bullet-}$ under this voltammetric timescale. In contrast, at slower scan rates where the BP/BP$^{\bullet-}$ feature

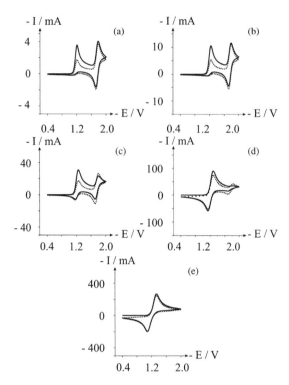

Fig. 7.19 Cyclic voltammograms with two cycles at varying scan rates for an ECE process. $E^0_{f,4-BBP} = -1.21$ V; $E^0_{f,BP} = -1.75$ V, $\alpha = 0.5$, $k^0 = 1$ cm s^{-1}; $[4 - BBP]_0 = 1$ mM; $A = 1$ cm^2; $D_{4-BBP} = D_{4-BBP^{\bullet-}} = 1.55 \times 10^{-5}$ cm^2 s^{-1}; $D_{BP} = D_{BP^{\bullet-}} = 1.65 \times 10^{-5}$ cm^2 s^{-1}. k_f [4-BBP$^{\bullet-} \rightarrow$ BP$^{\bullet-}$] $= 2400$s^{-1}; Scan rates: (a) $= 10$ Vs^{-1}; (b) $= 100$ Vs^{-1}; (c) $= 1000$ Vs^{-1}; (d) $= 10^4$ Vs^{-1}; (e) $= 10^5$ Vs^{-1}. The solid line shows the first voltammetric scan; the dotted line the second.

is additionally present, the current peak for the reduction of 4-BBP is much reduced on the second scan, indicating the kinetic loss of 4-BBP/4-BPP$^{\bullet-}$ which is not available to replenish the depleted concentration of BBP on the reverse scan of the first cycle via oxidation of BBP$^{\bullet-}$ to BBP. On the other hand, the BP/BP$^{\bullet-}$ feature is relatively consistent from the first to the second scan indicating the stability, at least on the timescales conducted, of both BP and BP$^{\bullet-}$.

7.8 ECE versus DISP

So far in this chapter we have considered relatively simple mechanistic schemes but even these have demonstrated the scope for non simple voltammetry! It suggests that as the degree of mechanistic complexity increases it may become difficult to discriminate between different reaction schemes.

This emerging dilemma is nicely illustrated by the difficulty in discriminating between the ECE mechanism and a mechanism involving disproportionation. It arises in a number of electrode reactions that behave as if two (or more) electrons are simultaneously transferred. Consider, for example, a reaction mechanism in which the first electron transfer is followed by a chemical reaction, the product of which is more easily reduced than the starting material, thus producing an overall two-electron process (see Table 7.1). The second electron transfer may occur either at the electrode (see Sec. 7.7 ECE Processes) via reaction (iii) or in bulk solution via disproportionation (DISP mechanism) through reaction (iv).

Within the DISP mechanism there are two further possibilities, DISP 1 and DISP 2, according to whether reaction (ii) or (iv) is rate-determining. The three limiting cases — ECE, DISP 1 and DISP 2 — arising out of steps (i)–(iv) are summarised in Table 7.1. The distinction is not unimportant because of the large number of electrode reaction embraced and the kinetic and synthetic implications when other reactions such as H atom transfer or dimerisation compete with the two-electron processes.

DISP 2 is readily identified from the forward peak potential dependence on scan rate and $[A]_{bulk}$ as shown in Table 7.1, but ECE and DISP 1 give identical responses. The mechanistic ambiguities can be resolved by double potential step chronoamperometry.[8] Specifically, a potential sequence of the following form can accomplish the sought resolution. Assuming a reductive process, as in Table 1, the potential starts at a value positive of both $E_f^0(A/B)$ and $E_f^0(C/Z)$. It is then stepped to a value negative of both these potentials during the ECE/DISP 1 process. Then,

Table 7.1. The general ECE – DISP mechanisms.

$A + e \rightleftarrows B$	$E_f^0(A/B)$	(i)
$B \rightleftarrows C$		(ii)
$C + e \rightleftarrows Z$	$E_f^0(C/Z)$	(iii)
$B + C \rightleftarrows A + Z$		(iv)

Mechanism	Step				Approximate Peak potential/mV decade	
	(i)	(ii)	(iii)	(iv)	$\partial E_p/\partial \log \upsilon$	$\partial E_p/\partial [A]_{bulk}$
ECE		rds		—	30	0
DISP 1		rds	—		30	0
DISP 2			—	rds	20	20

to complete the double step the potential is jumped to a value midway between $E_f^0(A/B)$ and $E_f^0(C/Z)$ so that the potential can drive the reduction of C to Z but now oxidise B to A. It follows that in the DISP 1 route:

$$A + e \rightleftarrows B$$

$$B \xrightarrow{k_1} 1/2A + 1/2Z.$$

The current flowing during the second step is necessarily oxidative due to the reconversion of B; negligible C is present. On the other hand, in the ECE limit,

$$A + e \rightleftarrows B$$

$$B \xrightarrow{k} C$$

$$C + e \rightleftarrows Z.$$

in the second step, there will be contributions to the currents not only from the re-oxidation of B but also from the reduction of C. Accordingly, the current responses are different and the sought distinction possible, at least in principle. Indeed, in favourable cases a characteristic current 'hump' appears during the second (anodic) step in the case of ECE and not of DISP as illustrated in Fig. 7.20.[8]

This approach has been used to show that the reduction of fluorescein.

where

proceeds via the DISP 1 route in aqueous solution at pH 10.

7.9 The CE Mechanism

The CE is a significant generic class of electrode reaction mechanisms which can be defined by the following general scheme

$$Y \rightleftarrows A$$

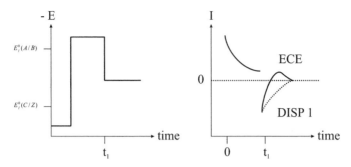

Fig. 7.20 Double potential step chronoamperometry. A double potential step as shown distinguishes ECE and DISP 1.

$$A + e \rightleftharpoons B,$$

in which chemical reaction (C) must precede electron transfer (E). Typical examples include the reduction of formaldehyde at mercury electrodes in aqueous solution. Formaldehyde exists predominately in the hydrated form which is electro-inactive; this form must dehydrate before reduction can take place:

$$C \qquad H_2C(OH)_2(aq) \rightleftharpoons H_2C = O(aq) + H_2O$$

$$E \qquad 2e^- + 2H^+ + H_2C = O(aq) \rightleftharpoons CH_3OH(aq).$$

A second example concerns the reduction of carboxylic acids such as acetic acid,

$$CH_3COOH(aq) \rightleftharpoons H^+(aq) + CH_3CO_2^-(aq)$$

$$H^+(aq) + e^- \rightleftharpoons 1/2H_2(g),$$

again in aqueous solution.

A third and particularly interesting example concerns the one-electron oxidation of 1,2,3-trimethylhexahydropyridazine in butyronitrile containing 0.1 M tetrabutylammonium perchlorate as supporting electrode.[9] Butyronitrile is an excellent low temperature electrochemical solvent. The molecule exists as two possible conformers ee or ae (e = equatorial, a = axial). The ae conformer is lower in energy. The molecule undergoes a one-electron oxidation at a gold electrode forming the corresponding cation radical. The ee conformer is more readily oxidised, by *ca* 0.25 V, than the ae conformer. At room temperature the oxidation occurs in a CE process as shown in Fig. 7.21.

A cyclic voltammogram shows a single oxidation peak at *ca*. 0.3 V. On cooling to −47°C two peaks are seen — one at +0.3 V and a larger one at 0.55 V. This latter

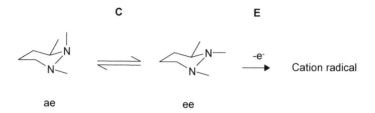

Fig. 7.21   The oxidation of 1,2,3-trimethylhexahydropyridazine.

peak corresponds to the direct oxidation of the ae form: at the lower temperature the rate of the conformational interconversion is slow and not all the material can react via the CE route. If the scan rate is increased the higher potential peak grows in size, as shown in Fig. 7.22, since there is less time for the conversion from the ae to the more easily oxidised ee form to take place on the voltammetric timescale.

7.10 The EC′ (Catalytic) Mechanism

The EC′ mechanism is defined by the following general kinetic scheme:

$$A + e \rightleftarrows B$$

$$B + Z \xrightarrow{k_2} A + Y.$$

The term 'catalytic' arises since the chemical reaction regenerates A from B. By analogy with the preceding sections it can be expected that the voltammetric wave-shapes will be controlled by three parameters: Λ, which dictates the extent of the electrochemical reversibility, a dimensionless rate constant

$$K_2 = \frac{k_2[Z]}{\upsilon}\left(\frac{RT}{F}\right)$$

and

$$\rho = \frac{[Z]_{bulk}}{[A]_{bulk}}.$$

It is illustrative to first consider the limit of a very large excess of Z: $\rho \rightarrow \infty$. Figure 7.23 shows the effect of changing the rate constant k_2 in the electrochemically reversible limit for 0.1 dm^3 mol^{-1} s^{-1} to 10^5 dm^3 mol^{-1} s^{-1} for a scan rate of 1 V s^{-1} and

$$\rho = 1000; \quad [A]_{bulk} = 10^{-3}\,\text{M}; \quad [Z]_{bulk} = 1\,\text{M}.$$

It is evident that at low rate constants the cyclic voltammogram is as expected for an unperturbed A/B reductive process. As k_2 increases however, the back peak for

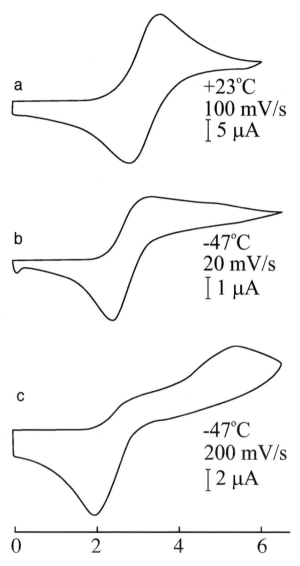

Fig. 7.22 Cyclic voltammetry curves for 1,2,3-trimethylhexahydropyridazine illustrating the appearance of a new peak at low temperature and its movement to higher potential at faster scan rates. Reprinted with permission from Ref. [9]. Copyright (1975) American Chemical Society.

the re-oxidation of B is lost, the forward peak current increases and for sufficiently large k_2, a plateau current is reached rather than the current passing through a maximum. All these characteristics are consistent with the following reaction consuming B and regenerating A. The current plateau is seen only for large p; effectively, there is a large amount of Z present to sustain the catalytic cycle. In the case of Fig. 7.23(g) the rate constant is so large that depletion of Z is only just seen so that the plateau current decreases slightly on the reverse scan.

Figure 7.24 shows the voltammetric responses for the same conditions as Fig. 7.23 except that

$$p = 1; [A]_{bulk} = 10^{-3} \, \text{M}; [Z]_{bulk} = 10^{-3} \, \text{M}$$

and the range of rate constants considered is now 10^2–10^8 dm^3 mol^{-1} s^{-1} so as to enhance the important observations.

Again, an increase in k_2 causes an increased forward current and a decreased back peak. However, since the amount of Z is limited, at high values of k_2, rather than a plateau, a splitting of the voltammogram is seen: the first peak corresponds to the catalysed process and the effective complete consumption of Z near the electrode surface whilst the second peak reflects the usual un-catalysed reduction of A to B.

A typical EC$'$ reaction involves the oxidation of cysteine (Z) mediated via the $Fe(CN)_6^{4/3-}$ redox cycle:

$$Fe(CN)_6^{4-} \rightleftarrows Fe(CN)_6^{3-} + e^-$$

$$Fe(CN)_6^{3-} + cysteine \xrightarrow{k_2} Fe(CN)_6^{4-} + cystine,$$

where k_2 has been reported as 8000 dm^3 mol^{-1} s^{-1} at pH 10.[10] A similar reaction has been proposed and patented as the basis of an amperometric hydrogen sulphide gas sensor:[11]

$$Fe(CN)_6^{4-} \rightleftarrows Fe(CN)_6^{3-} + e^-$$

$$Fe(CN)_6^{3-} + H_2S \longrightarrow 2Fe(CN)_6^{4-} + S + 2H^+.$$

7.11 Adsorption

Hitherto in this chapter we have assumed that all species participating in the electrode process are present in the solution phase. However, it is possible, and indeed not uncommon, that one or more species adsorb on the surface of the electrode and this can quite considerably complicate the voltammetry. Often experimentalists, unless interested in adsorption effects *per se*, will change ('optimise') the electrode

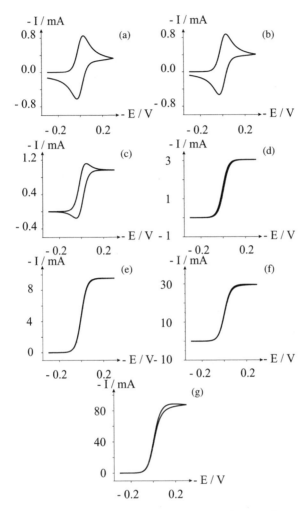

Fig. 7.23 Cyclic voltammograms at varying values scan rates for an EC′ process with a large excess of Z over A. $\Lambda = 100$, $[Z]_0 = 1\,M$; $E^0_{f,AB} = 0\,V$; Z/Y electrochemical process neglected; $\alpha = 0.5$, $k^0 = 1.973\,\text{cm s}^{-1}$; $[A]_0 = 1\,\text{mM}$; $A = 1\,\text{cm}^2$; $D_A = D_B = D_Z = D_{Y^-} = 10^{-5}\,\text{cm}^2\,\text{s}^{-1}$; Scan rate $= 1\,\text{Vs}^{-1}$. (a) $k_f = 0.1\,\text{dm}^3\,\text{mol}^{-1}\,\text{s}^{-1}$; (b) $k_f = 1\,\text{dm}^3\,\text{mol}^{-1}\,\text{s}^{-1}$; (c) $k_f = 10\,\text{dm}^3\,\text{mol}^{-1}\,\text{s}^{-1}$; (d) $k_f = 100\,\text{dm}^3\,\text{mol}^{-1}\,\text{s}^{-1}$; (e) $k_f = 1000\,\text{dm}^3\,\text{mol}^{-1}\,\text{s}^{-1}$; (f) $k_f = 10^4\,\text{dm}^3\,\text{mol}^{-1}\,\text{s}^{-1}$; (g) $k_f = 10^5\,\text{dm}^3\,\text{mol}^{-1}\,\text{s}^{-1}$.

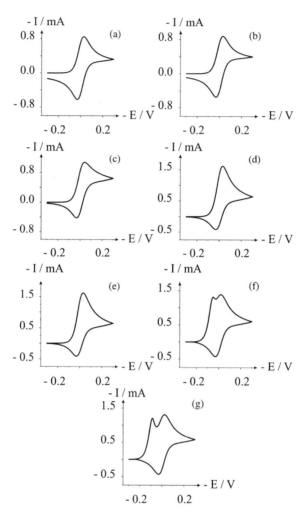

Fig. 7.24 Cyclic voltammograms at varying values scan rates for an EC′ process with comparable quantities of A and Z present. $E^0_{f,AB}$. $\Lambda = 100$, $[Z]_0 = 1$ mM; $E^0_{f,AB} = 0$ V; Z/Y electrochemical process neglected; $\alpha = 0.5$, $k^0 = 1.973$ cm s^{-1}; $[A]_0 = 1$ mM; $A = 1$ cm^2; $D_A = D_B = D_Z = D_{Y-} = 10^{-5}$ cm^2 s^{-1}; Scan rate = 1 Vs^{-1}. (a) $k_f = 100$ dm^3 mol^{-1} s^{-1}; (b) $k_f = 1000$ dm^3 mol^{-1} s^{-1}; (c) $k_f = 10^4$ dm^3 mol^{-1} s^{-1}; (d) $k_f = 10^5$ dm^3 mol^{-1} s^{-1}; (e) $k_f = 10^6$ dm^3 mol^{-1} s^{-1}; (f) $k_f = 10^7$ dm^3 mol^{-1} s^{-1}; (g) $k_f = 10^8$ dm^3 mol^{-1} s^{-1}.

material (Au, Pb, C,...) or the solvent (H$_2$O, EtOH/H$_2$O, MeOH/H$_2$O, ...) to minimise these effects.

We start by supposing that both chemical species A and B in the following redox process remain adsorbed, bound to the electrode surface throughout:

$$A(ads) + e^- \rightleftarrows B(ads).$$

We note that the amount of A and B on the electrode is quantified by the surface coverage, Γ measured in mol cm^{-2}. If the total surface coverage is constant throughout,

$$\Gamma_A(t) + \Gamma_B(t) = \Gamma_{Total},$$

where $\Gamma_A(t)$ is the coverage of species A and Γ_{Total} is the total coverage.

We hypothesize that the electrode kinetics of the A/B couple are fast so that a Nernstian equilibrium is attained in response to an applied potential:

$$\frac{\Gamma_A(t)}{\Gamma_B(t)} = \exp\left(-\frac{F}{RT}[E - E_f^0(A/B)]\right)$$

$$\frac{\Gamma_A(t)}{\Gamma_B(t)} = \exp\left(-[\Theta - \Theta_f^0(A/B)]\right),$$

where $E_f^0(A/B)$ is the formal potential of the adsorbed couple which is typically quite different from that of the potential for solution phase A and B. For simplicity, we have written

$$\Theta = \frac{FE}{RT} \quad \text{and} \quad \Theta_f^0 = \frac{F}{RT} E_f^0(A/B).$$

If follows that

$$\Gamma_A(t) = \frac{\exp\left\{-[\Theta - \Theta_f^0(A/B)]\right\}}{1 + \exp\left\{-[\Theta - \Theta_f^0(A/B)]\right\}} \Gamma_{Total}$$

and

$$\Gamma_B(t) = \frac{1}{1 + \exp\{-[\Theta - \Theta_f^0(A/B)]\}} \Gamma_{Total}.$$

The current is given by

$$\frac{I}{FA} = \frac{\partial \Gamma_B(t)}{\partial t} = -\frac{\partial \Gamma_A(t)}{\partial t},$$

where A is the electrode area.

For the cyclic voltammetry experiment,

$$E = E_1 + \upsilon t \quad (0 < t < t_{switch})$$

$$E = E_1 + \upsilon t_{switch} - \upsilon(t - t_{switch}) \quad (t_{switch} \geq t),$$

where υ is negative for a reduction. It follows that for the forward scan,

$$\frac{I}{FA} = \frac{\upsilon F}{RT} \Gamma_{Total} \frac{\exp\{-[\Theta - \Theta_f^0(A/B)]\}}{[1 + \{\exp -[\Theta - \Theta_f^0(A/B)]\}]^2}.$$

Figure 7.25 shows the predicted shape of the voltammogram. Note that the current response is symmetric about $E_f^0(A/B)$ and so the 'diffusional tail' where the current decays as $1/\sqrt{t}$ at potentials beyond the current peak, in the case of solution phase voltammetry is absent.

Note that the peak current is given by

$$I_p = \frac{F^2}{4RT} \upsilon A \Gamma_{Total}.$$

Furthermore, I_p is directly proportional to the voltage scan rate in contrast to the case of solution phase voltammetry where it scales with the square root of the scan

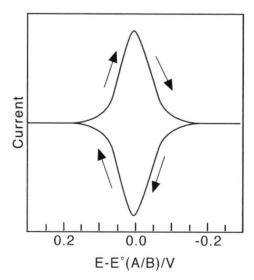

Fig. 7.25 Typical cyclic voltammogram for a reversible one-electron A/B couple where both A and B are adsorbed on the electrode surface.

rate. The width of the peak at half-height is given by

$$3.53 \frac{RT}{F} = 90.6 \text{ mV} \quad \text{at } 25°C.$$

On the reverse scan of the voltammogram the current inverts and the response is symmetrical about the zero current axis. Note that the potentials of the forward and reverse peaks are identical since there is no diffusional hysteresis.

Last, if the forward scan is integrated,

$$\int_0^{t_{switch}} I dt = FA\Gamma_{Total},$$

so that the total coverage can be evaluated simply for the area under the voltammetric trace provided the current has decayed to zero at the time t_{switch}.

It is interesting to question how the forward potentials for A/B with both species in solution and with both species adsorbed might be related. To do this we suppose that if the electrode surface is equilibrated with either A or B without any current flow,

$$\Gamma_A = b_A[A]; \quad \Gamma_B = b_B[B],$$

where the adsorption coefficients b_A and b_B reflect the standard Gibbs energies of adsorption of A and B respectively:

$$\Gamma_A = \exp\left(-\Delta G_A^0/RT\right)$$
$$\Gamma_B = \exp\left(-\Delta G_B^0/RT\right).$$

It follows that the formal potential for the absorbed species is given by

$$E_f^0(A/B)(ads) = E_f^0(A/B)(aq) - \frac{RT}{F} \ln\left(\frac{b_A}{b_B}\right).$$

The implication of this is that the voltammetric features for the absorbed A/B couple can appear at potentials either positive or negative of those relating to the solution phase species. This is illustrated in Fig. 7.26.

Note that where the product (B) is strongly adsorbed, there is a 'pre-peak' before the solution phase voltammetric peak (Fig. 7.26D), whereas when the reactant is strongly adsorbed the adsorption wave is seen after the solution phase peak. An example of a pre-peak concerns the voltammetry of the dye methylene blue in aqueous solution at a mercury drop electrode. The electrode process is a two-electron

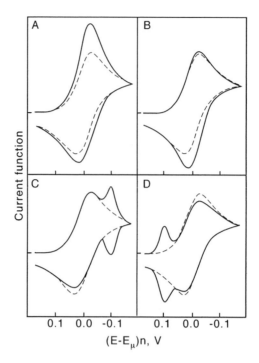

Fig. 7.26 Voltammetry with adsorption of reactants and products. A is where the reactant adsorbed weakly, B is where the product adsorbed weakly, C is where the reactant is adsorbed strongly and D is where the product adsorbed strongly. Dashed lines indicate behaviour for the uncomplicated Nernstian solution phase charge transfer. Reprinted with permission from Ref. [12]. Copyright (1967) American Chemical Society.

reduction forming leuco-methylene blue:

blue: $\xrightarrow{\text{+2e, +2H}^+}$

Figure 7.27 shows typical voltammetry with the pre-peak corresponding to the formation of strongly adsorbed leuco-methylene blue. The unreduced methylene blue is relatively weakly adsorbed.

An example of a strongly adsorbed reactant occurs in the oxidative voltammetry of hydroquinone on platinum electrodes in aqueous solution. In this case the solution voltammetry occurs at potentials less positive than required for the

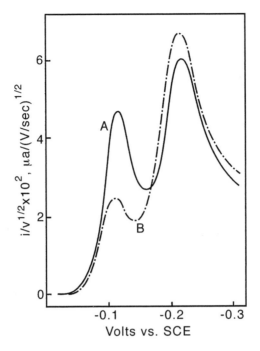

Fig. 7.27 Stationary voltammograms of methylene blue at two scan rates. Concentration = 1 mM, pH 6.5, 50% wt ethanol-water, (A) 82.4 mVs^{-1} and (B) 22.4 mVs^{-1}. Reprinted with permission from Ref. [13]. Copyright (1967) American Chemical Society.

oxidation of the adsorbed materials:

$$HO-\!\!\!\!\bigcirc\!\!\!\!-OH \; \xrightleftharpoons{\;-2e^-,\,-2H^+\;} \; O=\!\!\!\!\bigcirc\!\!\!\!=O \quad,$$

Figure 7.28 shows typical data. Note that in order to distinguish the two peaks the voltammetry was conducted in a 'thin layer cell' of volume just 4 microlitres and a very slow voltage sweep rate of 2 mVs^{-1} was employed.

Under these conditions the current–voltage characteristics observed above for the 'adsorbed' layer also applies to the thin layer cell solution phase voltammetry since the diffusion layer thickness ($\sim \sqrt{\pi D t}$) is large compared to the cell thickness. Accordingly, the solution phase voltammetry has the symmetrical shape observed under these 'thin layer' conditions.

Last, we emphasise that the discussion above has been simplistic in respect of the voltammetry of adsorbed species in two major respects. First, we have assumed electrochemical reversibility for the electrode kinetics of the adsorbed

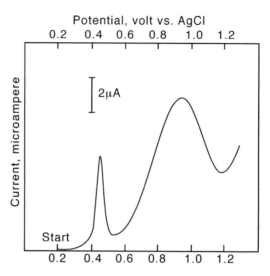

Fig. 7.28 Thin-layer voltammetric curve for hydroquinone at a polycrystalline platinum electrode. The solution contained 0.15 mM reactant and 1M $HClO_4$. The cell volume was 4.08 μL. Electrode area 1.18 cm^2, scan rate 2 mVs^{-1} and solution temperature 296 K. Reprinted with permission from Ref. [14]. Copyright (1982) American Chemical Society.

species whereas this need not be the case. Figure 7.29 shows the effect of varying the electrochemical rate constant: as with solution voltammetry an increase in the peak-to-peak separation is evident. Second, we have neglected the forces of attraction and repulsion between the adsorbed species; if those between the A molecules, the B molecules and between the A and the B molecules differ, then this causes derivation from the ideal voltammogram observed above for the reaction.

$$A + e \rightleftarrows B.$$

Since the relative amounts of adsorbed A and B changes during the voltammetric sweep this non-ideality cannot be handled simply by incorporation into the formal potential as we did in Chapter 2 in respect of solution phase non-ideality. If the forces between like species are parameterised by terms a_A and a_B, and those between unlike molecules as a_{AB}, with $a > 0$ corresponding to an attraction and $a < 0$ to a repulsion, then if

$$a_A + a_B > 2a_{AB},$$

the peak sharpens compared to the ideal case, whereas if

$$a_A + a_B < 2a_{AB},$$

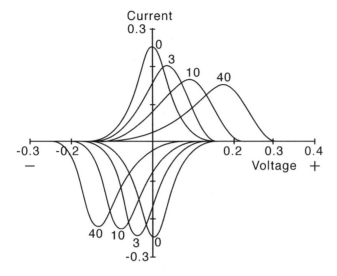

Fig. 7.29 Voltammetric peaks with changing rate of the electrochemical reaction for a surface catalysed reaction at $25°C$ with $\alpha = 0.6$. Reprinted from Ref. [15] with permission from Elsevier.

the peak becomes narrower and sharper. Obviously, the case

$$a_A + a_B = 2a_{AB}.$$

corresponds to the ideal limit described above. Figure 7.30 shows how the peak shapes responds to the value of

$$a = a_A + a_B - 2a_{AB}.$$

Interestingly, for large positive values of a the forward and back peak need no longer have the same potential, as shown in Fig. 7.31. These and other effects have been summarised and rigorously interpreted in a series of admirable papers by Laviron (see for example Ref. 15).

7.12 Voltammetric Studies of Droplets and Solid Particles

Most voltammetric studies are conducted on solution phase or on molecularly adsorbed species as discussed earlier in this Chapter. However, it is important to realise that this does not represent the limit of voltammetric substrates. For example, the study of immobilised microparticles and droplets has been shown to be viable with the groups of Scholz[16] and Marken[17] pioneering in the area. In typical

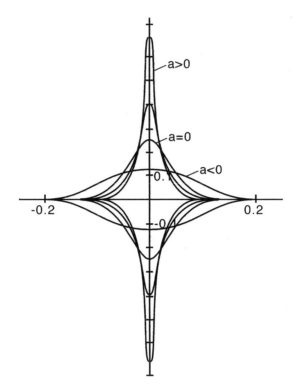

Fig. 7.30 Voltammetric peaks at 25°C in the case of a Frumkin isotherm. Reprinted from Ref. [15] with permission from Elsevier.

experiments 'arrays' of microparticles are formed through 'abrasive attachment' — rubbing the solid particles onto the surface of, say, a basal plane graphite electrode. Similarly, droplet arrays can be formed via the deposition of a solution of the water immiscible droplet medium (oil) on the electrode surface and then allowing the volatile carrier solvent (such as dichloromethane) to evaporate off. The droplet or microparticle array is then immersed in a suitable electrolyte, typically aqueous, which contains reference and counter electrodes. Note that it is not required that the solid or droplet is electrically conductive since electrolysis can occur at the three phase boundary formed between the electrolyte the oil or solid and the electrode, as shown in Fig. 7.32; we next present general illustrative examples.

First, Wain *et al.*[18] have studied the voltammetry of Vitamin K_1 microdroplets on a basal plane pyrolytic graphite electrode. The structure of Vitamin K_1, a quinone, is given in Fig. 7.33.

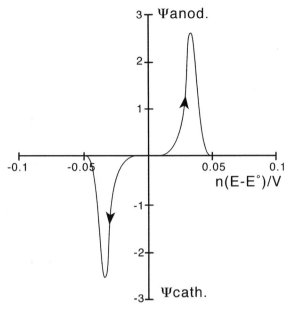

Fig. 7.31 Voltammetric peaks at 25°C in the case of a Frumkin isotherm.[15] For $a \gg 0$ a hysteresis appears in the surface confined voltammetry and the forward and backward peak occur at different potentials. Reprinted from Ref. [15] with permission from Elsevier.

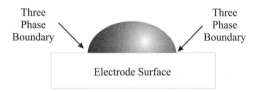

Fig. 7.32 Schematic diagram of a droplet immobilised at an electrode surface.

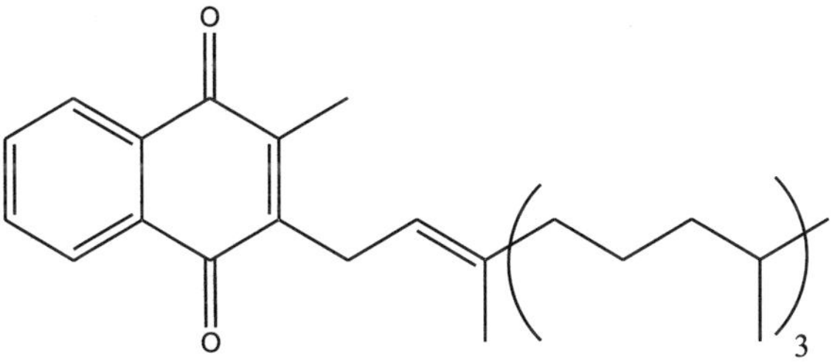

Fig. 7.33 The structure of Vitamin K_1.

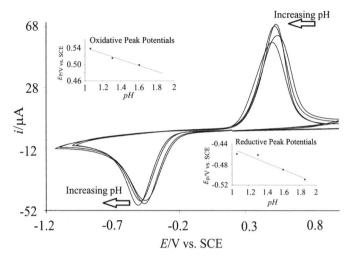

Fig. 7.34 Cyclic voltammograms of 5.3 nmol Vitamin K_1 immobilised on a 4.9 mm diameter basal plane pyrolytic graphite electrode immersed in aqueous HCl of various pH. The inserts illustrate the variation of the oxidation and reduction potential peaks on the solution pH. Reprinted from Ref. [18] with permission from Wiley.

Figure 7.34 shows that meaningful voltammetry can be obtained from the droplets when immersed in an aqueous acid solution and attributed to the two-electron, two-proton reduction of the quinone group:

$$VK_1(l) + 2H^+ + 2e^- \rightleftarrows VK_1H_2(l).$$

A second example[19] relates to the voltammetry of α-tocopherol (α-TOH) which is the most biologically active component of Vitamin E. The structure is shown in Fig. 7.35.

Figure 7.36 shows the oxidative droplet voltammetry in pH 2 aqueous buffer for the first and subsequent scans up to a maximum of ten in total. Two peaks are

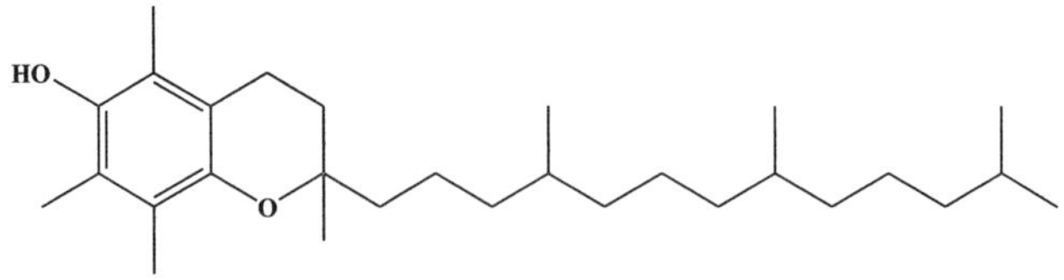

Fig. 7.35 The structure of α-tocopherol.

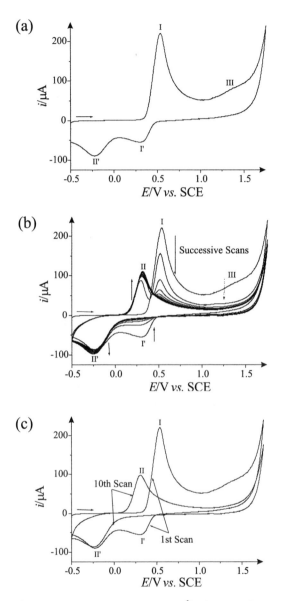

Fig. 7.36 Cyclic voltammograms (scan rate 200 mVs^{-1}) of 5.5 nmol α-tocopherol immobilised on a 4.9 mm diameter bppg electrode immersed into 0.12 M aqueous Britton–Robinson buffer solution at pH 2: (a) first scan, (b) the effect of consecutively scanning and (c) a comparison of the first and tenth scan in a repetitive cycling experiment. Reproduced from Ref. [19] with permission of the PCCP Owner Societies.

observed I/I′ and II/II′. The first of these, I/I′, has been attributed to the reaction.

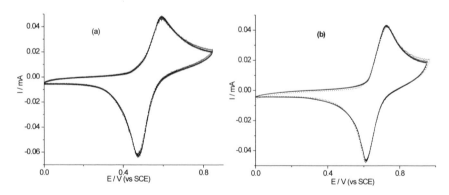

and the second II/II′ to the reduction and oxidation of the quinone group generated by the chemical reaction above.

A third example concerns a study of the oxidation of 1,3,5-tris[4-[(3-methylphenyl)phenylamino]phenyl]benzene, (TMPB), dissolved in toluene microdroplets in contact with aqueous solutions of sodium fluoride, perchlorate, nitrate and sulphate. Figure 7.37 shows typical voltammetry for the cases of perchlorate and nitrate where the oxidation was inferred to be accompanied by anion insertion into the toluene phase from the aqueous phase[20]:

$$TMPB(toluene) + X^-(aq) \rightleftarrows (TMPB^{\bullet+}X^-)(toluene) + e^-.$$

The peak potential was found to vary by ca 60 mV per order of magnitude difference in concentration of X- (= NO_3^- or ClO_4^-) consistent with the process proposed above. In the case of the strongly hydrated anions, SO_4^{2-} and F^-, no anion insertion was observed; rather, the cation $TMPB^{\bullet+}$ dissolved into aqueous solution:

$$TMPB(toluene) \rightleftarrows TMPB^{\bullet+}(aq) + e^-.$$

Finally, the oxidation of TMPD in the form of microparticles abrasively attached to a basal plane pyrolytic graphite electrode surface was studied. Figure 7.38 shows a

Fig. 7.37 Cyclic voltammogram recorded at 20 mVs^{-1} of TMPB in toluene deposited on a BPPG electrode in contact with (a) 1 M NaClO$_4$ (aq) and (b) 1 M NaNO$_3$ (aq). Reprinted from Ref. [20] with permission from Elsevier.

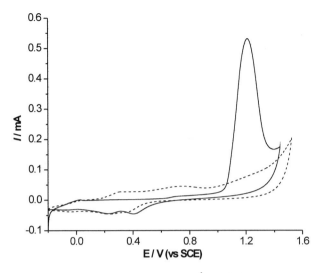

Fig. 7.38 Cyclic voltammogram recorded at 20 mVs^{-1} of solid TMPB on a BPPG electrode in contact with 0.1 M NaClO$_4$ (aq). Reprinted from Ref. [20] with permission from Elsevier.

typical result in aqueous sodium perchlorate solution. It was inferred that oxidation of the solid led to dissolution:

$$TMPB(s) \rightleftarrows TMPB^{\bullet+}(aq) + e^-.$$

References

[1] A.C. Testa, W.H. Reinmuth, *Anal. Chem.* **33** (1961) 1320.
[2] M.E. Peover, *J. Chem. Soc.* (1962) 4540.
[3] J.B. Conant, M.F. Pratt, *J. Am. Chem. Soc.* **48** (1926) 3178.
[4] R.G. Compton, J.C. Eklund, L. Nei, A.M. Bond, R. Colton, Y.A. Mah, *J. Electroanal. Chem.* **385** (1995) 249.
[5] R.G. Compton, R.G. Wellington, P.J. Dobson, P.A. Leigh, *J. Electroanal. Chem.* **370** (1994) 129.
[6] C. Amatore, C. Lefrou, *Portugaliae Electrochimica Acta* **9** (1991) 311.
[7] A.J. Bard, V.J. Puglisi, J.V. Kenkel, A. Lormax, *Faraday Discuss. Chem. Soc.* **56** (1973) 353.
[8] L. Nadjo, J. M. Savéant, *J. Electroanal. Chem.* **30** (1971) 41.
[9] S.F. Nelson, L. Echegoyen, D.H. Evans, *J. Am. Chem. Soc.* **97** (1975) 3530.
[10] O. Nekrassova, G.D. Allen, N.S. Lawrence, L. Jiang, T.G.J. Jones, R.G. Compton, *Electroanalysis* **14** (2002) 1464.
[11] P. Jeroschewski, K. Haase, A. Trommer, P. Gründler, *Electroanalysis* **6** (1994) 769.
[12] R.H. Wopschall, I. Shain, *Anal. Chem.* **39** (1967) 1514.

[13]   R.H. Wopschall, I. Shain, *Anal. Chem.* **39** (1967) 1527.
[14]   M.P. Soriaga, A.T. Hubbard, *J. Am. Chem. Soc.* **104** (1982) 2742.
[15]   E. Laviron, *J. Electroanalytical Chem.* **100** (1979) 263.
[16]   T. Grygar, F. Marken, U. Schröder, F. Scholz, *Coll. Czech. Chem. Commun.* **67** (2002) 163.
[17]   F. Marken, R.D. Webster, S.D. Bull, S.G. Davies, *J. Electroanalytical Chem.* **437** (1997) 209.
[18]   A.J. Wain, J.D. Wadhawan, R.G. Compton, *Chem. Phys. Chem.* **4** (2003) 974.
[19]   A.J. Wain, J.D. Wadhawan, R.R. France, R.G. Compton, *Phys. Chem. Chem. Phys.* **6** (2004) 836.
[20]   N.V. Rees, J.D. Wadhawan, O.V. Klymenko, B.A. Coles, R.G. Compton, *J. Electroanal. Chem.* **563** (2004) 191.

8

Hydrodynamic Electrodes

So far in this book we have considered the voltammetric response of stationary electrodes in a stationary solution in which transport to the electrode surface occurs solely by diffusion and in accordance with Fick's Laws (Chapter 3). In this chapter, we explore voltammetry in which defined diffusion is augmented by convection. There are two main reasons why convection is introduced. First, the additional means of transport gives higher current densities, especially where macroelectrodes are concerned. Indeed, we show later in this chapter that insonation of a macroelectrode can confer it with the mass transport properties of a micron-sized electrode with the benefits for the study of fast processes established in Chapter 5 for microelectrodes. Second, the presence of significant convection changes the voltammetric response from the characteristic peak shaped current–voltage curves typical of voltammetry under planar diffusion conditions into the steady-state response typical of a hydrodynamic electrode shown in Fig. 8.1.

Note that the current, rather than peaking, reaches a steady limiting current since the convective flow brings fresh electroactive material to the electrode from bulk solution. Furthermore, the hydrodynamic voltammogram is characterised by mass transport limited current, I_{LIM}, and the half wave potential, $E_{1/2}$, corresponding to the point where the current

$$I = I_{lim}/2.$$

8.1 Convection

There are two forms of convection: natural and forced. Natural convection occurs as a result of density gradients in solution causing a flow of material from the denser to

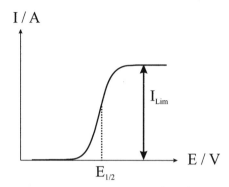

Fig. 8.1 A typical hydrodynamic voltammogram.

the less dense regions. Such effects can result directly from electrolysis occurring at the working and/or counter electrodes where the chemical transformations necessarily cause local density differences. Density gradients are also caused by localised thermal variations. Again, this might be intrinsic to the electrolytic processes: for example on strongly exothermic coupled homogeneous chemical reaction. Alternatively, imperfect thermostating can give rise to density gradients especially with larger voltammetric cells. Natural convection is generally irreproducible and undesirable in electrochemical experiments. Natural convection effects can usually be eliminated by ensuring that the time duration of voltammetric measurement does not exceed *ca.* 10–20 seconds. Sustained electrolysis for periods of time significantly in excess of this period is likely to lead to observable natural convection effects. Note that the implication of a maximum 'natural convection free' time window in which to conduct voltammetry provides a lower limit to the voltage scan rate used in current–potential experiments.

Forced convection is the deliberate stirring or agitation of the solution by mechanical means. It is often designed to be part of the experiment so as to dominate the mass transport in the system — except extremely physically close to the electrode surface where diffusional effects always dominate — and be of a well-defined and easily interpreted character. Hydrodynamic voltammetry is based on controlled convective mass transport. This is achieved either by stirring of the solution (as in sonovoltammetry), by rotation or vibration of the electrode (as in rotating disc voltammetry) or, as in the case of channel flow voltammetry, by the flow of solution over stationary electrodes.

Controlled movement of solution relative to the electrode leads to the dominance of forced over natural convection allowing the quantitative description of the electrode process operating, as will be seen below. Moreover, the rates of mass transport are not only controlled but can be readily altered, giving a wide range of reaction timescales that may be explored and giving a parameter through which kinetic and mechanistic information may be probed, for example by changing the flow rate in a channel electrode or the rotation speed of a rotating disc electrode. In addition, the fact that useful observations can be made under steady-state conditions precludes distortion of data as often happens in, for example, fast scan cyclic voltammetry, through charging of the double layer capacitance at the electrode surface.

8.2 Modifying Fick's Laws to Allow for Convection

It is readily appreciated that the convective equivalent of Fick's first law in one dimension, x,

$$j_{conv} = [B]V(x),\tag{8.1}$$

where $j_{conv}/$ mol cm^{-2} s^{-1} as the convective flux of species B, present at concentration $[B]$ at a point x where the local fluid velocity is $V(x)/$cm s^{-1}. To derive the equivalent of the second law we consider the one-dimensional transport shown in Fig. 8.2.

The temporal change in concentration of species B at the point x is obtained by considering the difference in flux entering the plane at x leaving the plane at $x + dx$. Since mass is conserved

$$[B](x, t + dt)Adx - [B](x, t)Adx = j(x, t)Adt - j(x + dx, t)Adt,\tag{8.2}$$

where A is the area shown in Fig. 8.1. Equation (8.2) rearranges to

$$\frac{\partial[B]}{\partial t} + \frac{\partial j}{\partial x} = 0.$$

Substitution of Eq. (8.1) gives

$$\frac{\partial[B]}{\partial t} = -V(x)\frac{\partial[B]}{\partial x}.\tag{8.3}$$

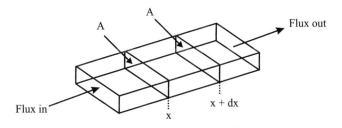

Fig. 8.2 Planar transport in the x-direction.

Equation (8.3) forms the basis for the convective–diffusion equation for transport in one dimension:

$$\frac{\partial [B]}{\partial x} = D\frac{\partial^2 [B]}{\partial x^2} - V(x)\frac{\partial [B]}{\partial x}. \tag{8.4}$$

The general three-dimensional versions of the above are, in terms of vector operators,

$$\vec{j} = -D\nabla[B] + [B]\,\vec{V} \tag{8.5}$$

and

$$\frac{\partial [B]}{\partial t} = D\nabla^2[B] - \vec{V}\,\nabla[B]. \tag{8.6}$$

An equation of the form (8.4) is sufficient to understand voltammetry at a rotating disc electrode provided the function $V(x)$ is known. A knowledge of the hydrodynamics prevailing at the electrode of interest is clearly a pre-requisite for developing an understanding of the voltammetry.

8.3 The Rotating Disc Electrode: An Introduction

The rotating disc electrode consists of a disc of the material of interest (e.g. Pt, Au, glassy carbon, etc.) embedded in a cylinder of insulating material (e.g. Teflon), as shown in Fig. 8.3.

The electrode is rotated in the solution under study; typically rotation speeds of $0 - 50$ Hz are employed although designs for a much faster electrode have been proposed.[1] Figure 8.4 shows the design of a gas-driven high-speed rotating disc (HSRD) capable of rotation at ca 650 Hz in aqueous solution.

The enhanced mass transport characteristics of this type of device are evident in Fig. 8.5 which shows the voltammetric oxidation of ferrocyanide in aqueous 0.1 M KCl at a gold electrode in stationary and rotating modes using the HSRD at ca 650 Hz.

Further illustration comes from the analytical detection of 1μM Arsenic (III) in 0.1 M nitric acid using anodic stripping voltammetry. The latter technique is more fully discussed in Chapter 9 of this book. In this case As (III) was reduced at -0.5 V to build up a deposit of As (0) on the electrode for a period of 60 seconds, after which a linear potential sweep in the positive direction was undertaken as shown in Fig. 8.6. A 'stripping' signal is observed at $\sim +0.16$ V:

$$1/3As(s) \longrightarrow 1/3As(III)(aq) + e^-.$$

This experiment was conducted comparatively under quiescent (no stirring) condition and using the high-speed rotating disc 'on' during the accumulation

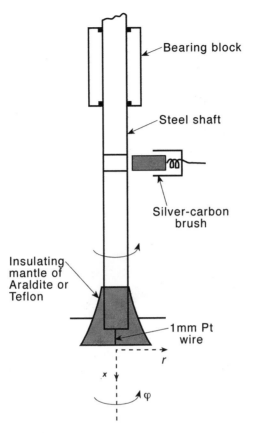

Fig. 8.3 A rotating disc electrode of the classical Riddiford design. The electrode is the polished end of a platinum wire of 1 to 3 mm radius. The coordinate system is also shown.

period. The charge (integral of current with time) under the stripping peak increases by a factor of *ca.* 16 and a small voltammetric signal is transformed with a large and easily quantifiable signature with the enhanced mass transport benefits due to rotation.

8.4 The Rotating Disc Electrode — Theory

When a rotating disc electrode is rotated in a large volume of solution, a well defined flow pattern is established: the electrode acts as a pump pulling the solution vertically upwards towards the disc, spinning it round, and then throwing it outwards, as illustrated schematically in Fig. 8.7.

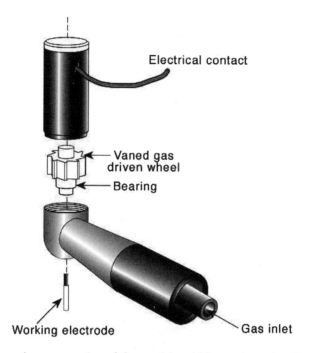

Fig. 8.4 Schematic representation of the gas-driven high-speed rotating disc. Reprinted with permission from Ref. [1]. Copyright (2005) American Chemical Society.

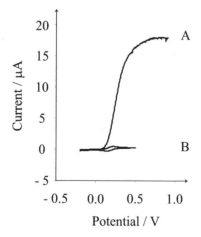

Fig. 8.5 Voltammetric oxidation of 1.47 mM ferrocyanide in 0.1 M KCl at a gold electrode in stationary (B) and rotating mode (A). Scan rate was 15 mV s^{-1}. Reprinted with permission from Ref. [1]. Copyright (2005) American Chemical Society.

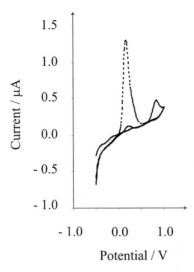

Fig. 8.6 Linear sweep voltammograms of $1\,\mu$M Arsenic (III) in 0.1 M nitric acid solution obtained under quiescent conditions (thick line) and rotating mode (dotted line) at the HSRD during the accumulation step. Parameters: -0.5 V for 60 second followed by a potential sweep at $50\,\text{mV s}^{-1}$ (vs. SCE). Reprinted with permission from Ref. [1]. Copyright (2005) American Chemical Society.

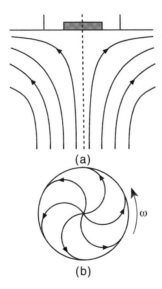

Fig. 8.7 Schematic diagrams of the flow patterns created by the rotating disc electrode. (a) View from the side showing how the solution is pumped towards the disc, then thrown outwards. (b) Solution flow close to the electrode surface, viewed from below.

This laminar flow is maintained provided that the Reynolds's number, defined by

$$R_e = \frac{\omega r^2}{\nu} \tag{8.7}$$

does not exceed a critical value of *ca.* 2×10^3. In Eq. (8.7), ω is the rotation speed measured in Hz, r is the radius of the electrode and ν is the kinematic viscosity $(\mathrm{m^2\,s^{-1}})$. Note that

$$\text{kinematic viscosity, } \nu, (\mathrm{m^2 s^{-1}}) = \frac{\text{viscosity } (\mathrm{Nm^{-2}\,s})}{\text{Fluid density } (\mathrm{kgm^{-3}})}.$$

Water at 25°C has a kinematic viscosity of the order of $10^{-6}\,\mathrm{m^2 s^{-1}}$ ($10^{-2}\,\mathrm{cm^2\,s^{-1}}$). Although this suggests that speeds of up to 1000 Hz might not cause turbulent flow, which sets in above the critical Reynolds number, experiments are almost invariably performed in the range 0–50 Hz except for the HSRD experiments as described in the previous section.

The distinction between laminar and turbulent flow originates in seminal work conducted by Reynolds in Manchester, UK in 1883. He conducted experiments in which a filament of dye was introduced into water flowing through a glass tube (Fig. 8.8).

In laminar flow, when the speed of flow was small, the dye followed a straight line path with only a slight blurring due to diffusion. At very fast flow rates the dye is blurred and apparently fills the entire pipe; this is the regime of turbulent flow. The

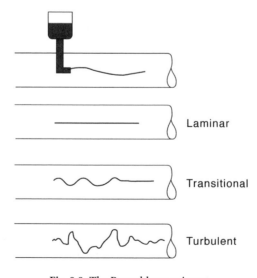

Fig. 8.8 The Reynolds experiment.

distinction can be made clear if instantaneous velocities of the dye molecules are measured at a point. Under laminar flow condition, there is a predominant velocity in the main flow direction whereas turbulent flow has a significant component of velocity in the direction normal to the predominant flow.

8.5 Osborne Reynolds (1842–1912)

Reynolds

Reynolds undertook an apprenticeship with the engineering firm Edward Hayes in 1861 before studying Mathematics at Cambridge and graduated in 1867. Reynolds was elected to a scholarship at Queens' College and spent a year as a civil engineer. In 1868, Reynolds became the first professor of engineering at Manchester and also the second only professor of engineering in England.

Reynolds held this post until he retired in 1905. His early work was on magnetism and electricity but he soon concentrated on hydraulics and hydrodynamics. He also worked on electromagnetic properties of the sun and of comets, and considered tidal motions in rivers. After 1873, Reynolds concentrated mainly on fluid dynamics and it was in this area that his contributions were of world leading importance. Around 1873, Reynolds researched mainly on fluid dynamics where he studied the flow along a pipe when it changes from laminar flow to turbulent flow (see above). He is famed for introducing the 'Reynolds number', a variable commonly used in modelling fluid flow. He was elected Fellow of The Royal Society in 1877 and won the Royal Medal in 1888. Reynolds retired in 1905. Many biographies of Reynolds suggest that he was not the best lecturer; his lectures were difficult to follow and frequently went through topics with little or no connection. An informative summary of Reynold's life can be found at Ref. [2].

8.6 The Rotating Disc Electrode — Further Theory

The hydrodynamics of the rotating disc under laminar flow conditions has been derived by Von Karman[3] and Cochran.[4] The equations are best described in terms of a hydrodynamic layer thickness

$$x_H = \sqrt{\nu/2\pi\omega}, \tag{8.8}$$

where ν is the kinematic viscosity and ω is the rotation speed in Hz. $2\pi\omega$ is therefore the rotation speed in radians per second.

The velocity components in each of the three cylindrical coordinates, x, r and ϕ (Fig. 8.3) are

$$V_r = 2\pi\omega r F(x/x_H)$$

$$= 2\pi\omega r \left[0.510 \left(\frac{x}{x_H} \right) - \frac{1}{2} \left(\frac{x}{x_H} \right)^2 + \frac{0.616}{3} \left(\frac{x}{x_H} \right)^3 + \cdots \right] \quad (8.9)$$

$$V_\phi = 2\pi\omega r G(x/x_H) = 2\pi\omega r \left[1 - 0.616 \left(\frac{x}{x_H} \right) + \frac{0.510}{3} \left(\frac{x}{x_H} \right)^3 + \cdots \right]$$

$$(8.10)$$

$$V_x = \sqrt{2\pi\omega\upsilon} H(x/x_H) = -\sqrt{2\pi\omega\upsilon} \left[0.510 \left(\frac{x}{x_H} \right)^2 - \frac{1}{3} \left(\frac{x}{x_H} \right)^3 + \cdots \right]$$

$$(8.11)$$

The functions F, G and H are shown in Fig. 8.9. It can be seen that at the disc surface where $x = 0$,

$$G(0) = 1; \quad F(0) = 0; \quad H(0) = 0.$$

In other words the solution is being spun round with the disc,

$$V_\phi = 2\pi\omega r$$

corresponding to the angular velocity of the disc at a radius r.

The angular velocity dies away as one moves away from the disc. The centrifugal velocity reaches a maximum at about

$$x \sim x_H.$$

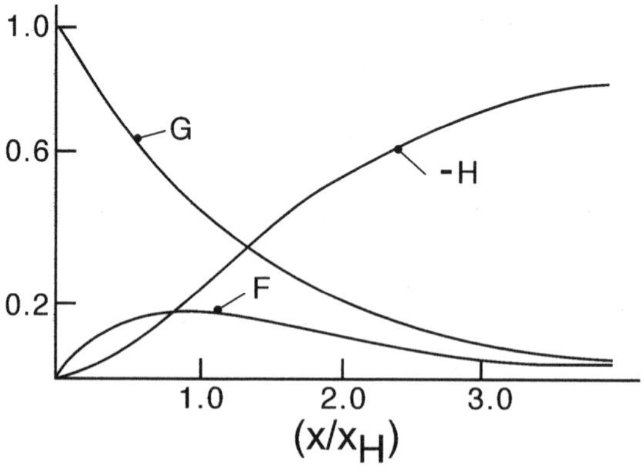

Fig. 8.9 Velocity distributions around the disc rotating in a stationary liquid. F is the radial, G the azimuthal, and H the axial component.

The solution that is thrown out radially is replaced by a steam of solution from below that is flowing towards the electrode. At distances $\sim 3x_H$ from the electrode both the angular and radial velocities are largely quenched and there is simply a steady, x-independent velocity towards the disc. The latter, to quote Albery,[5] 'acts as a pump sucking solution towards it, spinning it around, and flinging it out sideways.'

Two important features emerge from the above. First, the thickness of the hydrodynamic layer can be estimated from Eq. (8.8) given that for typical rotation speeds in the range 1–50 Hz, the scale of x_H is from 0.1 to 1 mm. Notice that this scale is more than an order of magnitude larger than a typical diffusion layer thickness.

Second, Eq. (8.11) shows that V_x, the velocity component normal to the electrode, is independent of r and ϕ, and only dependent on x. Thus, the solution velocity component bringing fresh solution to the electrode surface is uniform over the electrode surface. It is thus 'uniformly accessible', which greatly simplifies the theoretical treatment of the rotating disc in comparison with other hydrodynamic electrodes.

With a knowledge of the rotating disc hydrodynamics, it is possible to solve the mass-transport (convective–diffusion) equation to predict the current at the rotating disc electrode. The appropriate form of Fick's Second Law is

$$\frac{\partial[B]}{\partial t} = D\frac{\partial^2[B]}{\partial x^2} - V_x\frac{\partial[B]}{\partial x}, \tag{8.12}$$

where B is the species undergoing electrolysis in the presence of sufficient supporting electrolyte to suppress any migration effects. In experimental practice one is often interested in examining steady-state conditions for which

$$\frac{\partial[B]}{\partial t} = 0.$$

Experimentally, this corresponds to voltage scan rates which are slow in comparison with the time required to set up the steady-state concentration profile in the diffusion layer. It follows that

$$D\frac{\partial^2[B]}{\partial x^2} = V_x\frac{\partial[B]}{\partial x}, \tag{8.13}$$

where following Levich,

$$V_x \simeq -0.510\sqrt{2\pi\omega\upsilon}\left(\frac{x}{x_H}\right)^2$$
$$V_x \simeq -0.510\,(2\pi\omega)^{3/2}\,\upsilon^{-1/2}x^2. \tag{8.14}$$

The substitution

$$u = \left(\frac{0.510}{D}\right)^{1/3} (2\pi\omega)^{1/2} \, \upsilon^{-1/6} x \tag{8.15}$$

leads to the rewriting of Eq. (8.13) in the form

$$\frac{d^2[B]}{du^2} = -u^2 \frac{d[B]}{du}. \tag{8.16}$$

This equation can be integrated using the substitution $p = \frac{d[B]}{du}$ to give

$$\frac{d[B]}{du} = \left(\frac{d[B]}{du}\right)_{u=0} \exp\left(\frac{-u^3}{3}\right). \tag{8.17}$$

Integrating again gives

$$[B]_x - [B]_{x=o} = \left(\frac{d[B]}{du}\right)_{u=0} \int_0^{u(x)} \exp\left(\frac{-u^3}{3}\right) du. \tag{8.18}$$

For the case of $x \to \infty$

$$[B]_{bulk} - [B]_{x=o} = \left(\frac{d[B]}{du}\right)_{u=0} 3^{1/3} \Gamma\left(\frac{4}{3}\right), \tag{8.19}$$

since

$$\int_0^\infty \exp\left(-\frac{u^3}{3}\right) du = 3^{1/3}\Gamma\left(\frac{4}{3}\right) = 1.288, \tag{8.20}$$

where $\Gamma(x)$ is the gamma function and $\Gamma\left(\frac{4}{3}\right) = 0.893$. It follows that

$$[B]_{bulk} - [B]_{x=o} = 1.288 \left(\frac{D}{0.510}\right)^{1/3} (2\pi\omega)^{-1/2} \upsilon^{1/6} \left(\frac{\partial[B]}{\partial x}\right)_{x=0}. \tag{8.21}$$

Hence,

$$\left.\frac{\partial[B]}{\partial x}\right|_{x=0} = \frac{[B]_{bulk} - [B]_{x=0}}{\delta_D}, \tag{8.22}$$

where δ_D is the diffusion layer thickness,

$$\delta_D = 0.643\omega^{-1/2}\upsilon^{1/6}D^{1/3}, \tag{8.23}$$

where $0.643 = 1.288 \times 0.510^{-1/3} \times (2\pi)^{-1/2}$. Note that the ratio of the diffusion layer thickness to the hydrodynamic layer thickness

$$\frac{\delta_D}{x_H} = 1.61 \left(\frac{D}{\upsilon}\right)^{1/3}. \tag{8.24}$$

Typically, for experiments conducted in water,

$$D \approx 10^{-5} \text{cm}^2 \, \text{s}^{-1}; \upsilon = 0.01 \text{cm}^2 \, \text{s}^{-1},$$

so that

$$\frac{\delta_D}{x_H} \approx 0.16, \tag{8.25}$$

suggesting that the diffusion layer is significantly thinner that the hydrodynamic layer, thus validating the truncation of Eq. (8.11) at the first term in writing Eq. (8.14).

Figure 8.10 shows how the solution near the rotating disc electrode is effectively 'divided' into a well stirred bulk region and a stagnant 'diffusion layer' of thickness δ_D, the concentration of electroactive species differing appreciably from the bulk concentration only in the latter region.

Figure 8.10 suggests that the maximum current (largest concentration gradient at the surface of the electrode) will occur when

$$[B]_{x=0} = 0,$$

corresponding to the electrode potential being sufficiently large to electrolyse all B that can diffuse to it. The corresponding current is the transport limited

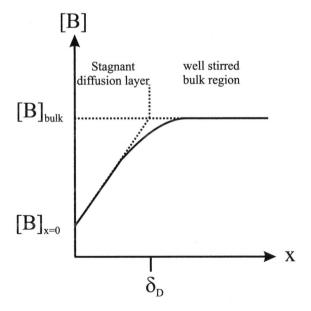

Fig. 8.10 Diffusion layer at a rotating disc electrode.

current,

$$I_{lim} = nFAD \left(\frac{\partial[B]}{\partial x} \right)_{x=0} \tag{8.26}$$

$$I_{lim} = 1.554nFAD^{2/3}\upsilon^{-1/6}[B]_{bulk}\omega^{1/2}, \tag{8.27}$$

where A is the electrode area and n the number of electrons transferred per molecule in the electrolysis of B. An important feature of Eq. (8.27) is the dependence on rotation speed: $\omega^{1/2}$. This provides the experimentalist with the means to vary the concentration profile at the electrode; the faster the rotation, the more compressed the diffusion layer becomes and the greater the transport limited flux. Figure 8.11 shows how a fourfold increase in rotation speed halves the diffusion layer thickness and hence doubles the concentration gradient at the electrode surface. Equation (8.27) is known as the Levich equation.

The predictions of the Levich equation have been well validated. By way of illustration, Fig. 8.12 shows the limiting current analysis of the two-electron reduction of the dye fluorescein, FH^-, at a pH close to 6:

$$FH^- + 2e^- + H^+ \longrightarrow L$$

The two-electron product is leucofluorescein:[6]

Figure 8.12 shows that the limiting current is proportional to $\omega^{1/2}$ over the rotation speed range 1–50 Hz. The slope of the plot is consistent with

$$n = 2 \text{ and } D = 3.2 \times 10^{-6}\,\text{cm}^2\,\text{s}^{-1}.$$

Finally, we point out that Levich's theory for the rotating disc has been developed beyond the level of truncation at the first term of Eq. (8.15). In particular, Newman[7]

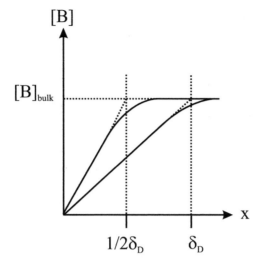

Fig. 8.11 A fourfold increase in rotation speed halves the diffusion layer thickness.

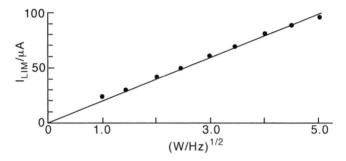

Fig. 8.12 Analysis of rotating disc voltammograms for the reduction of fluorescein at pH 5.88 shows two-electron Levich behaviour. Reproduced from Ref. [7] with permission of The Royal Society of Chemistry.

has shown that the expression

$$\delta_D = 1.61 \left(\frac{D}{v}\right)^{1/3} \left(\frac{v}{2\pi\omega}\right) 1/2 \left[1 + 0.2980 \left(\frac{D}{v}\right)^{1/3} + 0.14514 \left(\frac{D}{v}\right)^{2/3}\right]$$

(8.28)

predicts transport limited fluxes with an accuracy equal to or better than 0.1% for

$$Sc = \frac{v}{D} > 100,$$

(8.29)

where Sc is the Schmidt Number.

8.7 Chronoamperometry at the Rotating Disc Electrode: An Illustration of the Value of Simulation

It is illuminating to see how numerical simulation can be used to readily compute the form of the chronoamperometric transient at a rotating disc electrode resulting from a potential step from a potential corresponding to no current flow to one where the transport limited current is passed. The latter corresponds to a boundary condition at the electrode surface where the concentration of the electroactive species is zero.

The time dependent convective–diffusion equation is

$$\frac{\partial [B]}{\partial t} = D\frac{\partial^2 [B]}{\partial x^2} + Cx^2\frac{\partial [B]}{\partial x}, \tag{8.30}$$

where $C = 0.51 \times (2\pi)^{3/2} = 8.032$. Equation (8.30) can be written in dimensionless form by using the following variables

$$w = \left(\frac{C}{D}\right)^{1/3} x \tag{8.31}$$

$$t^* = \left(C^2 D\right) t \tag{8.32}$$

$$b = \frac{[B]}{[B]_{bulk}} \tag{8.33}$$

so that

$$\frac{\partial b}{\partial t^*} = \frac{\partial^2 b}{\partial w^2} + w^2\frac{\partial b}{\partial w}. \tag{8.34}$$

We now introduce the w^* coordinate in place of w, according to the Hale transformation:[8,9]

$$w^* = \frac{\int_0^w \exp\left(-\frac{1}{3}w^3\right) dw}{\int_0^\infty \exp\left(-\frac{1}{3}w^3\right) dw}. \tag{8.35}$$

Equation (8.34) now becomes

$$\frac{\partial b}{\partial t^*} = \frac{\exp\left(-\frac{2}{3}w^2\right)}{1.65894}\frac{\partial^2 b}{\partial w^{*2}}, \tag{8.36}$$

where

$$1.65894 = \left[\int_0^\infty \exp\left(-\frac{1}{3}w^3\right) dw\right]^2. \tag{8.37}$$

Note that the Hale transformation has reduced the two terms corresponding to diffusion and convection in Eq. (8.34) into just one expression. Moreover, because the x coordinate changes from zero (at the disc surface) to infinity, the transformed coordinate, w^*, changes from 0 to 1. Thus, the simulation of the transient response to a perturbation of some existing concentration distribution can be found by splitting this interval of unity into the N 'boxes' defined by the $N + 1$ grid points $\{0, 1, 2, 3, \ldots, N - 1, N\}$. Likewise, if the transient response over the time interval $t^* = 0$ to l is sought, then this interval is split into lM intervals such that

$$dt^* = \frac{1.0}{M}. \tag{8.38}$$

Choosing $b(m, n)$ to denote the concentration at the point $w^* = \frac{n}{N}$ and time $t^* = \frac{m}{M}$, the changes in concentration, $db(m, n)$ are computed by means of the following equation:

$$b(m + 1, n) = b(m, n) + db(m, n), \tag{8.39}$$

where

$$db(m, n) = \frac{db^* N^2 \exp\left(-\frac{2}{3} w^2\right)}{M 1.65894} \left[b(m, n - 1) - 2b(m, n) + b(m, n + 1) \right]. \tag{8.40}$$

Note that the term in square brackets is the finite difference form of the double derivative,

$$\frac{\partial^2 b}{\partial w^{*2}} = \frac{\partial}{\partial w^*} \left\{ \frac{\partial b}{\partial w^*} \right\}$$

$$\frac{\partial^2 b}{\partial w^{*2}} = \frac{1}{(1/N)} \left\{ \left[\frac{b(m, n + 1) - b(m, n)}{(1/N)} \right] - \left[\frac{b(m, n) - b(m, n - 1)}{(1/N)} \right] \right\}$$

$$\frac{\partial^2 b}{\partial w^{*2}} = N^2 \left[b(m, n - 1) - 2b(m, n) + b(m, n + 1) \right], \tag{8.41}$$

where $dw^* = (1/N)$.

Equations (8.38) and (8.39) permit the computation of the changing concentration distribution provided that some initial distribution is specified along with the boundary conditions that prevail at the electrode surface ($n = 0$) and in the bulk of the solution ($n = N$). In the specific case where the electrode potential is jumped from a value at which no current flows to one which corresponds to the transport limited current, the boundary conditions take the following form

$$m = 0 : \quad b(0, n) = 1$$
$$m > 0 : \quad b(m, 0) = 0; \quad b(m, N) = 1.$$

The current flowing at time t^* may be deduced as

$$I(t^*) = nFAD^{2/3}C^{1/3}[B]_{bulk} \frac{\exp\left(-\frac{1}{3}N^{-3}\right)}{(1.65894)^{1/2}} \left[\frac{b(m,1)}{(1/N)}\right], \qquad (8.42)$$

where A is the electrode area and the term in square brackets is

$$\frac{\partial b}{\partial w^*} = \frac{b(m,1) - b(m,0)}{(1/N)}$$

$$\frac{\partial b}{\partial w^*} = Nb(m,1),$$

since

$$b(m,0) = 0.$$

Note that provided a moderate number of boxes $(N \gg 10)$ are set up in Hale space the $(\exp -\frac{1}{3}N^{-3})$ term in Eq. (8.42) can always be taken as effectively unity. Figure 8.13 shows the current transient generated in the way outlined above[10] with

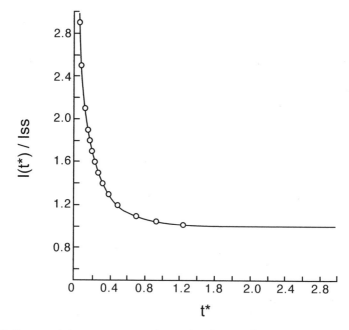

Fig. 8.13 Computed chronoamperometric transient for a single-potential-step at a rotating disc electrode when electrode reaction involves a simple electron transfer free from any kinetic complications. The current has been normalised with respect to the final steady-state value, I_{ss}. Reproduced from Ref. [10] with permission of The Royal Society of Chemistry.

M = 1000 and N = 20. Note that current has been normalised to its steady-state value.

8.8 The Rotating Disc and Coupled Homogeneous Kinetics

At the rotating disc electrode the rate of mass transport is varied by altering the disc rotation speed (ω/Hz). Let us consider how the use of controlled mass transport as a variable can provide a guide to the reaction mechanism of an electrode process and also give quantitative details. As an illustration we consider the behaviour of CE and EC processes (see Chapter 7). We assume that the process is a reduction and that the electron transfer steps (E) are electrochemically reversible (fast electrode kinetics). Figure 8.14 shows that the measured hydrodynamic voltammogram is defined by the half-wave potential, $E_{1/2}$, and the transport limited current, I_{lim}. It is the dependence of these two quantities on the disc rotation speed which permits

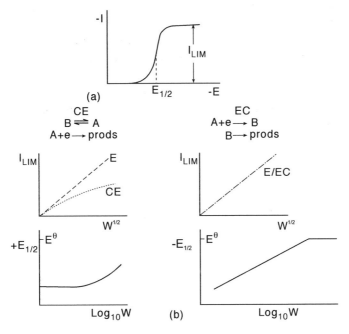

Fig. 8.14 CE and EC processes at a rotating disc electrode. (a) Current–voltage curves resulting from reversible electron transfer characterised by the half-wave potential $E_{1/2}$ and the mass transport limited current, I_{lim}. (b) the effect of coupled homogeneous kinetics (under the reaction layer approximation) on the parameters $E_{1/2}$ and I_{lim}. Note the sign change in the y-axis between the two $E_{1/2}$ versus $\text{Log}_{10} W$ plots.

the inference of mechanism, as shown in Fig. 8.14. For an electrode reaction free of any complications due to coupled homogeneous chemistry, I_{lim} varies as $\omega^{1/2}$ (see previous section) and $E_{1/2}$ is independent of ω.

For a CE process, at fast rotation speeds the limiting current is below that expected for a simple E reaction, since the precursor (B) of the electroactive material (A) spends insufficient time in the vicinity of the electrode surface for complete conversion to the electroactive form (A). At lower rotation speeds, the 'transit time' across the disc surface is increased so as to allow full reduction of the electroactive material and effectively 'one electron' behaviour is observed, as illustrated. Quantitative analysis of the complete $I_{lim}/\omega^{1/2}$ behaviour may provide confirmation of a CE mechanism.[11–13]

Considering next the EC process, it is evident from Fig. 8.14 that the chemical step has no influence of I_{lim}, since it occurs after the electron transfer. However, the following chemical step manifests itself in the $E_{1/2} - \omega$ behaviour; specifically for fast kinetics the reduction potential is shifted towards positive potentials by ca 30 mV/$\log_{10}\omega$ at 298 K when C is first order and by almost 20 mV/$\log_{10}\omega$ when C is a second-order irreversible process.[7] In contrast, it can be seen in Fig. 8.14 that in the CE case, where the chemical step precedes the electron-transfer, an additional Gibbs energy barrier is provided to the reduction process and $E_{1/2}$ is shifted to more negative potentials.

Last, we examine the ECE and DISP processes occurring at a rotating disc electrode. This can be illustrated with reference to the reduction of fluorescein at pH 9.5 to pH 9.7:

Electrode $\qquad F \underset{-e^-}{\overset{+e^-}{\rightleftarrows}} S^\bullet$

Solution $\qquad S^\bullet + H^+ \rightarrow SH^{\bullet+}$

Solution $\qquad S^\bullet + SH^{\bullet+} \rightleftarrows F + L$

where

$F =$

$S^\bullet =$

$R =$

$L =$

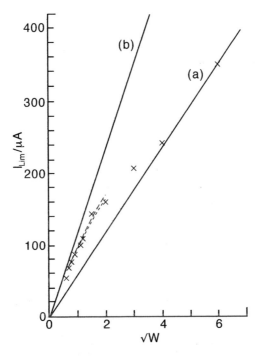

where the process has been written as a DISP1 reaction. Figure 8.15 shows experimental data obtained at a rotating disc electrode for this reaction.[14]

Also shown is the pure one-electron and two-electron behaviour calculated using the Levich equation and the diffusion coefficient of F measured at high pH (0.1 M NaOH) where the reaction suffers no kinetic complications, and a simple one-electron reduction of F to $S^\bullet$ is observed. Figure 8.14 shows a smooth transition from two- to one-electron behaviour as the rotation speed is increased. This transition is typical of DISP (and ECE) processes. At very fast rotation speeds the kinetics of the C step are outrun and the electrogenerated $S^\bullet$ is swept away from the electrode before there is any chance of chemical reaction to form $SH^{\bullet+}$.

Fig. 8.15 The experimentally determined limiting current versus rotation speed data (x), compared with the theoretically predicted behaviour for ECE $(\cdots)$ and DISP1$(—)$. The straight line shows the predicted behaviour for a single E process with (a) $n = 1$ or (b) $n = 2$. Reproduced from Ref. [14] with permission of The Royal Society of Chemistry.

Consequently, a one-electron process is seen. Conversely, at low rotation speeds the electrogenerated $S^\bullet$ spends much more time in the vicinity of the electron and there is sufficient time for reaction to take place, followed by disproportionation and further reduction of the F formed via this latter process. Accordingly, at low rotation speeds a two-electron process is observed. Detailed analysis permits the influence of the rate of the chemical step. Note that, as implicit in Fig. 8.15, the rotating disc technique is unable to distinguish ECE and DISP1 mechanisms even if an independent measurement of the homogeneous rate constant for the chemical step is available, for example from spectroscopic methods (see also discussion in Chapter 7).

8.9 The Channel Electrode: An Introduction

In the channel, electrode, solution flows through a cell such as that shown in Fig. 8.16 containing a rectangular electrode of length, x_e, and width w flush with one of the walls.

The flow cell is usually designed so that the flow pattern over the electrode is laminar and thus well-defined over the range of flow rates employed. The laminar or turbulent nature of the flow is characterised by the Reynolds number defined as

$$R_e = \frac{2hV_0}{\upsilon},\tag{8.43}$$

where V_0 (cm s^{-1}) is the solution velocity in the centre of the channel, h is the half-height of the cell and υ (cm^2 s^{-1}) is the kinematic viscosity of the solution. If the Reynolds number exceeds $\sim$2000 then the flow will become turbulent, and this is difficult to model. When $R_e < 2000$, local turbulence might still exist around imperfections in the joins of the cell, but this can be avoided if the cell is well constructed.

For suitably small R_e to ensure laminar flow, friction between the walls of the cell and the solution causes the velocity profile to progress from plug flow upon

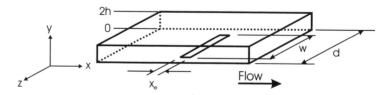

Fig. 8.16 A schematic diagram of a channel electrode with dimensions and coordinate system.

entering the cell to a parabolic shape, fully establishing the latter after an entry length of

$$l_e = 0.1hR_e, \qquad (8.44)$$

as illustrated in Fig. 8.17.

Under laminar flow steady-state the velocity components are given by

$$V_x = V_0 \left(\frac{h^2 - (y - h)^2}{h^2} \right); V_y = V_z = 0, \qquad (8.45)$$

where the coordinates x, y, z are defined in Fig. 8.16. Note that V_0, the velocity of the solution at the centre of the channel is related to the volume flow rate, V_f (cm^3 s^{-1}) by the expression

$$V_f = V_0 \int_0^d \int_0^{2h} \left(\frac{h^2 - (y - h)^2}{h^2} \right) dydz = \frac{4}{3} V_0 hd, \qquad (8.46)$$

where d is the channel width (Figure 8.16)

Knowledge of the cell hydrodynamics as above enables us to write the convective diffusion equation for the transport of B:

$$\frac{\partial[B]}{\partial t} = D_B \nabla^2[B] - \left(V_x \frac{\partial[B]}{\partial x} + V_y \frac{\partial[B]}{\partial y} + V_z \frac{\partial[B]}{\partial z} \right), \qquad (8.47)$$

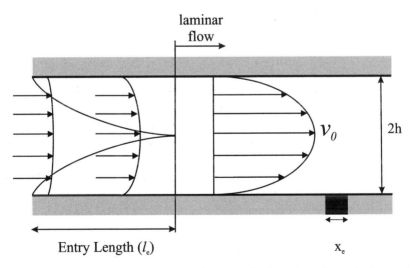

Fig. 8.17 Flow characteristics in the channel showing the formation of a parabolic velocity profile.

where ∇^2 is the Laplacian operator. Equation (8.47) can be simplified by noting the following.

(1) The cell geometry dictates that $V_y = V_z = 0$ as given in Eq. (8.45).
(2) If measurements are made at steady-state, then the derivative $\frac{\partial [B]}{\partial t} = 0$.
(3) For macroelectrodes (but not microelectrodes) and properly designed flow cells, axial and transverse diffusion can be neglected at typical flow rates,

$$V_x \frac{\partial [B]}{\partial x} \gg D_B \left(\frac{\partial^2 [B]}{\partial x^2} + \frac{\partial^2 [B]}{\partial z^2} \right).$$ (8.48)

Thus, Eq. (8.47) becomes

$$0 = D_B \frac{\partial^2 [B]}{\partial y^2} - V_x \frac{\partial [B]}{\partial x},$$ (8.49)

so that for a macroelectrode the only two forms of mass transport are convection axial and diffusion normal to the electrode surface.

Equation (8.49) can be simplified yet further using the Lévêque approximation[15] originally introduced in respect of the corresponding problem in heat transfer:

$$V_x = V_0 \left(\frac{h^2 - (y - h)^2}{h^2} \right) = V_0 \left(1 - \frac{(h - y)}{h} \right) \left(1 + \frac{(h - y)}{h} \right)$$ (8.50)

$$V_x \approx \frac{2 V_0 y}{h} \text{ for } y = 0.$$

Physically, the approximation is that the parabolic velocity profile is reduced to that of a linear profile close to the electrode ($y \sim 0$). Clearly, the approximation is better at fast flow rates as, under these conditions, the concentration perturbation of the substrate, B, due to electrolysis is small enough to be confined to the cell wall. The Lévêque approximation allows an analytical expression to be obtained for a simple electrode reaction (involving the transfer of n-electrons without homogeneous kinetic complications) for the transport limited current in terms of known experimental and cell parameters. The expression is the Levich equation:

$$I_{lim} = 0.925 n F [B]_{bulk} w (x_e D_B)^{2/3} \left(\frac{V_f}{hd} \right)^{1/3}.$$ (8.51)

The deviation of the equation is presented in the next section.

8.10 The Channel Electrode: The Levich Equation Derived

We start by combining Eqs. (8.49) and (8.50):

$$D_B \frac{\partial^2 [B]}{\partial y^2} = \frac{2V_0 y}{h} \frac{\partial [B]}{\partial x}.$$ (8.52)

We next make the substitution,

$$\eta = \left(\frac{V_0}{xhD} \right)^{1/3} y.$$ (8.53)

The solution to the problem of mass transport limited current at the channel electrode requires Eq. (8.53) to be solved subject to the boundary conditions

$$y \to \infty, \quad \eta \to \infty, \quad [B] \to [B]_{bulk}$$ (8.54)

$$y = 0, \quad \eta = 0, \quad [B] = 0.$$ (8.55)

The substitution,

$$p = \frac{d[B]}{d\eta},$$ (8.56)

in Eq. (8.53) leads to

$$\frac{1}{p} \frac{dp}{d\eta} + \frac{2}{3} \eta^2 = 0,$$ (8.57)

and so

$$p = p_{\eta=0} \exp \left(-\frac{2}{9} \eta^3 \right).$$ (8.58)

Thus,

$$\frac{[B]}{[B]_{bulk}} = \frac{\int_0^\eta \exp \left(-\frac{2}{9} \eta^3 \right) d\eta}{\int_0^\infty \exp \left(-\frac{2}{9} \eta^3 \right) d\eta}.$$ (8.59)

The integral in the denominator of Eq. (8.59) can be deduced, via the substitution

$$s = \eta^3,$$ (8.60)

to be

$$\int_0^\infty \exp \left(-\frac{2}{9} \eta^3 \right) d\eta = \frac{1}{3} \int_0^\infty s^{-2/3} \exp \left(-\frac{2}{9} s \right) ds$$

$$= \frac{\Gamma(1/3)}{3(2/9)^{1/3}}.$$ (8.61)

The average diffusional flux to the channel electrode is given by

$$J_{av} = \frac{1}{x_e} \int_0^{x_e} D_B \left. \frac{\partial [B]}{\partial y} \right|_{y=0} dx. \tag{8.62}$$

This may be evaluated by expanding the indefinite integral in Eq. (8.59) as a power series in η and integrating term by term, resulting in

$$J_{av} = \frac{1}{x_e} \frac{3 \left(\frac{2}{9}\right)^{1/3}}{\Gamma\left(\frac{1}{3}\right)} \int_0^{x_e} D_B [B]_{bulk} \left(\frac{V_0}{xhD}\right)^{1/3} dx \tag{8.63}$$

$$J_{av} = \frac{\left(\frac{9}{2}\right)^{2/3}}{\Gamma\left(\frac{1}{3}\right)} D_B^{2/3} x_e^{-1/3} \left(\frac{V_0}{h}\right)^{1/3} [B]_{bulk}. \tag{8.64}$$

It follows that

$$I_{lim} = nFx_e wJ_{av}$$

$$I_{lim} = \left(\frac{9}{2}\right)^{2/3} \frac{[B]_{bulk}}{\Gamma\left(\frac{1}{3}\right)} \left(\frac{V_0 D^2 x_e^2}{h}\right)^{1/3} nFw, \tag{8.65}$$

where $\Gamma(1/3) = 2.6789$. Substituting Eq. (8.46) gives the Levich equation (8.51).

8.11 Channel Flow Cells and Coupled Homogeneous Kinetics

Channel electrodes have proved attractive devices in the pursuit of electrochemical reaction mechanisms. Part of the reason for this is that the basic design is relatively readily adaptable to allow simultaneous spectroscopic measurements to permit the identity of reaction intermediates to be inferred. In other words the channel design is inherently more compatible with incorporation with various types of spectrometer than, say, a rotating disc electrode. Thus, channel cell designs have been reported which are compatible with *in-situ* UV/Visible, infra-red, florescence and ESR spectroscopies (for a review, see Ref. [16]). Figure 8.18(a) shows a typical practical channel cell for purely electrochemical measurement whereas Fig. 8.18(b) shows a silica channel flow cell design that has been widely and extremely successfully used for spectroelectrochemical studies.

Figure 8.19 shows a modified channel flow cell which is suitable for UV/Visible spectroscopy of electrogenerated species.

Given the power of the channel electrode approach in mechanistic studies we next illustrate the development of the voltammetric theory for channel electrode

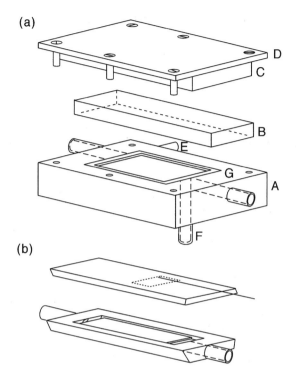

Fig. 8.18 (a) Perspex channel electrode cell. A: channel unit, B: cover plate, C: rubber block, D: metal plate, E: working electrode, F: possible reference electrode site, G: silicone rubber gasket. (b) A silica channel flow cell (unassembled) for spectroelectrochemical studies, showing the cover plate with electrode and lead-out wire and the channel unit. Reprinted from Ref. [16] with permission from Wiley.

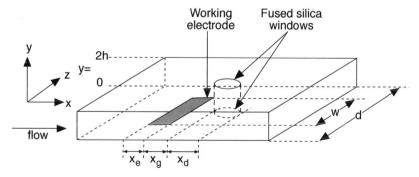

Fig. 8.19 A modified channel flow cell for *in situ* UV/Visible spectroscopy. The marked lengths represent the electrode length, the distance between the electrode and the window and the diameter of the window itself. Reprinted from Ref. [16] with permission from Wiley.

problems by addressing the case of an ECE reaction which can be written as a reduction:

$$A + e \rightleftarrows B$$

$$B \xrightarrow{k} C$$

$$C + e \rightleftarrows D.$$

We assume that C is more easily reduced than A so that depending on the value of the rate constant, k, the flow rate, V_f, and the dimensions of the cell and the electrode, between one and two times the limiting current for a simple one electrode process ($I_{1-electron}$) can flow (cf. Section 8.6). At faster flow rates, other factors remaining the same, there is less time for B to react to form C before the solution has flowed past the electrode and there is less possibility of C being reduced to D, hence relatively lowering the current, I. We can define an effective number of electrons transferred, N_{eff}, as

$$N_{eff} = \frac{I}{I_{1-electron}}; \quad 1 \le N_{eff} \le 2. \tag{8.66}$$

The steady-state transport equations for an ECE process, neglecting axial diffusion (see Section 8.7) and making the Lévêque approximation, are

$$D\frac{\partial^2 [A]}{\partial y^2} - \frac{2V_0 y}{h}\frac{\partial [A]}{\partial x} = 0 \tag{8.67}$$

$$D\frac{\partial^2 [B]}{\partial y^2} - \frac{2V_0 y}{h}\frac{\partial [B]}{\partial x} - k[B] = 0 \tag{8.68}$$

$$D\frac{\partial^2 [C]}{\partial y^2} - \frac{2V_0 y}{h}\frac{\partial [C]}{\partial x} + k[B] = 0, \tag{8.69}$$

where $V_0 = 3V_f/4hd$ and we have assumed $D_A = D_B = D_C = D$. If we introduce the following dimensionless variables

$$\chi = \frac{x}{x_e}; \quad \xi = \left(\frac{2V_0}{hDx_e}\right)^{1/3} y; \quad K_{norm} = k\left(\frac{h^2 x_e^2}{4V_0^2 D}\right)^{1/3}, \tag{8.70}$$

then the above equations simplify to

$$\frac{\partial^2 [A]}{\partial \xi^2} - \xi\frac{\partial [A]}{\partial \chi} = 0 \tag{8.71}$$

$$\frac{\partial^2 [B]}{\partial \xi^2} - \xi\frac{\partial [B]}{\partial \chi} - K_{norm}[B] = 0 \tag{8.72}$$

$$\frac{\partial^2 [C]}{\partial \xi^2} - \xi\frac{\partial [C]}{\partial \chi} + K_{norm}[B] = 0. \tag{8.73}$$

Hence, we can conclude that $[B]$ and $[C]$ depend *only* on ξ, χ, K_{norm}. Since the channel electrode current

$$I \propto \int_0^{\chi=1} \left(D \frac{\partial [A]}{\partial \xi}\bigg|_{\xi=0} + D \frac{\partial [C]}{\partial \xi}\bigg|_{\xi=0} \right) d\chi, \tag{8.74}$$

it follows that N_{eff}, being a ratio of currents, depends only on K_{norm}. It is therefore possible to summarise the response of a channel macroelectrode of arbitrary geometry at any flow rate by a 'working curve' which plots theoretical values of N_{eff} as a function of the dimensionless parameter K_{norm}. Figure 8.20 shows the working curve for an ECE process.

Similar working curves can be constructed for different electrode reaction mechanism. In each case they show N_{eff} as a function of a dimensionless rate constant, the exact form of which can vary from mechanism to mechanism. For example, in a DISP 2 process, the relevant dimensionless rate constant is

$$k_{DISP2} = k_2 K_{BC} [A]_{bulk} \left(\frac{4 x_e^2 d^2 h^4}{9 D V_f^2} \right)^{1/3}, \tag{8.75}$$

where k_2 and K_{BC} are defined in the following scheme:

$$A + e \rightleftarrows B$$

$$B \overset{K_{BC}}{\rightleftarrows} C \quad K_{BC} = \frac{[C]}{[B]}$$

$$B + C \overset{k_2}{\longrightarrow} D.$$

For DISP 1,

$$A + e \rightleftarrows B$$

$$B \overset{k_1}{\longrightarrow} C$$

$$C + B \overset{fast}{\longrightarrow} D.$$

The dimensionless rate constant is

$$k_{DISP1} = k_1 \left(\frac{4 x_e^2 d^2 h^4}{9 D V_f^2} \right)^{1/3}, \tag{8.76}$$

which is the same as for ECE except that the definition has been written in term of V_f rather than V_0. The working curve for the DISP1 process is also shown in Fig. 8.20.

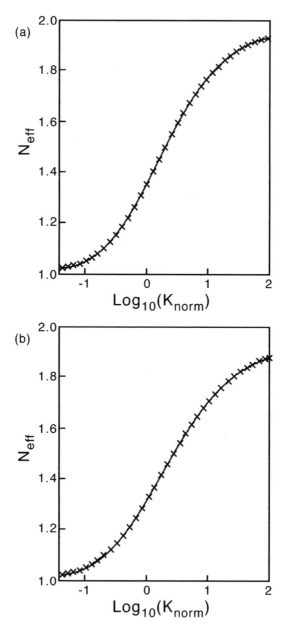

Fig. 8.20 Working curves of N_{eff} versus K_{norm} for (a) ECE and (b) DISP1 processes. Reprinted from Ref. [16] with permission from Wiley.

The working curves relating N_{eff} to the relevant K_{norm} permit the analysis of channel macroelectrode experiments. The value of K_{norm} at each flow rate is obtained from experimental current measurements in the working curves in the following manner. First, the limiting current is measured experimentally, and from this N_{eff} is determined by comparison with the current expected for a simple one-electron process. The working curves then give the corresponding values of K_{norm} for candidate mechanism, which are assessed by plotting K_{norm} against the flow rate raised to the appropriate power suggested by the definition of K_{norm}. For example, for an ECE process, K_{norm} is plotted against $V_f^{-2/3}$ (see Eq. (8.70)). A direct linear dependence consistent with Eq. (8.70) is supportive of the proposed mechanism. If the plot is not a straight line then alternative mechanisms need to be explored. This procedure is repeated until a 'fit' of experimental and simulated data is obtained and the kinetics and mechanisms are ascertained.

Next, we consider an EC mechanism

$$A + e \rightleftarrows B$$

$$B \longrightarrow products.$$

In this case the limiting current is independent of the coupled kinetics. Nevertheless, kinetic data and mechanistic insights are possible by considering the mass transport dependence of the half-wave potential, $E_{1/2}$. The pertinent convective–diffusion equations are

$$\frac{\partial^2 [A]}{\partial \xi^2} - \xi \frac{\partial [A]}{\partial \chi} = 0 \tag{8.77}$$

$$\frac{\partial^2 [B]}{\partial \xi^2} - \xi \frac{\partial [B]}{\partial \chi} - K_{norm}[B] = 0, \tag{8.78}$$

where ξ, χ, K_{norm} are all again given by Eq. (8.70). We next assume that the A/B redox couple has sufficiently fast electrode kinetics to be electrochemically reversible. Under these conditions the presence of the homogeneous kinetics is revealed by an anodic shift of $E_{1/2}$; the reduction of A to B occurs at less negative potentials than would be observed in the absence of the kinetic step. The shift is defined by

$$\Delta E_{1/2} = \left| E_{1/2} - E_f^0(A/B) \right|, \tag{8.79}$$

where $E_f^0(A/B)$ is the formal potential of the A/B couple. Assuming the Lévêque approximation and equality of diffusion coefficients, $\Delta E_{1/2}$ can be shown to be solely a function of K_{norm}. The relationship is represented by the working curve obtained by numerical simulation, shown in Fig. 8.21.

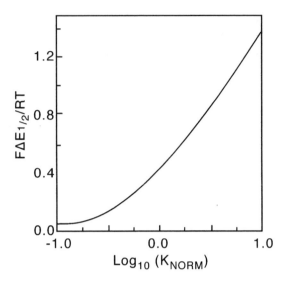

Fig. 8.21 Working curves for an EC reaction shown in the dimensionless potential shift as a function of K_{norm} under Lévêque conditions. Reprinted from Ref. [16] with permission from Wiley.

Working curves such as Fig. 8.21 permit the analysis of experimental data measured at channel macroelectrodes under Lévêque conditions, in terms of an EC mechanism. Typically, measured $\Delta E_{1/2}$ values would be used in conjunction with the working curve to generate corresponding K_{norm} values as a function of the flow rate, V_0 (or V_f). A direct dependence of the inferred values of K_{norm} on $V_0^{-2/3}$ suggests consistency with the proposed mechanism. Note that Fig. 8.21 implies that at very fast flow rates $\Delta E_{1/2}$ tends to zero consistent with the couple homogeneous kinetics being 'outrun' and, experimentally, allowing the determination of $E_f^0(A/B)$ which is required for the evaluation of $\Delta E_{1/2}$ at slower flow rates.

8.12 Chronoamperometry at the Channel Electrode

We briefly consider the basis for the description of potential step transients at the channel electrode and examine specifically the case where the electrode potential is stepped from one where no current flows to potentials corresponding to transport limited electrolysis.

The basis for the theoretical description is as follows. In a manner analogous to the normalised rate constant, K_{norm}, introduced in the previous section, a dimensionless time variable, τ, is defined. Thus, for the mass transport equation

$$\frac{\partial[A]}{\partial t} = D_A \frac{\partial^2[A]}{\partial y^2} - \frac{2V_0 y}{h} \frac{\partial[A]}{\partial x}, \tag{8.80}$$

the following variables

$$\chi = \frac{x}{x_e}; \quad \xi = \left(\frac{2V_0}{hD_A x_e}\right)^{1/3} y; \quad \tau = \left(\frac{4DV_0^2}{h^2 x_e^2}\right)^{1/3} t \tag{8.81}$$

reduce Eq. (8.80) to

$$\frac{\partial[A]}{\partial \tau} = \frac{\partial^2[A]}{\partial \xi^2} - \xi \frac{\partial[A]}{\partial \chi}. \tag{8.82}$$

This equation can be solved for a cell of any dimensions at any flow rate. The current, normalised to the steady-state value,

$$\frac{I(\tau)}{I(\tau \to \infty)}, \tag{8.83}$$

is solely a function of the parameter τ, when the potential step is as described above. Figure 8.22 shows the simulated behaviour.[17]

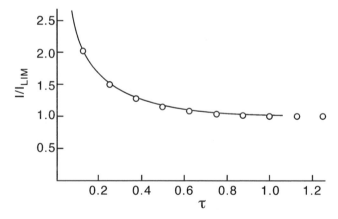

Fig. 8.22 A dimensionless current–time transient for a simple electron transfer process at a channel electrode, under Lévêque conditions, calculated using backwards implicit finite difference methodology. Reprinted with permission from Ref. [17]. Copyright (1991) American Chemical Society.

8.13 The Channel Electrode is not 'Uniformly Accessible'

In Section 8.4 we showed that the diffusion layer thickness of a rotating disc electrode is given by Eq. (8.23) and is a constant depending only on the diffusion coefficient, D, the kinematic viscosity, υ, and the disc rotation speed, ω. Since this is constant it follows that the current density (A m^{-2}) at all parts on the disc surface is the same. The electrode is said to be 'uniformly accessible'.

If we consider Eq. (8.58) with the electroactive species labelled A,

$$\frac{\partial[A]}{\partial \eta} = \left(\frac{\partial[A]}{\partial \eta}\right)_{\eta=0} \exp\left(-\frac{2}{3}\eta^3\right). \tag{8.84}$$

Integrating

$$[A]_{\eta=\infty} - [A]_0 = \left(\frac{\partial[A]}{\partial \eta}\right)_{\eta=0} \int_0^\infty \exp\left(-\frac{2}{3}\eta^3\right) d\eta \tag{8.85}$$

$$[A]_{\eta=\infty} - [A]_0 = \frac{\Gamma\left(\frac{1}{3}\right)}{3\left(\frac{2}{9}\right)^{1/3}} \left(\frac{\partial[A]}{\partial \eta}\right)_{\eta=0}, \tag{8.86}$$

it follows that

$$\frac{\partial[A]}{\partial y}\bigg|_{y=0} = \frac{3\left(\frac{2}{9}\right)^{1/3}([A]_{bulk} - [A]_{y=0})}{\Gamma\left(\frac{1}{3}\right) \cdot \left(\frac{xhD}{V_0}\right)^{1/3}} \tag{8.87}$$

and that the diffusion layer thickness,

$$\delta_d = \frac{\Gamma\left(\frac{1}{3}\right)}{\sqrt[3]{6}} \left(\frac{xhD}{\upsilon_0}\right)^{1/3}. \tag{8.88}$$

This is an interesting result: The diffusion layer thickness depends on $\sqrt[3]{x}$. At the upstream edge of the electrode the thickness is vanishingly small but the layer thickens as one moves downstream. It follows that the current density over the electrode is not uniform but is greatest at the upstream edge and least at the downstream edge. Under transport limited conditions the flux scales with $x^{-1/3}$, although in practice finite electrode kinetics will prevent the current density from becoming infinite at the upstream edge. Accordingly, the channel electrode is a 'non-uniformly accessible' electrode.

Further examination of Eq. (8.88) shows that as the flow rate increases the diffusion layer thickness decreases with $V_0^{-1/3}$. It follows that the transport limited current at a channel electrode,

$$I_{lim} \propto (flow\ rate)^{1/3}. \tag{8.89}$$

Also, Eq. (8.88) is interesting in that, unlike the corresponding expression for the rotating disc electrode, the diffusion layer thickness of the channel electrode does not depend on the kinematic viscosity. This is because the flow through the channel electrode is parabolic with each layer of solution moving at a steady velocity. In contrast, as a solution approaches a rotating disc it changes direction and speed, being spun round and flung out radially. Accordingly, the current response of a rotating disc electrode, but not a channel electrode, depends on the kinematic viscosity of the solution.

8.14 Channel Microelectrodes

When channel electrodes are made sufficiently small that axial diffusion, that is diffusion in the direction of the solution flow, becomes significant, they are referred to as microelectrodes. This both simplifies and complicates! The simplification is that the Lévêque approximation is very likely to hold since the concentration profiles extend a much smaller distance into the solution. On the other hand the term involving $\partial^2 [B]/\partial x^2$ in Eq. (8.47) can no longer be neglected.

To account for axial diffusion we next re-write Eq. (8.47) using the variables

$$\chi = \frac{x}{x_e}; \quad \Psi = \frac{y}{x_e} \text{ and } P_s = \frac{3x_e^2 V_f}{2h^2 dD}, \tag{8.90}$$

where P_s is the shear Peclet number which describes a balance between convection and diffusion with the Lévêque approximation made, the steady-state transport equation

$$\frac{\partial [A]}{\partial t} = D_A \frac{\partial^2 [A]}{\partial y^2} + D_A \frac{\partial^2 [A]}{\partial x^2} - \frac{2V_0 y}{h} \frac{\partial [A]}{\partial x} = 0 \tag{8.91}$$

becomes

$$\frac{\partial^2 [A]}{\partial \chi^2} + \frac{\partial^2 [A]}{\partial \Psi^2} - 3P_s \Psi \frac{\partial [A]}{\partial \chi} = 0. \tag{8.92}$$

If the electrode potential is sufficient that the concentration of A at the electrode surface is zero and the current is given by

$$I = nFwD_A [A]_{bulk} \frac{1}{x_e} \int_0^1 \frac{\partial \left(\frac{[A]}{[A]_{bulk}} \right)}{\partial \Psi} \Bigg|_{\Psi=0} d\chi, \tag{8.93}$$

then since Ψ appears in Eq. (8.93) as a constant and χ disappears on integration,

$$I = nFwD[A]_{bulk}f(P_s). \tag{8.94}$$

Therefore, the transport limit current is solely a function of P_s and hence, can be represented by a working curve of I_{lim} versus P_s.

In the limit that axial diffusion can be neglected (for macroelectrodes) the Levich equation (8.51) be re-written as

$$I_{Levich} = 0.925nFwD[A]_{bulk} \left(\frac{2}{3} \right)^{1/3} P_s^{1/3}. \tag{8.95}$$

In contrast, an analytical expression for the transport limited current in the limit of dominant axial diffusion has been derived by Akerberg et al.[18] This expression is valid at low shear Peclet numbers, $P_s \ll 1$:

$$I_{Ackerberg} = nFwD[A]_{bulk}\pi g(P_s)[1 - 0.04632P_s g(P_s),], \tag{8.96}$$

where

$$g(P_s) = \left\{ \ln\left[4(P_s)^{-1/2}\right] + 1.0559 \right\}^{-1}. \tag{8.97}$$

For other values of P_s numerical simulation is essential.[19] In practice there has been extensive simulation of the channel electrode both at steady-state and under transient conditions with voltammetric waveshapes characterised for a plethora of problems (see Cooper and Compton[16] for a review). Arguably, the channel electrode and its close relation, the tubular electrode, which is shown in Fig. 8.23, are today the best characterised hydrodynamic electrodes partly reflecting their importance in the increasingly important area of microfluidics with applications not least in analysis. The following two sections illustrate the power of the channel electrode approach.

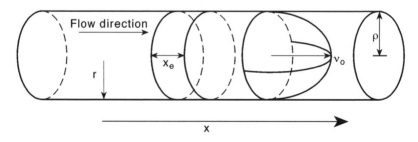

Fig. 8.23 The tubular electrode showing the laminar flow pattern.

8.15 Channel Microband Electrode Arrays for Mechanistic Electrochemistry

We have seen throughout this chapter that the mechanisms (ECE, DISP, EC′, etc.) of electrolyte reactions can be interrogated using variable mass transport. In the earlier part of this chapter, we saw that one way in which this has been achieved is through variable convection for example by changing the solution flow rate of a channel electrode or the rotation speed of a rotating disc electrode. An alternative approach is through the use of microelectrodes (Chapter 5) under diffusion-only conditions where altering the dimensions of the electrodes changes the dynamic range of the voltammetry. By using a channel electrode array with electrodes varying in size from the millimetre to the sub-micron scale, it is possible to use both variable convection and electrode size in the same experiment, so arguably conferring enhanced mechanistic discrimination[20] through a 'two-dimensional voltammetry' experiment. This has been accomplished using the channel microband array shown in Figs. 8.24 and 8.25. The array consists of 13 different-sized electrode ranging from 2 mm to less than 0.5 μm in the length (x_e).

The two-dimensional voltammetry experiment proceeds as follows. First, measurements of steady-state hydrodynamic voltammograms are made at each electrode in turn as a function of the solution flow rate giving rise to a three-dimensional plot involving limiting current (and hence N_{eff}) as a function of flow rate (V_f) and electrode length (x_e). Mass transport is greatest at high flow rates and small electrode lengths and so any mass transport sensitive data should display a systematic trend along the diagonal shown in Fig. 8.26.

Next, simulation (see previous section) can give the theoretical prediction of the expected current as a function of V_f and x_e for any chosen mechanism. Third, to test whether experimental and theory match, the ratio of experimental results and

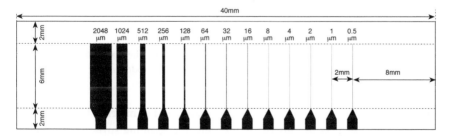

Fig. 8.24 The 13-electron array. The dotted lines delineate the position of the cell. The electrode lengths (x_e values) double from 0.5 μm to 2 mm. Reprinted with permission from Ref [20]. Copyright (1998) American Chemical Society.

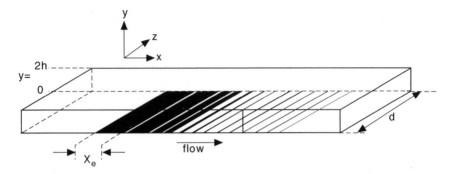

Fig. 8.25 Schematic diagram of the assembled channel electrode array, showing the coordinate system used to characterise the flow cell. Note that some authors, particularly in diffusion-only work, use 'width' to refer to the band's shortest dimension. Here we use the hydrodynamic convection: 'length' is the distance, x_e, in the direction of flow, and 'width' is in the z-coordinate. Reprinted with permission from Ref. [20]. Copyright (1998) American Chemical Society.

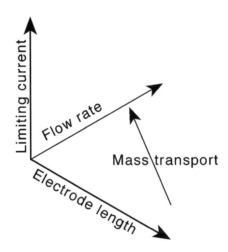

Fig. 8.26 The surface used to display 2-D voltammetry results. The diagonal line shows the increase in mass-transport at faster flow rates and smaller electrode size.

predictions is plotted as a surface against V_f and x_e: if they match, a flat featureless uniform plane will result; otherwise, typically, the resulting surface will show a systematic trend along the diagonal of the plot (Fig. 8.26). A flat plane is strong evidence in favour of a particular mechanism, and one which is markedly and systematically sloping is even stronger evidence against a particular mechanism.

Figure 8.27 shows the results of two-dimensional voltammetry applied to the reduction of 4-chlorobenzophenone in acetonitrile with tetra-n-butylammonium perchlorate as supporting electrolyte.

Table 8.1 shows the results of fitting the experiment transport-limited current data to a range of working surfaces. A measure of the fit is obtained through

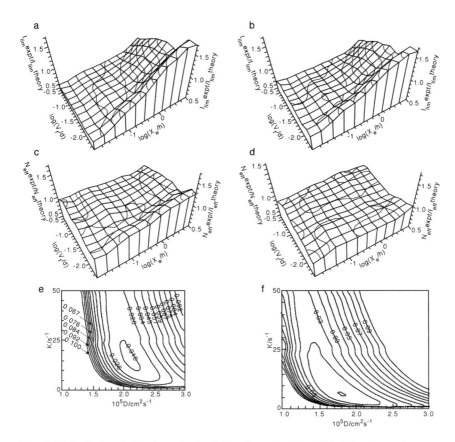

Fig. 8.27 Analysis of the data obtained for the reduction of 4-chlorobenzophenone. Surfaces showing (a) $I_{lim(expt)}/I_{lim(theory)}$ for an assumed E mechanism, contour interval = 0.04; (b) $I_{lim(expt)}/I_{lim(theory)}$ for an assumed EE mechanism, contour interval = 0.02; (c) $N_{eff(expt)}/N_{eff(theory)}$ for an assumed ECE mechanism, contour interval = 0.03; (d) $N_{eff(expt)}/N_{eff(theory)}$ for an assumed ECEE mechanism, contour interval = 0.025; (e) minimum MSAD corresponding to optimised diffusion coefficient and rates constant values for an assumed ECE mechanism; and (f) minimum MSAD for optimised diffusion coefficient and rate constant values for an assumed ECEE mechanism, contour interval = 0.01. Reprinted with permission from Ref. [20]. Copyright (1998) American Chemical Society.

Table 8.1 Analysis of data recorded for the reduction of 4-chlorobenzophenone in terms of candidate reaction mechanisms.

Mechanism	$D/cm^2\,s^{-1}$	k/s^{-1}	MSAD
E	3.578×10^{-5}		1.06×10^{-1}
EE	1.308×10^{-5}		8.39×10^{-2}
ECE	2.096×10^{-5}	15.7	1.66×10^{-2}
ECEE	1.827×10^{-5}	7.45	3.95×10^{-3}

evaluation of the mean-scaled absolute derivation (MSAD):

$$MSAD = \left(\frac{1}{N}\right) \sum_N \left| \frac{I_{lim}(expt) - I_{lim}(theory)}{I_{lim}(theory)} \right|, \qquad (8.98)$$

where N is the number of experimental points. By definition, the best fit of the experimental and theoretical values corresponds to the case of minimum MSAD. Table 8.1 shows the best-fit parameters for the data attained for the reduction of 4-chlorobenzopheneone (4-Cl-Ar) for E, EC, ECE, ECEE mechanisms; the parameters in question are the diffusion coefficient, D, of 4-Cl-Ar (assumed to be equal to other mechanistically significant species) and, for the last two mechanisms, the first order rate constant of the chemical step, k. The MSAD values in Table 8.1 point to the operation of an ECEE mechanism. This is supported by the comparative surfaces shown in Fig. 8.27. The three-dimensional plots shown all have the same vertical scale, and all but that for the ECEE mechanism, which is flat and featureless, shown a systematic variation over the surface in the direction of the diagonal of increasing mass transport. A possible ECEE mechanism is as follows:

$$4 - Cl - Ar + e^- \rightleftarrows [4 - Cl - Ar]^{\bullet -}$$

$$[4 - Cl - Ar]^{\bullet -} \longrightarrow Cl^- + Ar^\bullet$$

$$Ar^\bullet + HS \longrightarrow ArH + S^\bullet$$

$$ArH + e^- \longrightarrow [Ar - H]^{\bullet -}$$

$$S^\bullet + e^- \longrightarrow S^-,$$

where Ar-H denotes benzophenone and $S^\bullet$ either the $\bullet CH_2 CN$ radical or a species derived from the supporting electrolyte.

The surfaces showing the optimum diffusion coefficient and rate constant (25°C) for the ECE and ECEE analysis are shown in Figs. 8.27(e) and (f). The values of these parameters in the ECEE mechanism are $1.8 \times 10^{-5}\,cm^2\,s^{-1}$ and $7.5\,s^{-1}$.

8.16 The High Speed Channel Electrode

Channel flow cells have been used to investigate very fast electrode reactions, with standard electrochemical rate constants of 1 cm s^{-1} or more, by using a very high flow rate, achieved by forcing the electrolyte solution over the electrode using pressures of up to 500 atmospheres. This can give rise to a flow rate (V_0) at the centre of the cell of up to 270 km per hour! The cells used are rather different in design to conventional channel flow cells, having a much smaller height above the electrode (2h). They need thicker walls and are altogether more robust, as shown in Fig. 8.28.

The electrodes are used are microbands and the cell consists of two parts fused together with the thin electrode in between.

In order to correctly model the voltammetry it is essential to maintain laminar flow characteristics over the entire flow rate range studied as discussed earlier in this chapter. The Reynolds number, R_e, given by

$$R_e = \frac{3V_f}{2hd\upsilon},\tag{8.99}$$

where υ is the kinematic viscosity, can attain values as high as 9000 at the flow rates typically employed in the high speed channel apparatus. This would lead to turbulence under steady-state conditions. However the flow cell is designed so that the R_e numbers change from laminar to turbulent very quickly. Because a finite time is required for the transition to occur, the solution passes over the electrode before the flow collapses to become turbulent. The 'lead in' length, x_e, required for turbulence to develop can be estimated by

$$x_t = \frac{4h(3.04 \times 10^5)}{R_e(1 + \eta_t)},\tag{8.100}$$

where $\eta_t = (V_0 - \overline{V})/\overline{V}$, where V_0 is the axial centre line velocity and $\overline{V}$ the mean velocity.[21,22]

At the highest flow rates hitherto reported a 'lead in' length of ca 4 mm is required for turbulence to set in. Hence, turbulence does not develop until long after the solution has passed beyond the electrode, and is travelling down the exit tubing. The flow over the electrode is thus fully laminar and the limiting current scales with $V_f^{1/3}$ in accordance with the Levich equation.

A major use of the high speed channel electrode has been in the accurate measurement of fast standard electrochemical rate constants k_0 (e.g., Ref. [22]) up to *ca.* 3 cm s^{-1}. In this context the method may have merits over fast scan cyclic voltammetry (see Chapter 5) which has been widely used in the context but

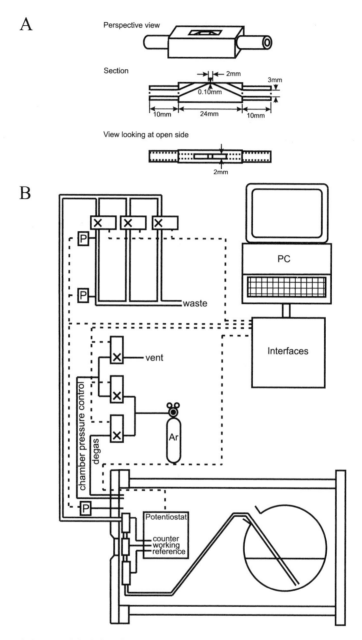

Fig. 8.28 (A) A modified fast-flow channel microelectrode for measuring very fast rate constants. (B) The equipment required to control and apply high pressure to the electrolytic solution. Reprinted with permission from Ref. [21]. Copyright (1995) American Chemical Society.

suffers from ohmic (IR) drop and especially capacitative charging. These effects can be partially countered by the use of atypically high concentrations of supporting electrolyte and potentiostat circuitry offering on-line IR drop compensation. At the scan rates required for the measurement of fast electrode kinetics, the Faradaic currents are often largely swamped by capacitative charging currents, and the IR drop compensation has to be set at exactly the right level in order to correctly elucidate the kinetic parameters. This requires a large degree of experimental skill to implement correctly and consistent repeat data is often hard to acquire. The attraction of the high speed channel electrode is that with it requiring only steady-state measurements the capacitative charging problem is entirely eliminated.

One example of the use of the high speed channel electrode is in the determination of the standard electrochemical rate constant for the oxidation of ferrocene and related derivatives[23]

$$Cp_2 Fe - e^- \longrightarrow Cp_2 Fe^+,$$

where Cp is cyclopentadienyl. Interestingly, the range of values for the standard electrochemical rate constant as measured by cyclic voltammetry spans from $0.02 \, cm \, s^{-1}$ to $12 \, cm \, s^{-1}$ reflecting the problems discussed above. Figure 8.29 shows typical data obtained at a high speed channel electrode from which a value of $k_0 = 1.0 \pm 0.2 \, cm \, s^{-1}$ was deduced.

8.17 Hydrodynamic Electrodes Based on Impinging jets

The wall-jet electrode is a disc-based arrangement. Unlike the rotating disc electrode the disc remains stationary, and a jet of solution is directed at the disc which is submerged in a stationary solution of the same composition as the jet. This results in a cylindrically symmetric 'umbrella shaped' flow profile as shown in Fig. 8.30. In terms of the rate of mass transport, the wall-jet is a considerable improvement over the rotating disc electrode and much higher rates of convection can be attained although at the price of losing the simplification of uniform accessibility. Provided the jet size is tiny compared with the electrode radius, the diffusion layer thickness scales with $r^{5/4}$, where r is the radial cylindrical coordinate from the centre of the disc.

The corresponding transport limited current is

$$I_{lim} = 1.35 n F D^{2/3} \upsilon^{-5/12} a^{-1/2} R^{3/4} V_f^{3/4} [A]_{bulk}, \tag{8.101}$$

where R is the disc radius, a is the diameter of the nozzle of the jet and V_f is the volume flow rate. Most uses of the wall-jet geometry have been confined to

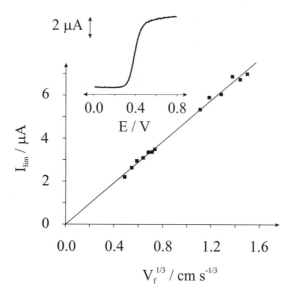

Fig. 8.29 Levich plot for the one-electron oxidation of 1.14 mM ferrocene in MeCN containing 0.1 M TBAP. Insert shows the steady-state linear sweep voltammogram at $V_f = 2.101 \, \text{cm}^3 \, \text{s}^{-1}$. I_{lim} is the steady-state limiting current and V_f is the volume flow rate. Reprinted from Ref. [23] with permission from Elsevier.

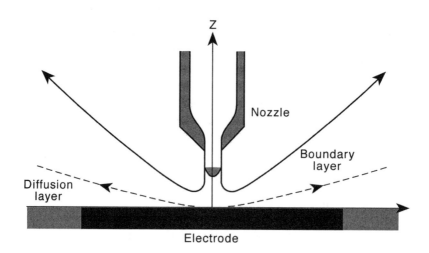

Fig. 8.30 Flow profile at a wall-jet electrode.

electroanalytical applications, although simulation methods have been developed to cover common mechanistic cases such as ECE.

Macpherson and Unwin[24] have developed a 'micro-jet electrode' in which a jet of solution is directed at a microdisc electrode. The radius of the jet is several times larger than that of the disc so that the convection is essentially one-dimensional and uniform across the electrode surface. In this way, the electrode can be made effectively uniformly accessible. This type of device offers high mass transport coefficients, much as the high speed electrode, although problems have been reported when using solvents such as acetonitrile, less viscous than water.[25]

8.18 Sonovoltammetry

The synergistic use of ultrasound with electrochemical processes and electroanalytical procedures in particular has recently been found to be significantly beneficial. The following chapter continues a section concerning the impact on electroanalysis in general; the present discussion seeks to introduce the physical effects arising from insonation and shows that a macroelectrode can be conferred with the mass transport properties of a microelectrode!

Power ultrasound is that within the frequency range 20 kHz to 100 kHz; the sound spectrum is shown in Fig. 8.31.

The latter shows the range of human hearing is from 16 Hz to *ca.* 18Hz (depending on age). Whilst high frequency ultrasound (1 MHz–10 MHz) is often utilised in medical applications it has found relatively little use in voltammetry. Instead

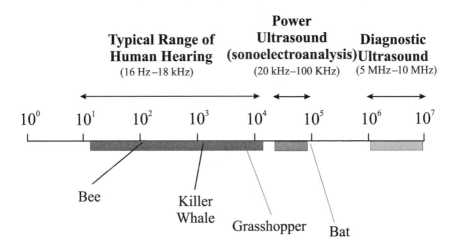

Fig. 8.31 Ultrasonic spectrum detailing the range of ultrasonic frequencies.

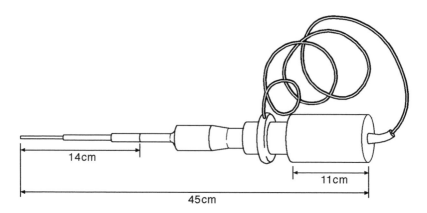

Fig. 8.32 A commercially available ultrasonic horn from either (a) Sonics and Materials, Inc. Newtown, CT 06470, USA or (b) Jencons-PLS, Bedfordshire, England.

electrochemical studies have overwhelmingly employed power ultrasound. This is introduced into a liquid by means of a sonic horn, such as shown in Fig. 8.32.

Immersion of the horn unit in a liquid leads to the efficient transmission of ultrasound. In the case of sonovoltammetry two possible electrode/horn configurations have proved effective as shown in Fig. 8.33.

The first experimental arrangement is the 'face-on' mode[26] in which the sonic horn is located opposite to a working electrode in an otherwise conventional voltammetric cell. In this case an important variable is the electrode-to-horn separation as well as the intensity of the power ultrasound. Second, as an alternative, a 'sonotrode' may be adopted.[27] This combines both the sonic transducer and the working electrode into a single unit as shown in Fig. 8.34; optimised versions are available commercially.

Figure 8.35 shows a typical sonovoltammogram alongside a conventional cyclic voltammogram run under exactly the same conditions but for the absence of insonation. The system studied in both cases is the oxidation of ferrocene at a 2 mm diameter platinum disc-electrode in acetonitrile solution containing supporting electrolyte.

The insonation data was obtained in face-on mode using the relatively low power of $50\,\text{W cm}^{-2}$ with a horn-to-electrode separation of 4mm. Comparison of the two voltammograms in Fig. 8.35 shows that the electrochemically reversible cyclic voltammogram recorded under silent conditions is transformed into a hydrodynamic voltammogram with no forward/back sweep hysteresis. Moreover, the maximum current under insonation is significantly larger than the silent signal. Note the limiting current is not perfectly flat as expected for a hydrodynamic

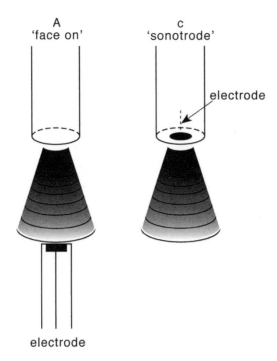

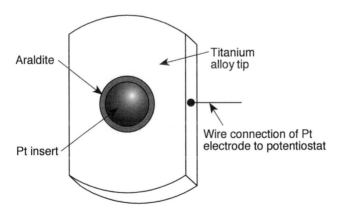

Fig. 8.33 The electrode geometries for sonoelectrochemistry. Reprinted from Ref. [26] with permission from Elsevier.

Fig. 8.34 A schematic design of a sonotrode. Reprinted from Ref. [27] with permission from Elsevier.

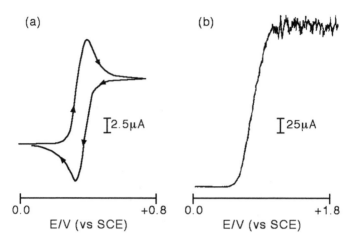

Fig. 8.35 Oxidation of 2 mM ferrocene in acetonitrile (0.1 M NBu$_4$ClO$_4$) at a 2mm diameter platinum electrode under silent conditions (a), and insonation of 50 W cm^{-2} (b), at a scan rate of 20 mVs^{-1} and a 40 mm electrode–horn separation. Reprinted from Ref. [27] with permission from Elsevier.

voltammogram measured as a channel or rotating disc electrode. Rather, there are irregular spikes present and the sizes of these are related to the power of the applied ultrasound.

The form of the sonovoltammogram can be understood by considering the modes of mass transport experienced by the electrode (Fig. 8.36). Under silent condition this is restricted to diffusion only, whereas with insonation two new modes of transport arise. The first is a strong turbulent convective flow, termed acoustic streaming, induced by the sonic horn in the liquid medium. The average speed in such a flow can typically be of the order of tens of centimetres per second. Consequently, an electrode either part of (sonotrode) or positioned below (face-on mode) the horn tip experiences a substantial convective flow leading to the hydrodynamic nature of the voltammogram seen.

Banks *et al.*[28] have shown that the limiting current arising from this flow is, in the face-on mode,

$$I_{lim} = C(h, \upsilon)D^{2/3}AC_{bulk}P_W^{1/2}, \tag{8.102}$$

where A is the electrode area, C_{bulk} is the concentration of electroactive species, D is the diffusion coefficient, P_W is the ultrasound power, and C is a function of the kinematic viscosity, υ, and of the electrode-to-horn separation, h, which falls off strongly as h increases.

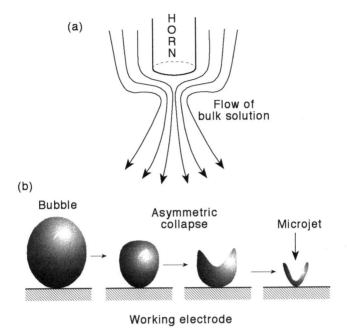

Fig. 8.36 Representation of acoustic streaming and bubble collapse at an electrode surface.

In addition to the acoustic streaming contribution, beyond a minimum ultrasound power threshold, cavitation (bubble formation) can occur[29] and this may take place preferentially on the electrode surface as opposed to bulk solution. Note that the cavitation threshold is dependent on the external applied pressure.

Microelectrodes have provided considerable insights into the nature of cavitation and of bubble dynamics.[30–34] For example, Fig. 8.37 shows the current–time response at a 30 μm platinum microdisc electrode for the ferri-/ferro-cyanide couple in aqueous solution when insonated at 20 KHz.

The electrode has been potentiostated at a value corresponding to the transport limited reduction of ferricyanide using a 'nanosecond' voltammetry potentiostat as described in Chapter 5, because of the timescales involved; note that the time axis of Fig. 8.35 shows microsecond (μs) as the units. The figure shows spikes superimposed on a steady background, and these are attributable to bubble formation at the electrode surface. The spikes can be very large, up to ca two hundred times larger than the steady-state current seen under silent conditions. Examination of Fig. 8.37 shows that different types of current signals are present. These range from 50 microseconds up to a few milli-seconds, embracing the range of transient cavitation (one acoustic cycle) through to stable cavitation (many acoustic cycles).

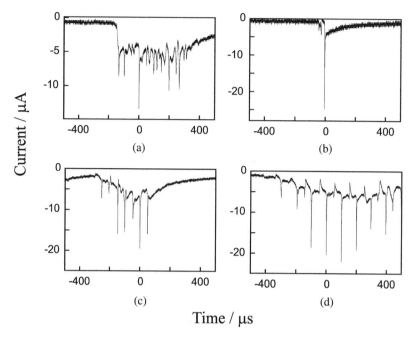

Fig. 8.37 Chronoamperometric currents observed for the reduction of 50 mM $K_3(Fe(CN)_6)$ in 0.1 M KNO_3 on a 29 μm platinum electrode (diameter) under insonation (8.9 W cm^{-2}). Horn-to-electrode distance: 1 cm (a,c,d) or 1.5 mm (b). Figures (c), (a) and (d) are different transients obtained under the same conditions. Reprinted with permission from Ref. [30]. Copyright (2001) American Chemical Society.

Notice also the periodicity of some of the spikes; frequencies of 10 and 20 KHz can be observed along with higher harmonics of the 20 KHz driving force.

As noted above the background current on which the spikes are imposed is much enhanced over the diffusion limited current expected for a microdisc electrode under silent conditions as a result of acoustic streaming. Interestingly, experiments with microelectrode of diameters spanning the range 25 to 400 microns show this current to scale with the electrode area rather than the radius, indicating a switch from the convergent diffusion regime under silent conditions to a ready uniform mass transport regime under insonation as described by Eq. (8.102). This can be attributed to the very strong convective flow to the electrode induced by acoustic streaming. Application of a simple Nernst diffusion layer model shows that even with gentle insonation using power ultrasound of just 10 W cm^{-2} a 30 μm diameter microelectrode has a diffusion layer thickness of only ca 8 μm. Moreover, with higher powers diffusion layers as small as 0.7 μm have been realised.[26]

It follows from the above that insonation of a macroelectrode can confer the mass transport characteristics of a microelectrode.[34] This can be illustrated with respect to the study of the reduction of 3-bromo benzophenone, RBr, in DMF using a 3 mm glassy carbon electrode. The process is thought to follow an ECE mechanism.

$$R - Br + e^- \rightleftarrows [R - Br]^{\bullet -}$$

$$[R - Br]^{\bullet -} \xrightarrow{k} R\bullet + Br^-$$

$$R\bullet + HS \longrightarrow RH + S\bullet$$

$$RH + e^- \rightleftarrows [RH]^{\bullet -},$$

where HS denotes the solvent/supporting electrolyte system, and RH is benzophenone.

A cyclic voltammogram is shown in Fig. 8.38 and compared with three corresponding sonovoltammograms measured at a fixed power but different electrode-to-horn separations.

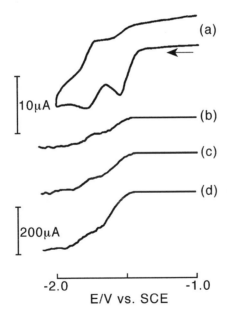

Fig. 8.38 (a) Cyclic voltammogram for reduction of 0.5 mM 3-bromobenzophenone in dimethylformamide (0.1 M NBu$_4$ClO$_4$) at a 3mm diameter glassy carbon electrode. (b)–(d) sonovoltammograms obtained with 25 W cm^{-2} power and (b) 27, (c) 15, (d) 8 mm horn-to-electrode distance. Scan rate: 50 mVs^{-1}. Reproduced from Ref. [34] with permission of The Royal Society of Chemistry.

The first chemically irreversible reduction process is followed by a second reduction process assigned to the reduction of benzophenone. Whereas peak-shaped responses are observed under silent condition, sigmoidal voltammograms with much higher currents result under insonation. The variation of the distance between the electrode and the horn allows the alteration of the diffusion layer thickness. The ratio of the limiting currents for the second and the first reduction process corresponds to the effective number of electrons transferred (see Section 8.6), N_{eff}, for an ECE process. For the case of a simple Nernst diffusion layer of thickness δ_d,[35]

$$N_{eff} = 2 - \frac{\tanh\left(\frac{\delta_d^2 k}{D}\right)^{1/2}}{\left(\frac{\delta_d^2 k}{D}\right)^{1/2}}, \qquad (8.103)$$

where k is the first order rate constant for the loss of Br^- from $[R - Br]^{\bullet-}$. Figure 8.39 shows a plot of the measured limiting currents of the reduction of 3-bromophenone versus δ^{-1} together with the expected values of $k = 600\ s^{-1}$ (ECE) and for simple one- and two-electron processes.

The data is therefore consistent with an ECE process tending to $N_{eff} = 1$ at small diffusion layer thickness and $N_{eff} = 2$ at thicker ones (as the horn-to-

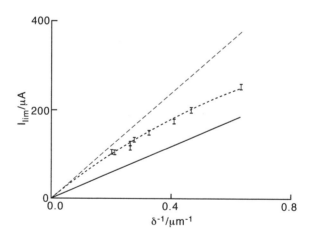

Fig. 8.39 Plot of sonovoltammetrically obtained limiting current vs. the reciprocal diffusion layer thickness for reduction of 0.5 mM 3-bromobenzophenone in dimethylformamide at a 3mm diameter glassy carbon electrode. The theoretically predicted curves for $k \to \infty$ (dashed line) and for $k = 6 \times 10^2\ s^{-1}$ (dotted line) are also shown. Reproduced from Ref. [34] with permission of The Royal Society of Chemistry.

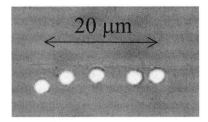

Fig. 8.40 An array of five microelectrodes sealed in epoxy each are 29 μm in diameter. Reprinted with permission from Ref. [30]. Copyright (2001) American Chemical Society.

electrode separation is increased). The rate constant of 600 s^{-1} is close to that of an independent measurement. We conclude that sonovoltammetry allow the determination of fast processes at *macro*electrodes placed in an ultrasonic sound field.

Microelectrode arrays can be used to 'size' cavitation bubbles. Figure 8.40 shows an array of five electrodes each of which can be addressed independently and simultaneously by a 5-channel 'nanosecond' potentiostat (overviewed by Banks and Compton).[36] It is thus possible to correlate signals seen on different electrodes.

Figure 8.41 shows chronoamperometry data recorded simultaneously on three electrodes, of diameter 29 μm, under 9 W cm^{-2} insonation with $h \sim 1$ cm using 50 mM potassium ferrocyanide in 0.1 M aqueous KCl with a potential corresponding to the mass transport limited reduction to ferrocyanide. In the case of Figs. 8.41(a)–(c) the inter-electrode distance was 104 μm whereas for (d)–(f) it was less than 5 μm.

In the first group of signals, (a)–(c), there is a strong correlation of the cavitational activity from one electrode to another whereas in the second group, (d)–(f), the signal seen on the central electrode (d) is voltammetrically visible on the two electrodes (e) and (f), located just 5 μm away! The conclusion follows that a very wide range of bubble sizes — between less than 1 μm and more than 400 μm — must be present at the electrode surface over a period of time. Furthermore, [30,31] the size of the cavitational current spikes seen over a range of fast chronoamperometric transients are all essentially of the same order of magnitude. This has implications on the physical nature of the cavitational current spiking; the microjetting mechanism illustrated in Fig. 8.36 in which acoustic collapse of the bubble at the interface leads to a fast moving microjet of liquid (moving at tens of metres per second) hitting the electrode surface may not, in defiance of some conventional wisdom, be reasonable. If microjetting were dominant, very different signals would be expected depending on whether the microelectrode was located within

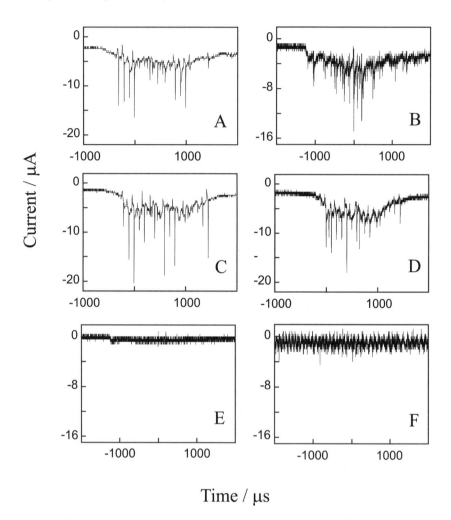

Time / μs

Fig. 8.41 Chronoamperometric currents recorded simultaneously (a, b and c or d, e and f) on three electrodes (diameter 29 μm) under 8.9 W cm^{-2} insonation using 50 mM ferricyanide in 0.1 M nitric acid. Horn-to-electrode distance: 1 cm. Inter-electrode distance: 104 um (a, b and c) or less than 5 μm (d, e and f). Reprinted with permission from Ref. [30]. Copyright (2001) American Chemical Society.

the microjet and the bubble, outside of the jet but within the bubble, or outside of the latter. Regardless of the mechanism of origin of the cavitational spikes, the effectiveness of ultrasound in cleaning and even ablating surfaces is well known and finds significant use in electroanalysis. We return to this topic in the following and final chapter of this book.

References

[1] C.E. Banks, A.O. Simm, R. Bowler, K. Dawes, R.G. Compton, *Anal. Chem.* **77** (2005) 1928.

[2] www-history.mcs.st-andrews.ac.uk/Mathematicians/Reynolds.html (accessed Jan 2007).

[3] T. von Karman, *Z. Angew. Math. Mech.* **1** (1921) 233.

[4] W.G. Cochran, *Proc. Camb. Phil. Soc. Math. Phys. Soc.* **30** (1934) 365.

[5] W.J. Albery, *Electrode Kinetics*, Oxford University Press, 1975.

[6] R.G. Compton, D. Mason, P.R. Unwin, *J. Chem. Soc. Faraday Trans.* **84** (1988) 483.

[7] J. Newman, *J. Phys. Chem.* **70** (1966) 1327.

[8] J. M. Hale, *J. Electroanal. Chem.* **6** (1963), 187.

[9] J. M. Hale, *J. Electroanal. Chem.* **8** (1964), 332.

[10] R.G. Compton, M.E. Laing, D. Mason, R.J. Northing, P.R. Unwin, *Proc. R. Soc. Lond.* **A418** (1988) 113.

[11] J. Koutecky, V.G. Levich, *Dokl. Akad. Nauk. ssse* **117** (1957) 441.

[12] J. Koutecky, V.G. Levich, *Zh. Fiz. Khim.* **32** (1958) 1565.

[13] W. Vielstich, D. Jahn, *Z. Elecktrochem.* **64** (1960) 43.

[14] R.G. Compton, R.G. Harland, P.R. Unwin, A.M. Waller, *J. Chem. Soc. Faraday Trans. 1* **83** (1987) 1261.

[15] M.A. Lévêque, *Ann. Mines. Mem. Ser.* **12** (1928) 201.

[16] J.A. Cooper, R.G. Compton, *Electroanalysis* **10** (1998) 141.

[17] A.C. Fisher, R.G. Compton, *J. Phys. Chem.* **95** (1991) 7538.

[18] R.C. Akerberg, R.D. Patel, S.K. Gupta, *J. Fluid Mech.* **86** (1978) 49.

[19] J. A. Alden, R.G. Compton, *J. Electroanal. Chem.* **404** (1996) 27

[20] J. A. Alden, M.A. Feldman, E. Hill, F. Prieto, M. Oyama, B.A. Coles, R.G. Compton, *Anal. Chem.* **70** (1998) 1707.

[21] N.V. Rees, R.A.W. Dryfe, J.A. Cooper, B.A. Coles, R.G. Compton, *J. Phys. Chem.* **99** (1995) 7096.

[22] A.D. Clegg, N.V. Rees, O.V. Klymenko, B.A. Coles, R.G. Compton, *J. Am. Chem. Soc.* **126** (2004) 6185.

[23] A.D. Clegg, N.V. Rees, O.V. Klymenko, B.A. Coles, R.G. Compton, *J. Electroanal. Chem.* **580** (2005) 78.

[24] J. V. Macpherson, S. Marcar, P.R. Unwin, *Anal. Chem.* **66** (1994) 2175.

[25] N.V. Rees, O.V. Klymenko, B.A. Coles, R.G. Compton, *J. Phys. Chem. B* **107** (2003) 13649.

[26] F. Marken, R.P. Akkermans, R.G. Compton, *J. Electroanal. Chem.* **415** (1996) 55.

[27] R.G. Compton, J.C. Eklund, F. Marken, T.O. Rebbitt, R.P. Akkermans, D.N. Waller, *Electrochemica Acta* **42** (1997) 2919.

[28] C.E. Banks, R.G. Compton, A.C. Fisher, I.E. Henley, *Phys. Chem. Chem. Phys.* **6** (2004) 3147.

[29] C. E. Banks, R.G. Compton, *Chem. Phys. Chem.* **4** (2003) 169.

[30] E. Maisonhaute, P.C. White, R.G. Compton, *J. Phys. Chem. B* **105** (2001) 12087.

[31] E. Maisonhaute, B.A. Brookes, R.G. Compton, *J. Phys. Chem. B* **106** (2002) 3166.

[32] E. Maisonhaute, F.J. Del Campo, R.G. Compton, *Ultrasonics Sonochem.* **9** (2002) 275.

[33] E. Maisonhaute, C. Prado, P.C. White, R.G. Compton, *Ultrasonics Sonochem.* **9** (2002) 297.

[34] R.G. Compton, F. Marken, T.O. Rebbitt, *Chem. Commun.* (1996) 1017.

[35] S. Karp, *J. Phys. Chem.* **7** (1968) 1082.

[36] C. E. Banks, R.G. Compton, *Chem. Anal. (Warsaw)* **48** (2003) 159.

9

Voltammetry for Electroanalysis

In this chapter we briefly address the use of voltammetric techniques in analysis; that is to say, the modification of the voltammetric techniques already presented so as to allow the detection of small concentrations of electroactive species. The chapter is written from the general outlook of the book in so far that it seeks to convey an *understanding* of the topic. We divert the reader interested in a detailed account of analytical electrochemistry to the inspiring and elegantly written general book by Joseph Wang[1] and the authoritative work written by K.H. Brainina and E. Neyman.[2]

9.1 Potential Step Voltammetric Techniques

Potential step techniques find wide application in analytical electrochemistry and are naturally compatible with digitally based potentiostats. Numerous methods have been devised including differential pulse voltammetry and square wave voltammetry as described in the following sections. The basis of the improved sensitivity of these methods is an enhanced response to the Faradaic current component *relative* to the capacitative current.

Figure 9.1 shows that the Faradaic response to a potential step is a pulse of current which decays with time as the electroactive species in the vicinity of the electrode is consumed.

Superimposed on this response is a capacitative constitution due to the double layer charging (movement of ions to and from the electrode, see Chapter 2). The latter dies away much more quickly than the Faradaic responses, often within 1 or

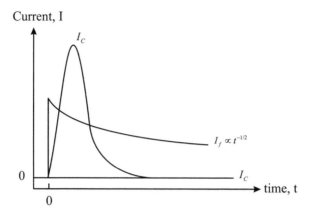

Fig. 9.1 Evolution of capacitative, I_C and Faradaic, I_f current following the application of a potential step.

2 milliseconds. After the capacitative component (I_C, see Fig. 9.1) has diminished, the Faradaic contribution, under transport limited linear diffusion conditions, is described by the Cottrell equation (Chapter 3) so that the current $I_F \propto t^{-1/2}$, and the charge, $Q \propto t^{1/2}$. Pulse and step techniques can be parameterised so as to sample the current after the capacitative current has died away and also to ensure that time is not allowed for natural convection (Chapter 8) to set in and perturb the expected diffusional response.

Differential pulse and square wave techniques find wide use for trace analysis and rank amongst the most sensitive means for the direct measurement of concentration. They can also provide information about the identification of oxidation states and the recognition of complexation effects.

9.2 Differential Pulse Voltammetry

Differential pulse voltammetry (DPV) is a technique typically capable of achieving detection limits of the order of 10^{-7} M. The potential waveform applied during the course of a DPV experiment is shown schematically in Fig. 9.2 as a waveform, of pulses superimposed on a staircase.

In the experiment the current, I_2, just before the end of each pulse (e.g. point 2 in Fig. 9.2) and the current, I_1 just before the pulse is applied (point 1) are measured and the difference

$$\Delta I = I_2 - I_1 \tag{9.1}$$

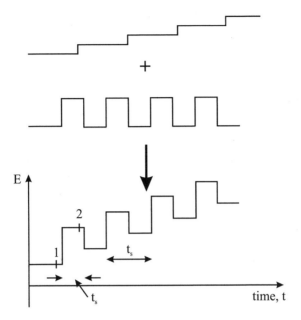

Fig. 9.2 Schematic waveform of pulses superimposed on a staircase where $t_S \sim$ 0.5–5 s and $t_P \sim$ 5–100 ms. The pulse height is typically $\sim$ 50 mV and the step height of the staircase $\sim$ 10 mV or less.

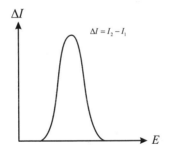

Fig. 9.3 Schematic illustration of the voltammetric profile resulting from a DPV scan.

is plotted against the staircase potential and leads to the peak-shaped profile as schematically shown in Fig. 9.3.

Note that the pulse duration t_P is at least ten times shorter than the period of the staircase waveform, t_S. DPV greatly reduces the contribution of the non-Faradaic processes, principally capacitive, to the recorded signal, mostly cancelling them out by subtracting I_1 from I_2. It is successful because these unwanted currents differ relatively little from the first point of sampling to the second.

9.3 Square Wave Voltammetry

Square wave voltammetry (SWV) was invented in 1952 by G.C. Barker and Jenkins[3] but at the time was limited in the extent of its application by the available instrumentation. Now suitable potentiostat electronics is available, it is often the method-of-choice in many analytical studies on account of its sensitivity.

The SWV waveform, as its name suggests, is a square wave which is superimposed on a potential staircase as shown in Fig. 9.4.

The square wave is characterised by a pulse height, or square wave amplitude, ΔE_P, the staircase height, ΔE_S, the pulse time, t_P, and the cycle period, t_S. Alternatively, the pulse time can be expressed in terms of the square wave frequency, $f = 1/2t_P$. The staircase shifts by ΔE_S at the start of each cycle so that the scan rate is

$$\frac{\Delta E_S}{2t_P} = f\,\Delta E_S \tag{9.2}$$

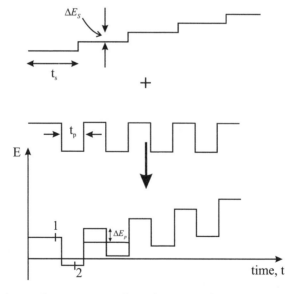

Fig. 9.4 Waveform and measurement scheme for SWV. It shows schematically the sum of a staircase and a square wave.

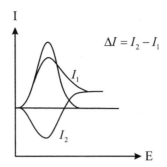

Fig. 9.5 Schematic voltammetric profiles of the current measured during the forward and reverse pulses and resultant difference, ΔI, plotted against the staircase potential E.

Usually in experimental practice[4] ΔE_S is much less than ΔE_P so that the latter parameter defines the resolution of voltammetric features. As in DPV the current is measured at two points in each cycle — for example points 1 and 2 in Fig. 9.4 corresponding to the end of the forward and reverse peaks respectively. The difference,

$$\Delta I = I_1 - I_2 \tag{9.3}$$

is plotted against the staircase potential as the principal result of the experiment. In reality, however, a single SWV potential scan generates three voltammograms showing I_1, I_2 and ΔI as a function of the staircase potential, as sketched in Fig. 9.5.

The major benefit of SWV is that of appreciable cancellation of the capacitative contributions to the current. This arises from the subtraction of I_1 from I_2 to generate ΔI and the fact that over the small potential range between forward and reverse pulses the interfacial capacitance is approximately constant. This also permits short pulses and faster scan rates than typically possible for other pulse techniques such as DPV. Detection limits of *ca.* 10^{-8} M, an order of magnitude lower than DPV, can be optimally attained. Not surprising SWV is available as an option as most modern conventional potentiostats, such as those manufactured by Eco-Chimie (Netherlands). Finally, a comment: the interpretation of SWV is not always obvious especially if irreversible electrode kinetics operate when two peaks may appear in the response of ΔI leading to the erroneous conclusion that two chemical species are present!

9.4 Stripping Voltammetry

Stripping voltammetry is a highly sensitive method for the electroanalytical detection of a wide range of metals and organic substances achieving low detection limits which can reach 10^{-10} M in favourable cases. The methodology is divided into three categories; anodic, cathodic or adsorptive stripping voltammetry, ASV, CSV or AdSV respectively. In each case there first is a 'pre-concentration step':

ASV: $\qquad\qquad\qquad\qquad M^{n+}(aq) + ne^- \rightarrow M(electrode),$

CSV:

Either $\qquad 2M^{n+}(aq) + mH_2O \rightarrow M_2O_m(electrode) + 2mH^+ + 2(m-n)e^-$

or $\qquad\quad M + L^{m-} \rightarrow ML^{(n-m)-}(ads) + ne^-,$

AdSV:

Either $\qquad\qquad\qquad A(aq) \rightarrow A(ads)$

or $\qquad\qquad M(aq) + L(aq) \rightarrow ML(ads),$

followed after suitable build up by a stripping step in which the potential from the working electrode is swept to induce Faradaic loss of the accumulated material from the electrode:

ASV: $\qquad\qquad\qquad M(electrode) \rightarrow M^{n+}(aq) + ne^-,$

CSV:

Either $\quad M_2O_m(electrode) + 2mH^+ + 2(m-n)e^- \rightarrow 2M^{n+}(aq) + mH_2O$

or $\qquad\qquad ML^{(n-m)-}(ads) + ne^- \rightarrow M + L^m$

AdSV:

Either $\qquad\qquad\qquad A(ads) \pm ne^- \rightarrow B(aq)$

or $\qquad\qquad ML(ads) \pm ne^- \rightarrow B'(aq).$

In ASV the stripping step, induced by scanning the working electrode towards positive potentials relative to that used for the deposition (pre-concentration step), produces a characteristic peak allowing quantification of the target trace ion. An example of a typical anodic stripping peak is shown in Fig. 9.6 for the anodic stripping voltammetry of cadmium using a boron-doped diamond electrode. Note the use of SWV to improve the sensitivity (see previous section) and that the potential at which the stripping peak occurs is intrinsic to cadmium, with other metals being stripped at different potentials.

In the example shown, additions of cadmium (II) to a 0.5 M acetate buffer were explored using a 60-second accumulation period during which metallic cadmium was plated onto the electrode, [5] as shown in Fig. 9.7.

Figure 9.6 shows well-defined square wave anodic stripping signatures with a detection limit of 2.5×10^{-8} M easily achievable. Significantly, lower detection limits for metals are similarly readily attainable if a (liquid) mercury electrode is employed rather than a solid electrode. This is because the metal usually dissolves in the mercury to form an amalgam rather than nucleates and grows as metal

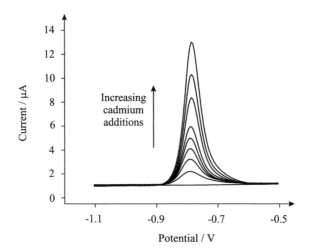

Fig. 9.6 Square-wave anodic stripping voltammetry in 0.5 M acetate buffer for increasing additions of cadmium (II). Parameters: 60 seconds deposition at $-1.8\,V$ vs. SCE, followed by scanning from $-1.1\,V$ to $-0.5\,V$. Reprinted from Ref. [5] with permission.

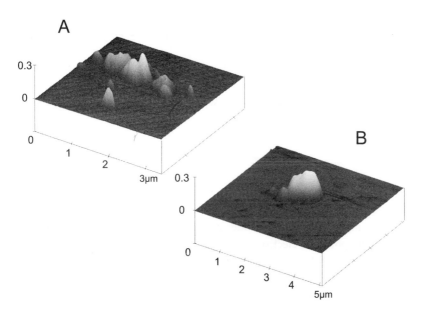

Fig. 9.7 AFM images after the deposition of cadmium from a solution containing 2.1 mM $CdNO_3$ in 0.5 M acetate buffer using a 60 second deposition procedure at $-1.8\,V$ (vs. SCE). (A) no surfactant present, (B) in the presence of a 3 μM solution of the neutral surfactant Trition X-100. Reprinted from Ref. [5] with permission.

particles on the electrode surface. The kinetic barriers to the latter are usually larger than for the amalgamation process. That said, the use of mercury is unattractive in the light of current and expected environmental legislation.

Examples of stripping voltammetry include the following.

(A) The detection of As(III) in some drinking water supplies using a gold electrode and ASV:

Pre-concentration: $\qquad As(III)(aq) + 3e^- \longrightarrow As(electrode)$

Stripping: $\qquad As(electrode) \longrightarrow As(III)(aq) + 3e^-.$

In this case after pre-concentration at a suitable reducing voltage, the working electrode is swept to positive potential and the resultant peak is used to quantify trace levels of As(III).

(B) The analysis of lead in river sediments utilises CSV in conjugation with a boron-doped diamond electrode. The lead is pre-concentrated as lead dioxide, PbO_2:

Pre-concentration: $\quad Pb^{2+}(aq) + 2H_2O(aq) \longrightarrow PbO_2(electrode) + 4H^+ + 2e^-$

Stripping: $\qquad PbO_2(electrode) + 4H^+ + 2e^- \longrightarrow Pb^{2+}(aq) + 2H_2O(aq).$

(C) The analysis of TNT (tri-nitrotoluene) is important in the detection of explosives and terrorist weapons for security surveillance. It can be analysed using AdsSV with glassy carbon electrodes modified with multi-walled carbon nanotubes:

Pre-concentration: $\qquad TNT(aq) \longrightarrow TNT(ads)$

Stripping: $\qquad TNT(ads) + ne^- \longrightarrow Reduced - TNT(aq).$

Whilst the attractions of mercury electrodes for ASV have been mentioned above it is worth adding a comment that the use of ASV for quantitative analytical measurements using solid electrodes can be prone to some generic limitations of which the potential electroanalyst needs to be aware[5] even if the effects are subsequently hopefully 'calibrated out'.

First, there are issues regarding electrode homogeneity and morphology so that metal deposition may not be uniform over the surface of the working electrode. By way of illustration, Fig. 9.8 shows microscopic images of the deposition of lead on boron-doped diamond and glassy carbon electrodes where the deposit density can vary widely over distances of hundreds of microns.

A) B)

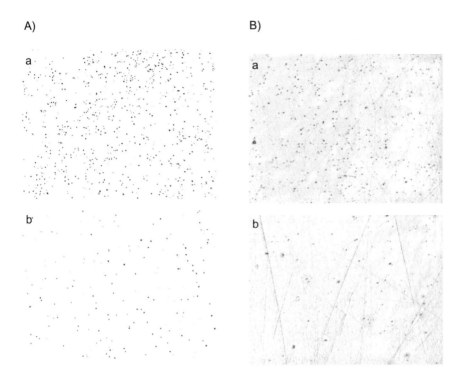

Fig. 9.8 Microscopic images of lead deposited from a solution containing 5 mM PbNO$_3$/0.1M HNO$_3$ on (A) BDD and (B) GC, showing in each case (a) the area of highest deposition density and (b) area of lowest deposition density (see text). Deposition conditions: −0.48 V (vs. SCE) for 60 seconds. Images show a total area of 540 × 400 μm. Reprinted from Ref. [5] with permission.

Similarly, Fig. 9.9 shows silver and lead deposition at different locations on the same boron-doped diamond; again, appreciable heterogeneity is evident.

Also, in respect of glassy carbon, a tendency has been noted for electro-deposition to occur preferentially at 'scratch' marks on manually polished samples: Fig. 9.10 shows clear atomic force microscopic evidence for this.

Second, there is the phenomenon of 'incomplete stripping' during ASV. When performing an ASV experiment, it is necessarily assumed that the stripping current is directly representative of the amount of material accumulated on the electrode surface during the deposition stage. However, this is not always the case.

Incomplete stripping is especially evident at high concentrations. However, even under conditions similar to those of analytical ASV it can be apparent. For example, Fig. 9.11 shows atomic force microscopic images of a boron-doped

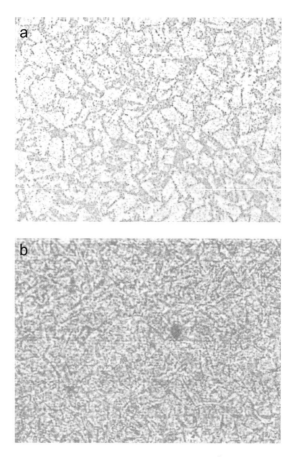

Fig. 9.9 Microscopic images of (a) silver deposited from a solution containing 5 mM AgNO$_3$/0.1 M HNO$_3$ at −0.3 V (vs. Ag) for 5 seconds and (b) lead deposited from a solution containing 5 mM PbNO$_3$/0.1 M HNO$_3$ at −0.6 V (vs. SCE) for 5 second on BDD. Image area 540 × 400 μm. Reprinted from Ref. [5] with permission.

diamond electrode (a) after 10 minutes of silver deposition from a 50 μM aqueous solution of AgNO$_3$ in 0.1 M HNO$_3$ and (b) after 'stripping' at a suitable anodic potential: approximately 20% of the silver remains on the surface.

Lead ASV using low, 50 μM, concentrations of Pb(NO$_3$)$_2$ shows a similar effect as depicted in Fig. 9.12.

Third, only in the case of a uniformly accessible electrode, or for fully irreversible electrode kinetics should it be assumed that the electrode surface is uniformly stripped of the deposited metal during the quantification step. Simulations

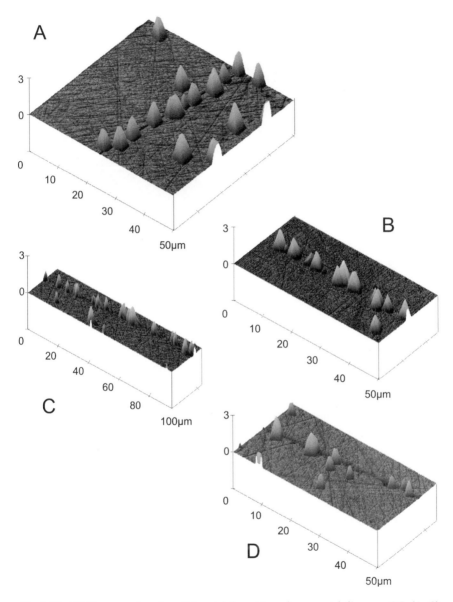

Fig. 9.10 AFM images of preferential metal deposition along scratch lines on GC. (A–C) deposition from a 5 mM AgNO₃/0.1M HNO₃ solution at −0.05 V (vs. Ag) for 120 seconds; (D) 5 mM PbNO₃/0.1M HNO₃ solution at −0.5 V (vs. SCE) for 120 seconds. Reprinted from Ref. [5] with permission.

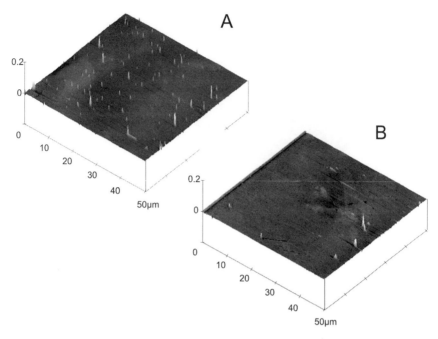

Fig. 9.11 Successive AFM images of a similar region of a BDD electrode. (A) after deposition from a 50 μM AgNO$_3$/0.1 M HNO$_3$ solution at -0.5 V (vs. Ag) for 10 min. (B) after a linear sweep from -0.15 V to $+0.2$ V at 20 mVs^{-1}. Reprinted from Ref. [5] with permission.

have shown that for non-uniformly accessible electrodes, for example the channel and wall-jet electrodes (see Chapter 8), and for electrochemically reversible or near reversible M^{n+}/M redox couples stripping occurs in a spatially non-uniform fashion with the flow inducing the upstream edge of the electrode, in the case of the channel, or the centre of the disc electrode in the case of a wall-jet electrode, to be depleted before, and at less positive potentials than the downstream edge (channel) or the radial extremities (wall-jet). Moreover, in the case of the wall-jet[6] with flow and sweep rates similar to those employed in analytical practice, it is seen that material oxidised from the electrode centre can become re-deposited at radial distances closer to the electrode edge where the diffusion layer is thicker, before being re-oxidised later on in the potential sweep at more positive potentials.

9.5 Sono-electroanalysis

Voltammetric stripping methods as discussed in the previous section can, with care, show good accuracy and reproducibility with easy to use instrumentation

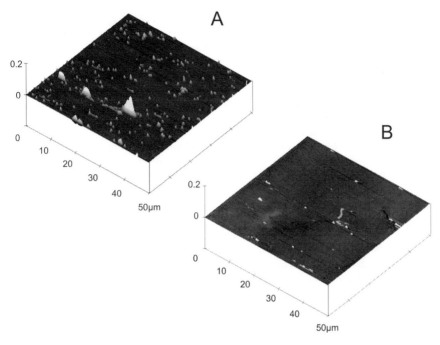

Fig. 9.12 Successive AFM images of a similar region of BDD electrode. (A) After deposition from a 50 μM PbNO$_3$/0.1 M HNO$_3$ solution at -0.8 V (vs. Ag) for 10 min. (B) After a linear sweep from 0 V to $+0.4$ V at 20 mVs^{-1}. Reprinted from Ref. [5] with permission.

which is cheap and simple in comparison to alternative methods. Nevertheless, despite many appealing studies with 'model systems', real world analytical usage of stripping voltammetry has not yet realised it full potential. The limitations are two-fold. First, in real world rather than model samples there are inevitably a wider range of molecules present some of which may be surface-active. For example many biological samples will contain proteins and environmental samples surfactants. These can lead to significant interference and electrode passivation problems. Whilst they can often be overcome by suitable sample pre-treatments, they give rise to a methodology which is slow and cumbersome. Second, as noted earlier, stripping voltammetry has been largely dependent on mercury electrodes to achieve the best limits of detection but environmental concerns and legislation strongly discourage the use of this material. Table 9.1 emphasis the reality of these pressures and Fig. 9.13 shows the corresponding impact on mercury usage in the USA.

Table 9.1. US Domestic Legislation, Regulation and Agreement Related to Mercury.

Year	Event
1971	Mercury designated as a hazardous pollutant.
1972	Insecticide, Fungicide, Rodenticide Act banned many pesticides containing mercury. Water Pollution Control Act authorized EPA to regulate mercury discharges into waterways.
1973	Mercury designated as a toxic pollutant. Standards for mercury ore processor and chloro-alkali plants enacted. Dumping mercury and mercury compounds into the ocean was prohibited.
1978	Resource Conservation and Recovery Act (RCRA) established regulations for the disposal of mercury waste.
1992	EPA banned land disposal of high mercury content wastes generated by chlor-alkali facilities.
1993	EPA cancelled the registrations for the last 2 mercury-containing fungicides at the manufacturer's request.
1994	Congress suspended the sale of National Defense Stockpile of mercury because of EPA's concerns with environmental problems related to the toxin.
1995	EPA's new regulation on municipal waste combustors issued. Regulations are designed to reduce mercury emissions from these facilities by 90% from 1990 emission levels.
1996	The Mercury Containing and Rechargeable Battery Management Act: prohibited batteries being sold without recyclability or disposal labels and phased out most batteries containing mercury.
1997	EPA issue new standards for medical waste incinerators which will reduce mercury emissions from these facilities by 94% from 1990 levels once fully implemented in 2002. The U.S./Canadian Great Lakes Bi-National Toxics Strategy is created. This agreement sets a goal to significantly reduce human use and release of mercury in Great Lakes Basin by 2006. The Chlorine Institute and the USA mercury cell chloro-alkali producers voluntarily commit to a 50% reduction goal in mercury use by 2005 and to providing EPA with an annual progress report.
1998	The 1998 Protocol on Heavy Metals of the Convention on Long Range Transboundary Air Polution: Involves the U.S., Canada and all European nations.
1999	EPA's new standards for hazardous waste combustors are designed to reduce mercury emissions from these facilities by 50% from 1990 emission levels.
2000	EPA lowered the threshold level for reporting mercury emission to the Toxics Release Inventory. Phase II North American Regional Plan on Mercury, under North American Agreement on Environmental Cooperation, Involves US, Canada and Mexico.

(Continued)

Table 9.1. *(Continued)*

Year	Event
2002	The Chlorine Institute reports chlor-alkali has achieved the 50% reduction goal three years early and pledges to continues its mercury reduction efforts.
2003	EPA issues its final rule for limiting the emissions from mercury cell chloro-alkali plants under the Clean Air Act. This new rule requires additional reduction in the emission limit from the existing rule.
2004	EPA developing emission standards for small sources of air toxin including mercury. The Chlorine Institute reports the overall mercury usage reduction to date over an eight-year period is 76%.
2005	EPA is scheduled to issue its final rule to regulate mercury emission from power plants.

Adapted from 'Mercury in Prospective', a US government report.

We have proposed the joint use of ultrasound and stripping voltammetry both to overcome electrode passivation effects and to open up the under usage of solid electrodes as alternative electrode materials to mercury with insonation being used to significantly enhance the sensitivity. Typically, ultrasound is applied during the accumulation protocol, then terminated when the potential is swept to quantify the target. The utility of the method can be exemplified by the use of sono-ASV to detect copper in untreated raw industrial effluent simply diluted with acid containing 1 M KCl to remove problems due to variable levels of chloride in the samples which can otherwise influence the stripping signal due to complexation of Cu(I) formed as an intermediate.

Figure 9.14 shows that the ASV signal at a glassy carbon electrode under conventional silent conditions produces a tiny, irreproducible stripping peak, whereas

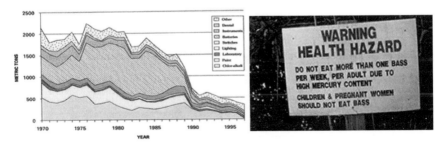

Fig. 9.13 (Left) US industrially reported consumption of mercury (1970–1997). (Right) A sign resulting from local mercury contamination. Adapted from 'Mercury in Prospective', a US government report.

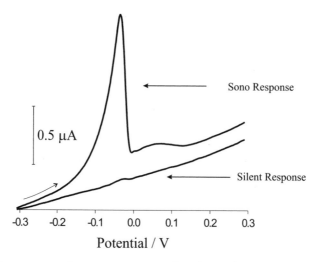

0.5 μA

Sono Response

Silent Response

Potential / V

Fig. 9.14 Linear sweep voltammograms comparing silent and sono response for stripping of copper in raw diluted industrial effluent using a bare glassy carbon electrode with a 30 second deposition time. Copyright (2001) from 'Sonoelectrochemical detection of copper within industrial effluent: a critical assessment'. Reproduced by permission of Taylor & Francis Group, LLC. See Ref. [7].

applying ultrasound during the copper deposition transforms this response into a large and reproducible peak ideally suited to quantitative analysis.[7] Table 9.2 shows the results of analysis in two effluent samples, both containing high levels of diverse organic materials including pesticides, herbicides, methanol, xylene, and acetone. As can be seen excellent agreement is obtained with independent blind analysis confirming the quantitative success of the sono-ASV approach.

Similar experiments have been reported in which sono-ASV was used to detect copper in beer.[8] A glassy carbon (and so mercury-free) electrode was used with no

Table 9.2. Comparison of copper analysis via sonoelectro-analysis with Atomic Absorption Spectroscopy in industrial effluent.

	Samples-Cu (ppm)	
	1	2
Sono ASV	2.3	8.6
Existing method used on site	3.0	9.0
Independent analysis (AAS)	2.5	8.1

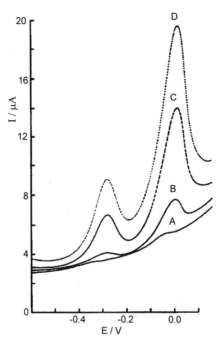

Fig. 9.15 Square wave ASV of aqueous solution of beer using a 240 s deposition with a horn to electrode distance of 5 mm: (A) 'blank' sample, silent deposition; (B) 48 μg L^{-1} standard addition of copper, silent deposition; (C) 143 μg L^{-1} standard addition of copper insonated deposition 200 W cm^{-2}; (D) 191 μg L^{-1} standard addition of copper insonated deposition 200 W cm^{-2}. Reproduced from Ref. [8] with permission of The Royal Society of Chemistry.

sample pre-treatment other than dilution with acid. Figure 9.15 shows the sono-square wave ASV; again, the excellent insonation during the accumulation step is seen to transform an almost negligible and analytical useless signal into the basis for quantitative electroanalysis. The method was successfully validated against blind, independent analysis.

It is evident from the above that insonation during the accumulation step can be highly beneficial in ASV. Why is this? We saw in the previous chapter that insonation of the electrode leads to a strong convective flow arising from acoustic streaming and, at insonation powers above the cavitation threshold, bubble formation on the electrode surface. The former leads to a greater amount of material deposited during the accumulation step and so very usefully enhances the sensitivity of the stripping method. The latter provides a mechanism for the *in situ* cleaning of the electrode surface so removing possible passivating materials and allowing the metal deposit in ASV to grow without inhibition. The mechanisms for this surface

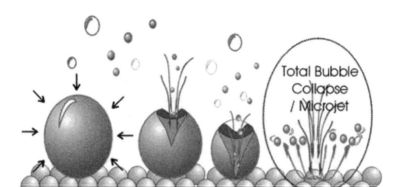

Fig. 9.16 A schematic depicting the bubble collapse at the electrode surface classically thought responsible for *in situ* activation via microjetting. Reproduced with permission from Ref. [9].

cleaning may involve microjetting as illustrated in Fig. 9.16 and/or via oscillating bubbles generating a strong shear flow across the electrode surface, as shown in Fig. 9.17.

The power of ultrasound for the *in situ* activation (cleaning) of working electrodes can be illustrated by an experiment[11,12] in which an egg (!) is ultrasonically homogenised after mixing with an equal volume of aqueous electrolyte. In the presence of ultrasound, linear sweep voltammetry could be successfully conducted provided insonation was maintained throughout the scan. Figure 9.18 shows the voltammetry of dopamine at a glassy carbon electrode in the egg/electrolyte

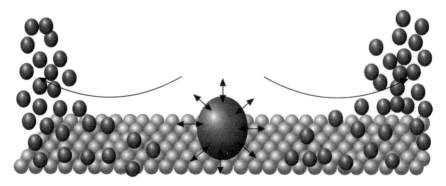

Fig. 9.17 Cleaning via oscillating (20 kHz) bubbles generating a strong shear flow across the electrode surface. Reproduced from Ref. [10] with permission of The Royal Society of Chemistry.

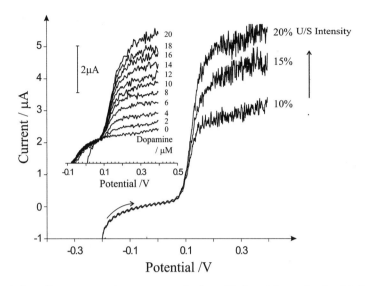

Fig. 9.18 Sono linear sweep voltammograms for the oxidation of dopamine (2 μM aliquots) on a carbon electrode. Scan rate 50 mVs^{-1}. Reprinted from Ref. [13] with permission from Elsevier.

medium; concentrations as low as 20 μM could be detected by linear sweep voltammetry.

Needless to say, voltammetry in 'ultrasonically scrambled egg' under silent conditions is pointless. The approach has been used to quantify nitrite in egg: the molecule 1,3,5-trihydroxybenzene is added to the sonicated mixture and, in the presence of acid, reacts with nitrite to form a nitroso compound which is electroactive and permits the quantitative determination of nitrite in egg (Fig. 9.19).

Finally, the use of new solid electrodes in sono-stripping voltammetry is nicely exemplified by the CSV detection of manganese (II) using boron-doped diamond electrodes. The wide potential window available in aqueous solution using the

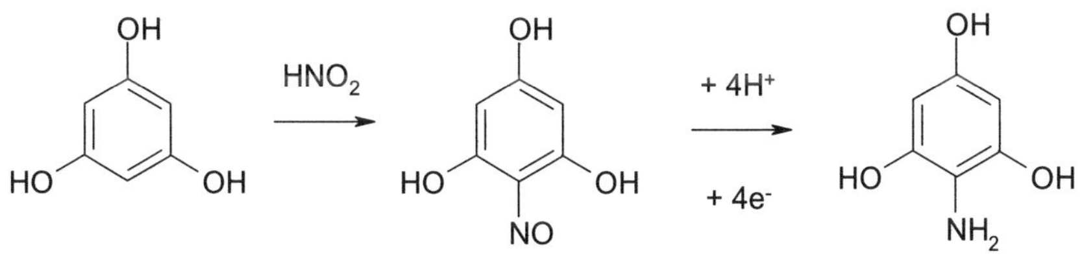

Fig. 9.19 Electrochemical pathway for the detection of nitrite.

latter enables the pre-concentrations of the metal as MnO_2. In this way the sono-CSV detection of manganese (II) with the very low detection limit of 10^{-11} M has been proved possible. The stripping step is

$$MnO_2 + 2e^- + 4H^+ \longrightarrow Mn^{2+} + 2H_2O.$$

The method has been used for environmental analysis.[13]

References

[1] J. Wang, *Analytical Electrochemistry*, Wiley-VCH, 2nd ed., 2000.

[2] K.H. Brainina, E. Neyman, *Electrochemical Stripping Methods*, Wiley, 1993.

[3] G.C. Barker, I.L Jenkins, *Analyst* 77 (1952) 685.

[4] J.L. Hardcastle, DPhil. Thesis, University of Oxford, 2002.

[5] M.E. Hyde, C.E. Banks, R.G. Compton, *Electroanalysis* 16 (2004) 345.

[6] J.C. Ball, R.G. Compton, C.M.A. Brett, *J. Phys. Chem.* B 102 (1998) 162.

[7] J. Davis, M.F. Cardosi, I. Brown, M.J. Hetheridge, R.G. Compton, *Anal. Letters* 34 (2001) 2375.

[8] C. Agra-Gutierrez, J.L. Hardcastle, J.C. Ball, R.G. Compton, *Analyst* 124 (1999) 1053.

[9] Reproduced with permission from J. Davis, Nottingham Trent University.

[10] C.E. Banks, R.G. Compton, *Analyst* 129 (2004) 678.

[11] J. Davis, R.G. Compton, *Anal. Chim. Acta.* 404 (2000) 241.

[12] E.L. Beckett, N.S. Lawrence, Y. C. Tsai, J. Davis, R.G. Compton, *J. Pharm. Biomed. Anal.* 26 (2001) 995.

[13] A. Goodwin, A.L. Lawrence, C.E. Banks, F. Wantz, D. Omanovic, Š. Komorsky-Lovrić, R.G. Compton, *Analytical Chimica Acta* 533 (2005) 141.

10

Voltammetry in Weakly Supported Media: Migration and Other Effects

Previously we have considered mass transport in solution resulting from either diffusion (Chapter 3) and/or convection (Chapter 8). In this chapter we explore the movement of ions resulting from a non-uniform electrical potential, ϕ. That is when an electrical field, or potential gradient is present where

$$\text{electrical field} = -\nabla\phi, \tag{10.1}$$

or, in one dimension, x,

$$\text{electrical field} = -\frac{\partial\phi}{\partial x}. \tag{10.2}$$

Such fields drive the movement of ions in the direction of the field or against it according to their charge.

10.1 Potentials and Fields in Fully Supported Voltammetry

In Section 2.5 we noted that voltammetry is usually conducted in the presence of a large concentration of so-called supporting electrolyte, considerably in excess of the concentration of the electroactive species being studied. Thus, for example, in typical non-aqueous voltammetry in, say, acetonitrile, the species of interest to be studied might be present at the *ca.* millimolar levels whereas the supporting electrolyte, for example tetra-n-butylammonium tetrafluoroborate, would be present at a concentration of at least *ca.* 0.1 M, approaching or exceeding some two orders of magnitude in excess. Under these conditions the availability of ions from

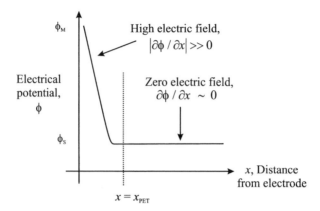

Fig. 10.1 The potential distribution for 'supported' voltammetry where the location of electron transfer, $x_{PET} \sim 10\text{--}20$ Å.

the supporting electrolyte results in these being attracted to, or repelled from, the working electrode according to charge, and so the electrical potential drops from the value characteristic of the (metal) electrode ϕ_M to that of the bulk solution, ϕ_S, over a very short distance of no more than *ca.* 10–20 Å. As a result there exists a very large electric field in this narrow interfacial region but outside of this, where the potential has the constant value of ϕ_S, the electric field is zero. Figure 10.1 shows this situation. Note that the high field at the interface can be as large as of the order 10^8–10^9 Vm^{-1}.

Under the conditions of Fig. 10.1, a molecule being electrolysed at the electrode ($x = 0$) can diffuse to a distance $x = x_{PET}$ of it without experiencing any electric field. Accordingly the only means of transport from bulk solution ($x \sim \infty$) to x_{PET} is by diffusion. If we suppose that x_{PET} corresponds to the plane of electron transfer, that is to a location close enough to the electrode to allow electron transfer to or from the electrode by means of quantum mechanical tunnelling, then the full drop in potential ($\phi_M - \phi_S$) is available at x_{PET} to 'drive' this electron transfer.

Note that because tunnelling is only effective over short distance $x_{PET} \sim$ 10–20Å (see Chapter 2), if a low concentration of supporting electrolyte is used then there are fewer ions in the solution to be attracted or repelled from the electrode surface. As a result the potential changes at the electrode solution interference from ϕ_M to ϕ_S over a much larger distance than in the 'fully' supported cases, as shown in Fig. 10.2. The consequences of this are twofold. First, at distances consistent with efficient electron tunnelling, $x \le x_{PET}$, only a fraction of the maximum possible drop in potential, $\phi_M - \phi_S$, will be available to drive the electrode reaction. Second, when the electroactive species is transported from bulk solution to the

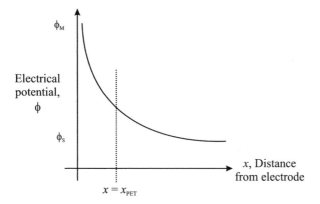

Fig. 10.2 The potential distribution for less than fully 'supported' voltammetry.

location of electron transfer it experiences a finite electric field and so, if the species carries an electrical charge, it is itself attracted or repelled from the electrode by virtue of the electric field it experiences. Paradoxically for the uninitiated, it is only under less than fully 'supported' situations that a species undergoing electrolysis at an electrode will be influenced by the charge on the electrode; under the usual conditions of fully supported voltammetry the molecules undergoing electrolysis will diffuse to x_{PET} and undergo electron transfer without any attraction or repulsion by the electrode. It is for this reason that is possible for, say, positively charged species to undergo oxidation or reduction at an electrode with an absolute positive charge (potential) if the thermodynamics are appropriate. Thus for example the reduction

$$Fe^{3+}(aq) + e^- \rightleftharpoons Fe^{2+}(aq)$$

takes place at positive potentials in aqueous solution. The standard electrode potential is $E^0(Fe^{3+}/Fe^{2+}) = +0.77\,V$. A voltammogram for the reduction of Fe^{3+} is shown in Fig. 10.3. Note that the potential scale is reported relative to the Saturated Calomel Electrode (SCE). The latter has a potential of 0.242 V on the hydrogen scale.[a]

In the following sections we address first the distribution of ions around an electrode, and second, the transport of (charge) ions in solution as driven by an electric field.

---

[a] Standard electrode potential are, of course, potential values *relative* to the standard hydrogen electrode. Whilst *absolute* potentials cannot be measured (see Chapter 1), they can be estimated by means of a thermodynamic cycle. Trasatti, on behalf of IUPAC, recommends a value of $4.44 \pm 0.02\,V$ for the absolute potential of the standard hydrogen electrode at 298 K.

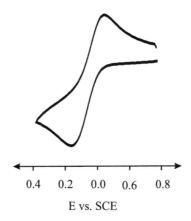

0.4   0.2   0.0   0.6   0.8

E vs. SCE

Fig. 10.3 Cyclic voltammogram for the reduction of 10^{-3} M *Fe*(III) in 1.0 M H_2SO_4 at a platinum electrode.

10.2 The Distribution of Ions Around a Charged Electrode

We have noted that if we apply a potential to an electrode immersed in a solution containing ions then, assuming no electrolysis takes place, ions will be attracted or repelled from the electrode according to its potential. Figure 10.4 shows that the application of a negative potential to the electrode results in the attraction of cations and the repulsion of anions from the interface so that local to the electrode is an excess of the former over the latter.

The electrical potential — defined as the work in hypothetically transferring a unit positive charge from infinity to the position in question — is seen in Fig. 10.4 to vary smoothly between ϕ_S in bulk solution and ϕ_M at the electrode where $\phi_M < \phi_S$ corresponding to a negative charge on the electrode. Figure 10.5 shows the corresponding situation for positive potential applied to the electrode resulting in the attraction of anions and the repulsion of cations. In this case the solution local to the electrode carries an excess of negative charge.

In general, the charge density (charge per unit volume) in the solution $\rho(x)$ can be defined by the expression

$$\rho(x) = \sum_i Z_i F c_i(x) \tag{10.3}$$

where the summation extends over all the ions, in the solution, $Z_i F$ (Coulombs per mole) is the charge on one mole of the ion i of concentration c_i (moles per unit volume). Figure 10.6 shows how $\rho(x)$ varies with the distance, x, away from a planar electrode in each of the cases described in Figs. 10.4 and 10.5. Note that the

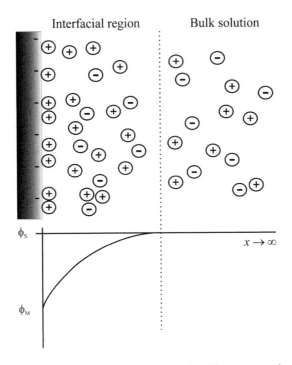

Fig. 10.4 Distribution of ions at an electrode with a negative charge.

total excess charge in the solution near the electrode will be balanced by an exactly equal charge of appropriate sign on the surface of the electrode.

The distribution of ions shown in Figs. 10.4 and 10.5 is, of course, not a static one; rather the ions are moving about in solution since typically (see below), at least sufficiently far away from the electrode, the applied potential can be assumed to be relatively small compared to the energy of the thermal motions of the ion. Consequently the ion distribution pictures in Figs. 10.4 and 10.5 should be thought of as representing a time average. Assuming that these time average distributions obey the Boltzmann distribution law, we can write

$$c_i(x) = c_i(x \to \infty) \exp[-Z_i F(\phi_x - \phi_S)/RT] \qquad (10.4)$$

where $c_i(x \to \infty)$ and ϕ_S are, respectively, the concentration of ion i and the potential in bulk solution. It follows that

$$\rho(x) = \sum_i Z_i F c_i(x \to \infty) \exp[-Z_i F(\phi_x - \phi_S)/RT]. \qquad (10.5)$$

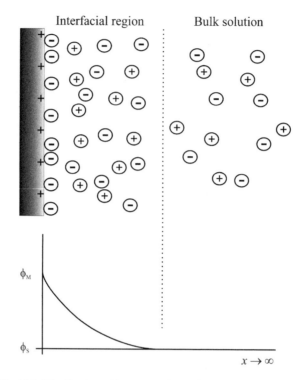

Fig. 10.5 Distribution of ions at an electrode with a positive charge.

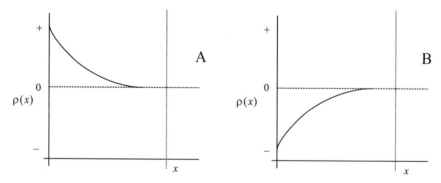

Fig. 10.6 The charge density in solution corresponding to the potential and ion distributions shown in Figs. 10.4 (A) and 10.5 (B). Note that the electrode ($x < 0$) will carry a charge equal and opposite to the total excess charge in solution. This charge will reside at or very close to the surface of the electrode.

The physics relates the charge density $\rho(x)$ and the potential ϕ_x through the Poisson equation:

$$\frac{\partial^2 \phi_x}{\partial x^2} = -\frac{\rho}{\varepsilon_0 \varepsilon_r} \tag{10.6}$$

where ε_0 is the permittivity of vacuum ($8.854 \times 10^{-12}\,\mathrm{C^2\,J^{-1}\,m^{-1}}$) and ε_r is the dielectric constant (relative permittivity; see Section 2.15). Substituting Eq. (10.5) into Eq. (10.6) gives

$$\frac{\partial^2 \phi(x)}{\partial x^2} = -\frac{1}{\varepsilon_0 \varepsilon_r} \sum_i Z_i F c_i (x \to \infty) \exp[-Z_i F(\phi_x - \phi_S)/RT]. \tag{10.7}$$

Let us simplify the problem, but with no loss of physical insight, by assuming a 1:1 electrolyte, $M^{z+} X^{z-}$. Equation (10.7) becomes in dimensionless form

$$\frac{\partial^2 \Theta}{\partial \chi^2} = \sinh(\Theta) \tag{10.8}$$

where

$$\Theta = \frac{ZF}{RT}(\phi_x - \phi_S) \tag{10.9}$$

$$\chi = \frac{x}{\kappa^{-1}}$$

$$\sinh \Theta = \frac{1}{2}[\exp(\Theta) - \exp(-\Theta)] \tag{10.10}$$

and the parameter

$$\kappa^{-1} = \frac{1}{ZF}\left(\frac{\varepsilon_0 \varepsilon_r RT}{2c(x \to \infty)}\right)^{1/2} \tag{10.11}$$

is known as the Debye length and

$$c(x \to \infty) = c_M(x \to \infty) = c_X(x \to \infty). \tag{10.12}$$

For water at 25°C,

$$\kappa \simeq 10^{-8}[c(x \to \infty)]^{1/2} \tag{10.13}$$

if c is measured in $\mathrm{mol\,m^{-3}}$ (or mM or 10^{-3} M). So as c changes from 1 mM to 1 M, κ varies from *ca.* 100 Å to ~3 Å.

The solution of Eq. (10.8) is as follows:

$$\tanh(\Theta) = \tanh(\Theta_0)\exp(-\chi) \tag{10.14}$$

where

$$\tanh \Theta = \frac{\sinh \Theta}{\cosh \Theta} = \frac{\exp(\Theta) - \exp(-\Theta)}{\exp(\Theta) + \exp(-\Theta)}$$

and

$$\Theta_0 = \frac{ZF}{RT}(\phi_M - \phi_S). \tag{10.15}$$

Recasting Eq. (10.14) in dimensional form and assuming that $F(\phi_M - \phi_S)$ is not too large compared with RT, we obtain the approximate relationship

$$\phi = (\phi_M - \phi_S)\exp(-\kappa x) \tag{10.16}$$

which shows that the potential falls from ϕ_M to ϕ_S as κ increases from zero over a distance scale of the order of κ^{-1}, as can be seen from Fig. 10.7 which shows how ϕ varies with x for three different concentration values assuming an aqueous solvent.

The approximate exponential nature of the fall off is evident. Figure 10.8 shows the distributions of the cation and anion for the case where the bulk concentration is 10^{-2} M.

The theory above is based on the independent work of Gouy[2,3] and Chapman.[4] It provides an explanation of why, when a supporting electrolyte is added to the solution studied in a voltammetric experiment, the interpretation of the experiment is much facilitated. First, for large electrolyte concentrations (> 0.1 M) the potential drop between the electrode and bulk solution, $\phi_M - \phi_S$ will occur over a distance of just a few Angstroms so that this full thermodynamic driving force is

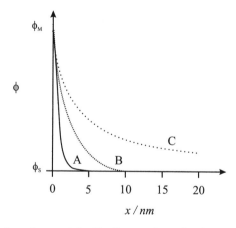

Fig. 10.7 The variation of potential with distance from the electrode for three different concentration for the case of an aqueous solution and $\phi_M - \phi_S = 100$ mV. $A = 0.1$ M, $B = 0.01$ M and $C = 0.001$ M.

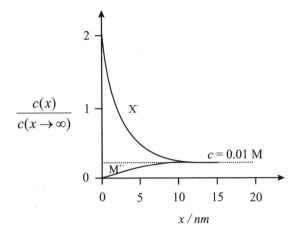

Fig. 10.8 The concentration profiles for the cation M^+ and the anion X^- for the case of a bulk electrolyte concentration of 10^{-2} M and $\phi_M - \phi_S = 100$ mV assuming an aqueous solvent.

available to drive an electrode reaction of a species located at this distance from the electrode, since tunnelling of electrons to and from the electrode is efficient only over distances of a few Angstroms. In contrast, if only a diluted solution of electrolyte is present, the fall off between $\phi_M - \phi_S$ occurs over a larger distance so that when the species is transported to a location close to the electrode consistent with efficient tunnelling to facilitate electron transfer, only a small fraction of $\phi_M - \phi_S$ is available to 'drive' the reaction.

Second, for the fully supported case, the species undergoing electrolysis is transported up to the site of electron transfer purely and solely by diffusion. This is because the electric field outside of the region within the tunnelling distance is essentially zero so that movement of the species, even if it is charged by the electric field, does not occur.

We consider each of these effects in more detail in the rest of the chapter. First, however, we examine the structure of the interfacial layer between the electrode and the solution in more detail.

10.3 The Electrode–Solution Interface: Beyond the Gouy–Chapman Theory

The pictures of the interfacial region presented in Figs. 10.4 and 10.5 are incomplete. The Gouy–Chapman theory assumes that the electrode simply attracts or repels ions in solution so that there is a build-up of either cations or anions and a depletion of the other ion at all potentials at which the electrode carries a charge (potentials other than the so-called 'Potential of Zero Charge'). In practice the theory needs modification first to recognise that the attracted ions have a finite size which reflects their level of solvation. Second, they can, in many cases, interact 'specifically' with the electrode by which is meant chemical bonding usually after partial or full desolvation. Note that anions are more prone to loss of hydration since they interact more weakly with water molecules than do cations. Third, the electric field at the interface can be sufficient to orientate solvent molecules which have a dipole moment so that rather than rotating relatively freely they take up a preferential orientation at the interface. Figure 10.9 shows the different possible cases. In (A) the anions approach as closely to the electrode as their solvation shells and the forces of electrical (only) attraction allow. The plane of closest approach is the so-called 'Outer Helmholtz Plane' (OHP). Beyond the OHP is the 'diffuse layer' described by Gouy–Chapman theory. In (B) there is specific adsorption; some anions are desolvated and bind chemically directly to the electrode surface. The plane of closest approach of these desolvated anions is the 'Inner Helmholtz Plane' (IHP). Further away from the electrode is the OHP and the diffuse layer. Figure 10.9(C) shows the case of strong specific adsorption; note again the presence of an OHP and IHP. In the latter case it is sometimes possible that more anions can adsorb on the electrode than is required to 'balance' the charge on the latter so that the ions at the OHP and in the diffuse layer carry the same charge as present on the electrode ('charge reversal')! Such effects are thought to occur for mercury electrodes and KCl or KBr electrolyte. Indeed, bromide ions are thought to interact so strongly with mercury they specifically adsorb even at potentials where the electrode carries a negative charge!

It should be pointed out that the diagrams in Fig. 10.9 omit the solvent molecules both in the interfacial region and in bulk solution. For the case of water the molecule can be oriented, depending on the electrode potential, as shown in Fig. 10.10(a). Figure 10.10(b) shows a general and more complete image of the interfacial region and identifies various types of species at the metal-aqueous electrolyte interface including water in different orientations in the 'primary' and 'secondary' solvent layers.

Lastly, note that except for liquid electrodes, such as mercury of other molten metals, it is unlikely that the metal surface will be atomically flat but rather that surface roughness will feature. Moreover, for the case of polycrystalline metals, heterogeneity may result as a consequence of different crystal faces being exposed at different locations on the surface.

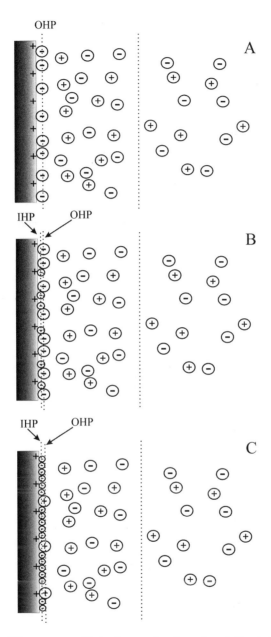

Fig. 10.9 Three different types of behaviour at the electrode-solution interface: (A) non-specific adsorption; (B) weak specific adsorption; (C) strong specific adsorption. Note the solvent molecules are not shown.

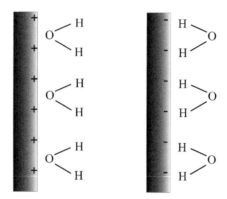

Fig. 10.10 (a) Orientation of the water layer next to an electrode depends on the electrode charge.

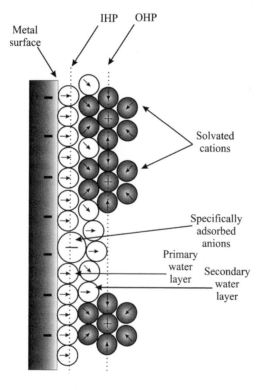

Fig. 10.10 (b) Possible structure of a metal electrolyte interface.

10.4 Double Layer Effect on Electrode Kinetics: Frumkin Effects

In Section 10.2 we saw that if the solution was not 'fully supported' then effects on the electrode kinetics observed voltammetrically would be seen. Considering an irreversible reduction, $A + e \rightarrow B$, and with the notation of Section 2.3, we recall that for the fully supported case,

$$j \propto [A]_0 \exp\left(-\frac{\alpha F}{RT}(\phi_M - \phi_S)\right). \tag{10.17}$$

Where the solution is less than fully supported we can expect two effects to modify the electrode kinetics. First, the driving force for the potential seen by the species A at distances compatible with electron transfer is rather less than the full amount $\phi_M - \phi_S$ seen under the corresponding fully supported conditions, as discussed in Section 10.2. Second, the concentration $[A]_0$ in the case that A is charged (i.e. A is not an electrically neutral molecule) will be different under weakly supported conditions because A will be attracted or repelled by the electrode since the levels of electrolyte present are insufficient to fully 'screen' the electrode charge from the approaching A molecule. These effects together constitute so called 'Frumkin effects' on the electrode process. If the adsorption behaviour at the electrode of interest is understood then attempts can be made to quantify these on the basis of the two physical effects noted. Albery gives characteristically insightful accounts.[5]

The two effects can be illustrated with reference to the reduction

$$S_2O_8^{2-}(aq) + 2e^- \rightarrow 2SO_4^{2-}(aq)$$

which was extensively studied by Frumkin *et al.*[6,7] Figure 10.11(A) shows the reduction of 10^{-3} M peroxysulphate in the presence of various concentration of K_2SO_4 at a mercury amalgam rotating disc electrode.

It can be seen under fully supported conditions, where 1.0 M K_2SO_4 is present, that a steady constant limiting current is attained at sufficiently negative potential for the reduction to take place. However, as the concentration is progressively lowered then the sigmoidal reduction wave becomes more and more distorted. Note that the distortion occurs at potentials negative of the Potential of Zero Charge (PZC). Figure 10.11(B) shows the same experiment conduction with only 10^{-3} M K_2SO_4 supporting electrolyte but using rotating disc electrodes of different metals and hence of different PZC. Note that in each case the onset of deviation from the limiting current corresponds to the potential becoming negative of the PZC. Thus as the potential becomes negative the initial fall in the current can be

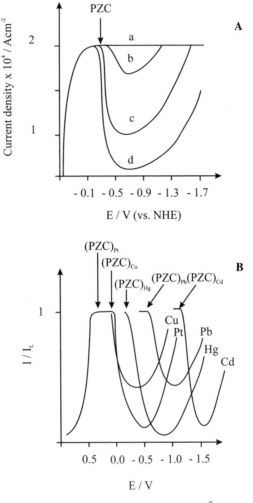

Fig. 10.11 Rotating disc experiments for the reduction of $S_2O_8^{2-}$ (A) at a mercury amalgam electrode with the following concentrations of K_2SO_4: (a) 1 M, (b) 0.1 M, (c) 0.08 M, (d) 0 M, and (B) in the presence of 10^{-3} M K_2SO_4 for different metals as indicated.

attributed to the repulsion of the negative $S_2O_8^{2-}$ ions from the electrode. The rise in current at more extreme negative potentials reflects the increasing compression of the diffuse layer as the electrode becomes more and more negatively charged.

The complexity of the current–potential curves shown in Fig. 10.11 is such as to encourage all experimentalists to ensure that they work with 'fully supported' conditions. If in doubt, always add more supporting electrolyte in your experiment and see what happens. On the other hand, the interpretation of voltammetry in weakly supported conditions is sometimes necessary and can offer considerable insight into the structure of the electrode-solution interface. We will return to the topic later in this chapter after some necessary theory is explained.

10.5 A.N. Frumkin

Alexander Naumovich Frumkin was without doubt a leading electrochemist of the twentieth century. He was born on 24th October 1895 in Kishinev (now Moldova, then part of the Russian Empire), the son of an insurance agent. He was, however, schooled in Odessa and further educated in Strasbourg and Bern before returning to the intellectual centre of Odessa, obtaining his first degree in 1915 from the Faculty of Mathematic and Physics in Novosossiya University (currently Odessa University). He subsequently worked with Professor A.N. Sakhanov, leading in 1919 to his seminal work, 'Electrocapillary Phenomena and Electrode Potentials', which was to inform and inspire generations of electrochemists. Among the ideas introduced for the first time were those of the Potential of Zero Charge (PZC) and the use of the Gibb's equation to derive surface excesses from electro-capillary curves. Parts of Frumkin's thesis were published in English (*Philosophical Magazine*, 1920) and German (*Z. for Physikalische Chemie*, 1923) but since the doctorate degree was abolished with the Russian Revolution of 1917 he received no academic degree for it.

(A) (B)

Fig. 10.12 A.N. Frumkin as (A) a young man and (B) an older man. Pictures are subject to copyright and kindly supplied from: http://www.elch.chem.msu.ru

In 1920 Frumkin worked in the Institute for People's Education in Odessa before moving to Karpov Institute, Moscow, in 1922. The latter was concerned with the scientific investigation of industrial problems and was founded in 1918. In Moscow he continued work on the metal–electrolyte interface but also began study on the air–solution interface and the measurement of the Volta potential between two electrolytes. At this time he also developed the adsorption isotherm which carries his name.

In 1930 Frumkin moved to Moscow University and in 1933 established there the department of electrochemistry. He headed this department until his death on 27th May 1976.

10.6 Transport by Diffusion and by Migration

In Chapter 1 we introduced the electrochemical potential, $\overline{\mu_i}$, of a species i,

$$\overline{\mu_i} = \mu_i + Z_i F \phi, \tag{10.18}$$

where Z_i is the charge on the species i and ϕ is the electrical potential. The term μ_i describes the chemical potential of i whilst the quantity $Z_i F \phi$ represents its electrochemical energy. Both have units of energy *per mole.*

Species will tend to move from high to low electrochemical potential; an electrochemical potential gradient is thus a driving force causing displacement of the species:

$$Force = -\frac{1}{N_A} \frac{\partial \overline{\mu_i}}{\partial x}, \tag{10.19}$$

where N_A is the Avogadro constant. Such a force imposed constantly would, in the absence of anything else, cause the species to accelerate and possess an ever increasing velocity. In practice, for the case of ions and molecules in solution, a frictional forces opposes the motion. The frictional force is thought to be proportional to the velocity, v, of the moving ion or molecules:

$$Frictional\ Force = -fv, \tag{10.20}$$

where f is a frictional coefficient. The existence of this force retarding the acceleration of the moving species has the consequence that the species i reaches a steady state velocity v_i, when the driving force resulting from the electrochemical potential gradient equals the frictional retarding force. Under these conditions

$$v_i = -\frac{1}{f N_A} \frac{\partial \overline{\mu_i}}{\partial x}. \tag{10.21}$$

We can expand

$$\overline{\mu_i} = \mu_i^0 + RT \ln \gamma_i [i] + Z_i F \phi$$

with the notation of Section 1.8 and where γ_i is the activity coefficient of i. Hence

$$v_i = -\frac{1}{fN_A}\left[RT\frac{\partial \ln \gamma_i[i]}{\partial x} + Z_iF\frac{\partial \phi}{\partial x}\right] \quad (10.22)$$

which is a form of the Nernst–Planck equation, of which we can usefully consider two limiting cases. First we assume that the electrical potential, ϕ, is constant so that the last term on the right-hand side of Eq. (10.22) disappears. It follows that the flux of i is given by the product of v_i and $[i]$:

$$j_i = v_i[i] = -\frac{[i]RT}{f_iN_A}\frac{\partial \ln \gamma_i[i]}{\partial x}. \quad (10.23)$$

If the activity coefficient is constant, for example because it is provided by an excess of supporting electrolyte (Section 2.5), then

$$j_i = -\frac{RT}{f_iN_A}\frac{\partial [i]}{\partial x}. \quad (10.24)$$

This is Fick's first law of diffusion (Section 3.1) with

$$D_i = \frac{RT}{f_iN_A}. \quad (10.25)$$

It can be seen therefore that Eq. (10.19) is a generalisation of Fick's laws of diffusion.

In the second limiting case we assume that the concentration (strictly activity) of species i is uniform, independent of x. In this case,

$$v_i = \frac{Z_iF}{f_iN_A}\frac{\partial \phi}{\partial x}, \quad (10.26)$$

which expresses the velocity of the ion i as a result of *electrical migration* induced by the electric field $-\frac{\partial \phi}{\partial x}$. The quantity

$$u_i = \frac{|Z_i|F}{f_iN_A} \quad (10.27)$$

is known as the mobility of the ion, i.

The frictional coefficient, f_i, appears in both Eqs. (10.26) and (10.27). Elimination thus gives

$$D_i = \frac{RTu_i}{|Z_i|F}, \quad (10.28)$$

known as the Einstein relation. It enables us to rewrite the Nernst–Planck equation as

$$j_i = -D_i\left(\frac{\partial [i]}{\partial x} + \frac{Z_iF[i]\partial \phi}{RT\partial x}\right) \quad (10.29)$$

where again the activity coefficient has been assumed constant.

10.7 Measurement of Ion Mobilities

We have seen in earlier chapters that voltammetric measurements allow the determination of diffusion coefficients. The Einstein relation, Eq. (10.28), shows that the measurement of ion mobilities is equivalent. The latter have been traditionally measured by means of conductivity experiments.

Conductivity measurements treat the bulk electrolyte as obeying Ohms law so that the electrical resistance, R_r, is given by the equation

$$R_r = \frac{\Delta\phi}{I}, \tag{10.30}$$

where $\Delta\phi$ is the voltage drop and I is the current flowing. Resistance, for an ohmic conductor, depends on geometric size and therefore an extensive quantity. For a uniform solution of cross-sectional area A and length L, the resistivity is given by

$$\rho = \frac{R_r A}{L}. \tag{10.31}$$

Resisitivity, which has unit Ωm, is an intensive quantity. Since we will ultimately be interested in ion mobilities and hence the ease, rather than difficulty, of current flow in the electrolyte, it is useful to introduce the quantity

$$K = \frac{1}{\rho} = \frac{L}{R_r A}, \tag{10.32}$$

where K is the 'conductivity' of the electrolyte solution.

The experimental measurement of R_r and hence K (for example by means of a Wheatstone bridge circuit) is often discussed in introductory physical chemistry textbooks.[8] The resulting values, for the case of fully dissociated electrodes, are seen to scale linearly to a good degree of accuracy with concentration, c. Accordingly it is helpful to introduce the quantity, the molar conductivity,

$$\Lambda = K/c, \tag{10.33}$$

where $\Lambda = K/c$ has units of $\Omega^{-1} \, m^2 \, mol^{-1}$.

Accurate conductivity measurement shows, in fact, that Λ has a weak dependence on concentration:

$$\Lambda = \Lambda^0 - A\sqrt{c}, \tag{10.34}$$

where A is a small constant. The quantity Λ^0 is the molar conductivity extrapolated to infinite dilution. For any electrolyte Λ^0 is the sum of two independent contributions, one characteristic of the cation, and the other of the anion:

$$\Lambda^0 = \Lambda_+ + \Lambda_- \tag{10.35}$$

Values of Λ_+ and Λ_- can be found in Table 1.1 (Chapter 1). The implication of this is that the ions move essentially independently of one another.

Equation (10.35) is a statement of Kohlrausch's law of the independent migration of ions which underpins almost all of the ideas about transport in this chapter (and book). The Einstein relationship implies the independent diffusion of ions:

$$D_i = \frac{RT\Lambda_i N_A}{Z_i F^2}. \tag{10.36}$$

Note however that ion-ion effects are included in the $\sqrt{c}$ term in Eq. (10.34). In terms relevant to voltammetry, D_i values would be reported for a specific electrolyte compensation recognising that they too will have a corresponding (weak) concentration dependence.

Lastly we note that an implication of Kohlrausch's law is that the current passing through the bulk of an electrolyte is carried to a different extent by the cations and anions. For example, in an aqueous solution of LiCl both the lithium cations and the chloride anions contribute to carrying the current. The value of molar conductivities are

$$Li^+ : \Lambda_+ = 38.7\ \Omega^{-1}\,\mathrm{cm}^2\,\mathrm{mol}^{-1}$$

$$Cl^- : \Lambda_- = 76.3\ \Omega^{-1}\,\mathrm{cm}^2\,\mathrm{mol}^{-1}.$$

It is evident that a larger part of the current is carried by the chloride anion.

It is useful to introduce the concept of the transport numbers, t_+ and t_-, which describe the fraction of the current carried by the cation and by the anion, respectively:

$$t_+ = \frac{\Lambda_+}{\Lambda_+ + \Lambda_-} \qquad t_- = \frac{\Lambda_-}{\Lambda_+ + \Lambda_-}.$$

For the case of LiCl it is apparent that $t_+ = 0.34$ and $t_- = 0.66$. In contrast, for a solution of KCl where

$$K^+ : \Lambda_+ = 73.5\ \Omega^{-1}\mathrm{cm}^2\,\mathrm{mol}^{-1},$$

it is evident that $t_+ = 0.49$ and $t_- = 0.51$ so that the two ions carry the current almost equally between themselves.

The inequality of transport between ions in solution is at the heart of the origin of liquid junction potentials as introduced qualitatively in Section 1.6. In the next section we provide a more rigorous and hopefully insightful re-examination of these potentials.

10.8 Liquid Junction Potentials[9]

At the boundary between two electrolyte solutions where ionic species have different transport numbers, a charge separation arises because the various ions diffuse at different rates. The charge separation, as noted in Section 1.6, creates an electric field in solution and this results in ion migration which accelerates the transport of the slower species but retards that of the faster moving ion(s).

Lingane classified liquid junctions into types 1, 2 and 3.[10] Type 1 liquid junctions are between two electrolyte solutions which are identical except for having different ionic concentrations. Type 2 is a junction between two solutions containing different ions but common concentrations. Type 3 covers all other junctions. Figure 10.13 illustrates types 1 and 2.

The traditional view of a liquid junction is illustrated by the schematic in Fig. 10.14 and was developed by Nernst and Planck[11−13] who reasoned that the electric field will develop in time until, considering a type 1 situation, the fluxes of the two differing ions are equal with any difference in their intrinsic diffusion rate (reflected in their diffusion coefficient) exactly balanced by their migrational

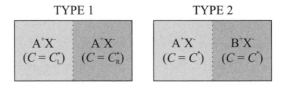

Fig. 10.13   Lingane's type 1 and type 2 liquid junctions.

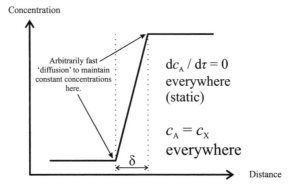

Fig. 10.14   Schematic showing the traditional 'static' view of the liquid junction. Reprinted with permission from Ref. [9]. Copyright (2010) American Chemical Society.

attraction or repulsion. Once the steady-state is established no further charge separation occurs across the junction and the potential difference across the junction is constant.

In the classical work the steady-state is thought of as being confined to a boundary layer of finite thickness with constant concentration boundaries as illustrated in Fig. 10.14. Within the boundary layer, concentration profiles and the potential all vary in a linear fashion, the solution is electro-neutral and the Nernst–Planck equation is at steady-state. Under these conditions, for a monovalent electrolyte A^+X^-, the liquid junction potential is given by

$$\Delta E_{LJP} = (t_A - t_X)\frac{RT}{F}\ln\frac{C_L^*}{C_R^*} \tag{10.37}$$

where C_L^* and C_R^* are defined in Fig. 10.13. In the type 2 for monovalent electrolytes A^+X^- and B^+X^- the potential is

$$\Delta E_{LJP} = \frac{RT}{F}\ln\left(\frac{D_A + D_X}{D_B + D_X}\right). \tag{10.38}$$

To develop a *dynamic* theory of liquid junction potential we consider a planar junction normal to the x-coordinate between the two solutions of binary monovalent electrolyte (Fig. 10.13).[9] The flux of any ion, $i = A, B$ or X is given by the Nernst–Planck equation

$$Flux = J_i = -D_i\left(\frac{\partial c_i}{\partial x} + \frac{Z_i F}{RT}c_i\frac{\partial \phi}{\partial x}\right) \tag{10.39}$$

where c_i is the concentration. Generalising the arguments in Section 3.2, mass conservation gives

$$\frac{\partial c_i}{\partial t} = -\frac{\partial j_i}{\partial x} \tag{10.40}$$

so that

$$\frac{\partial c_i}{\partial t} = D_i\left[\frac{\partial^2 c_i}{\partial x^2} + \frac{Z_i F}{RT}\frac{\partial}{\partial x}\left(c_i\frac{\partial \phi}{\partial x}\right)\right]. \tag{10.41}$$

As in Section 10.2 the potential must also obey the Poisson equation:

$$\frac{\partial^2 \phi}{\partial x^2} = -\frac{\rho}{\varepsilon_r\varepsilon_o} \tag{10.42}$$

where

$$\rho = F\sum_i Z_i c_i. \tag{10.43}$$

Equations (10.41)–(10.43) can be solved numerically.[9] The results are most simply reported using dimensionless variables:

$$\Theta = \frac{F}{RT}\phi \tag{10.44}$$

$$\chi = \frac{x}{\kappa^{-1}} \tag{10.45}$$

$$D_i' = D_i/D_X \tag{10.46}$$

and

$$\tau = \frac{D_X t}{2(\kappa^{-1})^2} \tag{10.47}$$

where κ^{-1} is the Debye length defined in Eq. (10.11).

The results of the numerical calculations confirm the results in Eqs. (10.36) and (10.37) for type 1 and type 2 systems, respectively. Further, however, they allow insights into the temporal and spatial evolution of the junction potential. For example, the dynamics of liquid junction formation were investigated for a concentration discontinuity in an aqueous solution of HCl from 1 mM to 10 mM. Figure 10.15 shows the (dimensionless) liquid junction potential, $\Delta\Theta_{LJP}$, evolving

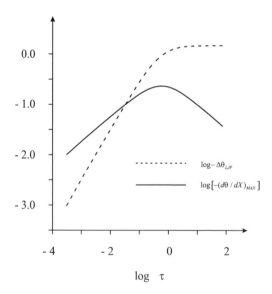

Fig. 10.15 Dynamic evolution of the liquid junction potential for a type 1 junction of 1 and 10 mM HCl. Reprinted with permission from Ref. [9]. Copyright (2010) American Chemical Society.

as a function of (dimensionless) time, $\tau = \frac{D_x t}{2(\kappa^{-1})^2}$ on a logarithmic scale. Also shown in Fig. 10.15 is the maximum electric field, $(-\frac{\partial \Theta}{\partial x})_{max}$. Note that the latter is not the field at $x = 0$ corresponding to the original location of the junction but is found to diffuse with the species themselves.

The conversion of Fig. 10.15 to dimensional form shows that the electric field resulting from the unequal mass transport of Li^+ and Cl^- ions achieves a maximum at $\tau \sim 0.5$ corresponding to a field of ca. $1.3\,MVm^{-1}$ at a time $\sim 5\,ns$ after contact between the solutions. Prior to this time the potential difference increases proportionally to τ and the maximum field proportionally to $\tau^{1/2}$. The potential difference approaches the limiting value predicted by Eq. (10.36) at long times but the electric field after passing through a maximum relaxes at $\tau^{1/2}$. The maximum electric field scales inversely with κ^{-1}.

Figure 10.16 (A–E) shows the concentration profiles of H^+ and Cl^- at different times relative to that for the creation of the maximum field τ_{trs}.

Concentration profiles are shown for a logarithmic range of time from $10^{-2}\tau_{trs}$ to $10^2\tau_{trs}$. Figure 10.17 shows the evolution of the associated electric field. It is apparent that the location of the maximum field is mobile and varies away from the initial position of the junction. At the same time the concentration profiles become more and more asymmetric.

Simulations such as Fig. 10.17 present a dynamic picture underpinning the liquid junction concept. The steady liquid junction potential arises from charge separation resulting from unequal rates of ionic diffusion. The charge separation creates an electric field that changes the rates of transport and allows the liquid junction to begin to discharge towards this state of electro-neutrality throughout the system. However the latter is only attained at infinite time; continuing diffusion (from high to concentration) causes the junction to grow at a rate equal and opposite to its discharge so that a steady potential difference arises. The steady potential arises on the time scales of 10–1000 ns after junction formation for typical aqueous systems. At this point the zone of the liquid junction has expanded to 10–1000 nm and continuously expands at a rate proportional to $\sqrt{\tau}$. Figure 10.18 summarises the physical picture.

10.9 Chronoamperometry and Cyclic Voltammetry in Weakly Supported Media

In Section 3.5 we saw that if a potential step was applied to a macro-electrode from a potential of no current flow to one corresponding to diffusion-controlled electrolytes, then for the case of a simple oxidation or reduction (with no

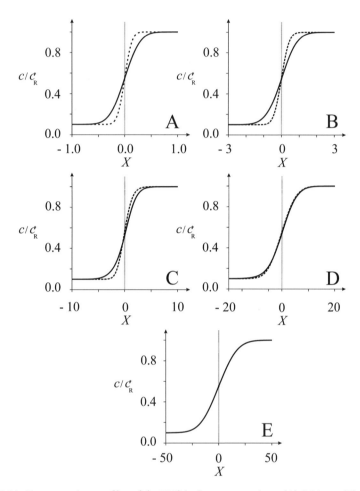

Fig. 10.16 Concentration profiles of the HCl (aq) systems at times (A) $0.01\tau_{trs}$. (B) $0.1\tau_{trs}$, (C) τ_{trs}, (D) $10\tau_{trs}$ and (E) $100\tau_{trs}$. Note the increasing asymmetry of the profile with time up to τ_{trs}. After this time symmetry begins to return, along with an *apparent* return to electro-neutrality In fact a finite charge separation exists and maintains a steady potential difference as the junction grows. The point of iso- concentration continuously diffuses away from $x = 0$. Reprinted with permission from Ref. [9]. Copyright (2010) American Chemical Society.

coupled homogeneous chemistry) the current decreased continuously so that it was inversely proportional to the square root of the time (the Cottrell equation). On the other hand, for a microelectrode the initial decrease continued not to zero but until a steady-state current was attained (Section 5.1), reflecting transition from

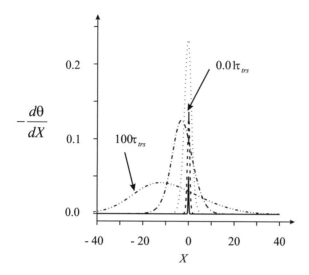

Fig. 10.17 Evolution of the electric field for the HCl (aq) systems of Fig. 10.16. Reprinted with permission from Ref. [9]. Copyright (2010) American Chemical Society.

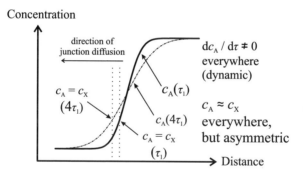

Fig. 10.18 The dynamic model of the liquid junction. Reprinted with permission from Ref. [9]. Copyright (2010) American Chemical Society.

planar to convergent diffusion. These behaviours were deduced on the assumption that transport was by diffusion only and hence it was implicit that the solution contained sufficient electrolyte to be 'fully supported'.

It is of interest to consider the effects of partial or weak support on chronoamperometry.[14,15] Figure 10.19 shows three chronoamperograms recorded using a 300 μm gold hemisphere electrode for the oxidation of ferrocene in acetonitrile:

$$Cp_2Fe - e^- \longrightarrow Cp_2Fe^+.$$

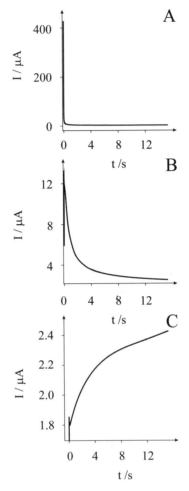

Fig. 10.19 Chronoamperograms for potential steps from open circuit to +500 mV (versus Ag/Ag$^+$) for the oxidation of ferrocene (3 mM) in acetonitrile with various support ratios: (A) SR = 33.3, (B) SR = 0.33, (C) SR = 0.033. Reprinted with permission from Ref. [14]. Copyright (2009) American Chemical Society.

Varying quantities of tetra-n-butylammonium perchlorate were used as a supporting electrolyte and under 'fully supported' condition a formal potential of 98 mV (versus Ag/Ag$^+$) was measured. The chronoamperograms were recorded by stepping from the open circuit (no current) to a potential of +500 mV (Ag/Ag$^+$) using differing support ratios, SR, of supporting electrolyte concentration to ferrocene concentration of 33.3, 0.333, and 0.033, respectively.

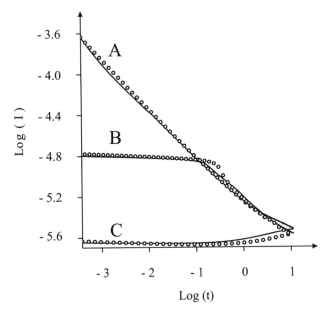

Fig. 10.20 Comparison between simulation (circle) and experiment (line) for the potential steps shown in Fig. 10.19: (A) SR = 33.3, (B) SR = 0.33 and (C) SR = 0.033. Note that the curves are plotted in log–log form. Reprinted with permission from Ref. [14]. Copyright (2009) American Chemical Society.

Figure 10.19(A) shows the expected behaviour for a high support ratio. The current depends inversely on the square root of time except at long times where the approach to a steady state limiting current begins to set in.

However, the other transients, measured at lower support ratios, show significant migration effects and hence deviation from the expected diffusion-only behaviour. Figure 10.20 shows that a model based on the Nernst–Planck equation can quantitatively describe this.[14,15] The simulations permit insights into the causes of the behaviour for the transient. In particular, Fig. 10.21 shows how the current relates to the driving force $(\phi_M - \phi_S)$ at a point adjacent to the electrode. Specifically curve (A) shows the diffusion only behaviour seen for SR = 33.3. For the lower values of SR the driving force is initially too small to induce electrolysis since a diffuse layer cannot be instantly created. After some time (*ca.* 0.1 s and 3 s for SR = 0.33 and 0.033, respectively) transport of the charge species has led to an excess of ClO_4^- and a depletion of tetra-n-butylammonium cation near the electrode. Corresponding to this a driving force for electron transfer is developed and eventually a diffusion-like transient is attained.

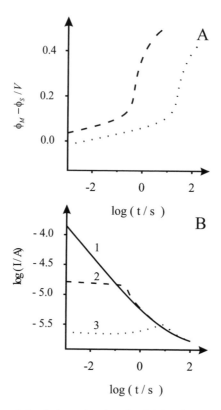

Fig. 10.21 The results of simulations showing the driving force ($\phi_M - \phi_S$) for electron transfer as a function of time (A) along with the transients considered in Figs. 10.19 and 10.20 (SR = 33, 0.33 and 0.033). Reprinted with permission from Ref. [14]. Copyright (2009) American Chemical Society.

Thus for curve (B) (SR = 0.33) there is a period at short times (< 0.15) where the response is 'ohmic drop' controlled. For curve (C) (SR = 0.033) this period is longer as a greater time is required to build up interfacial excesses and depletions of ions.

Figure 10.22 shows the cyclic voltammetric responses for the three systems hitherto examined by potential step transients. The effect of 'ohmic drop' is apparent and considerable. Analogous to the delayed onset of diffusional behaviour in the chronoamperometry the voltammetric peak appears at much more positive potentials for low values of SR corresponding to the increased time required to develop a suitable driving force at the electrode-solution interface. Simulations have been performed to estimate the support ratio required to obtain accurate voltammetry

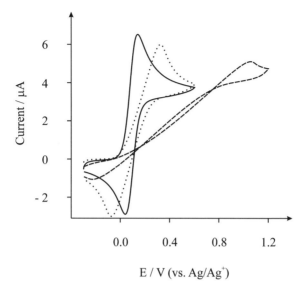

Fig. 10.22 Cyclic voltammetry for ferrocene oxidation in acetonitrile at different support ratios: SR = 33.3 (solid line), SR = 0.33 (dashed line) and SR = 0.033 (dotted line). Reprinted with permission from Ref. [14]. Copyright (2009) American Chemical Society.

in the strongly supported limit for reversible voltammetry (fast electron kinetics).[16] It was concluded that for transient cyclic voltammetry, macroelectrode systems require SR > 100 to avoid detectable peak broadening for ohmic drop. The much lower currents drawn by microelectrodes make these much less susceptible to ohmic drop.

References

[1]   S. Trasatti, *Pure Appl. Chem.* **58** (1986) 955.
[2]   L.-G. Gouy, *Compt. Rend.* **149** (1909) 654.
[3]   L.-G. Gouy, *J. Phys.* **9** (1910) 457.
[4]   D.L. Chapman, *Phil. Mag.* **25** (1913) 475.
[5]   W.J. Albery, *Electrode Kinetics*, Clarendon Press, Oxford, 1975.
[6]   A.N. Frumkin, *Z. Phys. Chem.* **164A** (1933) 121.
[7]   N.V. Nikolmeua-Federouch, B.N. Rybakou, K.A. Rudyushkiun, *Soviet Electrochemistry* **3** (1967) 967.
[8]   P.W. Atkins, *Physical Chemistry*, 3rd edn., Oxford University Press, Oxford, 1986; W.J. Moore, *Physical Chemistry*, 5th edn., Longman, London 1972.
[9]   E.J.F. Dickinson, R.G. Compton, *J. Phys. Chem. B.* **114** (2010) 187.
[10] J.J. Lingane, *Electroanalytical Chemistry*, 2nd edn., Wiley, New York, 1958.

[11] W.H. Nernst, *Z. Phys. Chem.* **4** (1889) 165.

[12] M. Planck, *Wied. Ann.* **39** (1890) 161.

[13] M. Planck, *Wied. Ann.* **40** (1890) 561.

[14] J.G. Limon-Petersen, I. Streeter, N.V. Rees, R.G. Compton, *J. Phys. Chem. C.* **113** (2009) 333.

[15] I. Streeter, R.G. Compton, *J. Phys. Chem.* **112** (2008) 13716.

[16] E.J.F. Dickinson, J.G. Limon-Peterson, N.V. Rees, R.G. Compton, *J. Phys. Chem. C.* **113** (2009) 11157.

11

Voltammetry at the Nanoscale

In Section 5.10 we considered the fabrication of 'nanodes' — electrodes with characteristic dimensions in the nanometre range. A full account of methods for making such electrodes has been given by Arrigan.[1] 'Nano electrode arrays' have also been made by supporting pre-formed nanoparticles on an electrode surface.[2] Such structures are now widely used in electroanalysis, the supporting electrode serving to make electrical contact with the nanoparticles but the electrolysis at the potential of interest, at least, being confined to the surface of the nanoparticles. We consider first the case of well-separated (diffusionally isolated) nanoparticles.

11.1 Transport to Particles Supported on an Electrode

Figure 11.1 shows well-supported spherical or hemispherical particles supported on an electrode together with the coordinate system used for their description. We assume that the planar electrode is conductive and makes electrical contact with the particles but that the electrode is electrolytically inactive at the potentials of interest so that the electron transfer is only kinetically feasible on the surface of the particles.

We assume also that the particles are sufficiently far apart as to be diffusionally independent.

For the case of the simple electrode reaction

$$A + ne^- \rightleftarrows B$$

the diffusion-limited current at a hemispherical particle is

$$I = 2\pi nFDr[A]_{Bulk} \qquad (11.1)$$

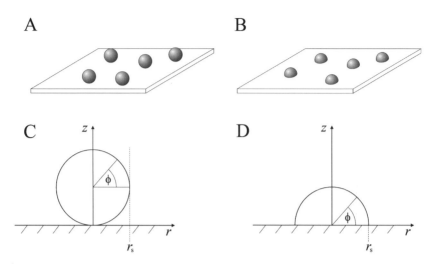

Fig. 11.1 Well-separated particles on an electrode: (A) spherical, (B) hemispherical and schematic diagram of (C) a spherical particle and (D) a hemispherical particle sitting upon a supporting planar surface. Reprinted with permission from Ref. [3]. Copyright (2007) American Chemical Society.

where r is the radius of the hemispherical particle and D is the diffusion coefficient of species A. Equation (11.1) has been derived assuming only Fickian diffusional transport (but see below). The corresponding result for a spherical particle is[3]

$$I = 8.71 nFDr[A]_{Bulk}. \tag{11.2}$$

For an isolated spherical electrode (see Section 5.1) of course,

$$I = 4\pi nFDr[A]_{Bulk} \tag{11.3}$$

corresponding to twice the value for a hemisphere of the same radius. Note that

$$8.71 < 4\pi$$

since the underlying electrode 'shields' the sphere from the full extent of the diffusion that would be seen in the absence of the electrode (see Fig. 11.2).

The limiting currents quantified above are only seen at sufficiently slow voltage scan rates. For faster scan rates, by analogy with the behaviour of microdisc electrodes (Section 5.6), we would expect peak-shaped voltammetry to develop. The transition from steady-state convergent to transient, almost planar diffusion is controlled by a dimensionless scan rate,

$$\sigma = \left(\frac{F}{RT}\right)\left(\frac{\upsilon r^2}{D}\right) \tag{11.4}$$

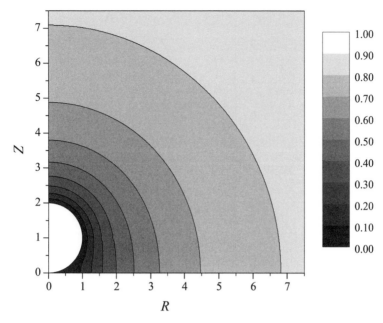

Fig. 11.2 Simulated concentration profile at a spherical particle on an electrode under diffusion-limiting conditions. Z and R are dimensionless distances of the cylindrical coordinates z and r, relative to the sphere radius. Reprinted with permission from Ref. [3]. Copyright (2007) American Chemical Society.

where υ is the scan rate (Vs^{-1}). Figure 11.3 shows voltammetry simulated for three different values of σ : 10^{-3}, 1 and 10^3. The transition between the two expected limits is apparent.

Figure 11.4 shows the concentration profiles at the voltammetric peak potentials in the two cases. The change from convergent to planar diffusion is implicit in these figures.

Expressions for the Fickian diffusion-only transported controlled current to other shapes of particles, notably distorted spheres and hemispheres, have been given.[3]

Consideration of the above suggests that the particle radius will have a qualitatively similar effect on the voltammetric response of an isolated particle as does the radius of a microdisc electrode, as detailed in Chapter 5. It is therefore to be expected that size-dependent diffusion effects will be observed in nanoparticle voltammetry.

An example of the effect of particle size on the voltammetric response can be illustrated with reference to the reduction of hydrogen peroxide under acidic

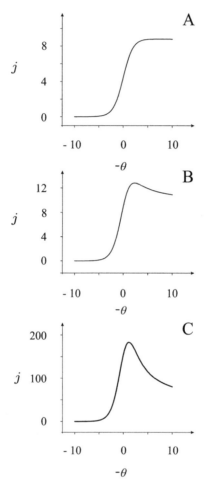

Fig. 11.3 Simulated voltammetry for a reversible electrode transfer at a spherical particle supported on a planar electrode. The following scan rates are used (A) $\sigma = 10^{-3}$; (B) $\sigma = 1$; (C) $\sigma = 10^3$. Reprinted with permission from Ref. [3]. Copyright (2007) American Chemical Society.

conditions at silver nanoparticles.[4]

$$H_2O_2 + H^+ + e^- \xrightarrow{slow} {}^\bullet OH(ads) + H_2O$$

$${}^\bullet OH(ads) + H^+ + e^- \xrightarrow{fast} H_2O$$

This reaction has been studied using both macro-silver electrodes and at well-separated nanoparticles of size ranges 25–40 nm, 55–75 nm and 80–120 nm. At

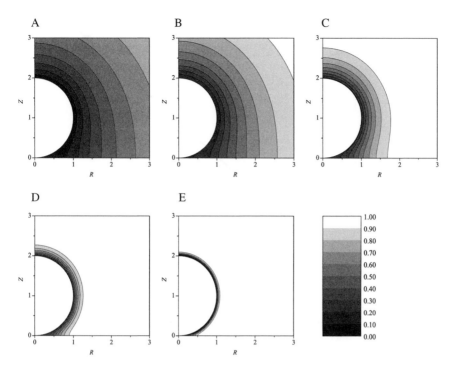

Fig. 11.4 Simulated concentration profiles at the spherical particle on a plane: (A) $\sigma = 0.1$; (B) $\sigma = 1$; (C) $\sigma = 10$; (D) $\sigma = 100$; (E) $\sigma = 1000$; R and Z are defined in the legend to Fig. 11.2. Reprinted with permission from Ref. [3]. Copyright (2007) American Chemical Society.

the macroelectrode a transfer coefficient of $\alpha = 0.25$ was observed. Figure 11.5 shows how the measured peak potential varies with radius for the latter nanoparticles.

It can be seen that

$$\frac{\partial E_p}{\partial \log_{10} r} = 233 \, \text{mV}. \tag{11.5}$$

Assuming that the nanoparticles behave as isolated hemispheres, it can be predicted that

$$\frac{\partial E_p}{\partial \ln r} = \frac{RT}{\alpha F}; \quad \frac{\partial E_p}{\partial \log_{10} r} = \frac{2.3RT}{\alpha F}, \tag{11.6}$$

giving a value of α consistent with the macroelectrode data and suggesting that for nanoparticles, of this size at least, the electrode reaction mechanism is unchanged from the macro- to the nanoscale.

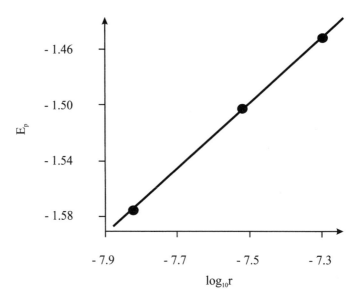

Fig. 11.5 Reduction of H_2O_2 at silver nanoparticles modified electrode: E_p versus $Log_{10}r_{(avg)}$ for isolated nanoparticles; $\alpha = 0.253$. Reprinted with permission from Ref. [4]. Copyright (2009) American Chemical Society.

Size-dependent peak potentials have also been predicted theoretically, and observed experimentally, for the case of the stripping voltammetry of diffusionally isolated nanoparticles,[5] for example in the case of silver nanoparticles

$$Ag(np) - e \longrightarrow Ag^+(aq).$$

Particle size effects are expected for both electrochemically reversible and irreversible kinetics with more positive peak potentials seen for larger particles at a fixed scan rate given the more the amount of metal to be oxidised, the larger the scan has to sweep before exhaustive oxidation takes place.

When the coverage of particles is sufficient to allow the overlap of adjacent diffusion fields then, depending on the voltage scan rates, categories of behaviour akin to those described for arrays of microelectrodes (see Section 6.2) are observed. Figure 11.6 illustrates the different situations[6] the dimensionless scan rate is as defined in Eq. (11.4).

It can be seen that if the centre-to-centre spacing between adjacent particles is four times their radius, for $\sigma = 10^3$ the voltammetric response will be that of near linear diffusion to isolated microelectrodes and peak-shaped voltammograms will result (Category 1). If the scan rate is decreased to $\sigma = 10$ then

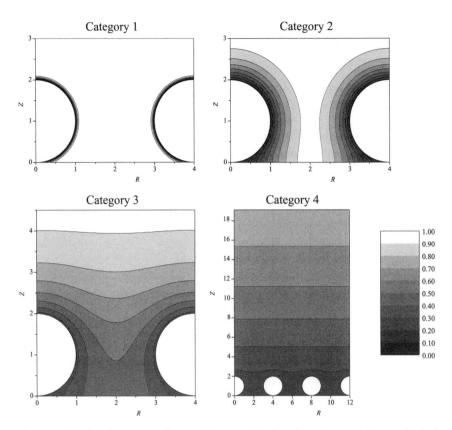

Fig. 11.6 Simulated concentration profiles at a diffusion domain containing a spherical particle. Category 1: $\sigma = 1000$. Category 2: $\sigma = 10$. Category 3: $\sigma = 1$. Category 4: $\sigma = 0.01$. For all categories $R_0 = 2$. Concentration profiles were taken at the linear sweep's peak potential. Reprinted with permission from Ref. [6]. Copyright (2008) American Chemical Society.

approximate convergent diffusion to the isolated particles will be observed and nearly sigmoidal-shaped voltammograms will be seen with well-defined limiting currents (Category 2). As the scan rate is decreased further to 1 (Category 3) or 0.01 (Category 4) diffusional overlap occurs and in the Category 4 limit linear diffusion to the electrode supporting the entire array of electrodes is seen with voltammograms characteristic of linear diffusion with the peaked response expected of a macroelectrode. The waveshape expected for the Category 3 situation is intermediate between the sigmoidal steady-state response of isolated particles and the linear diffusional behaviour of Category 4 viz, a reduced back peak and a weaker fall-off of current after the current peak as compared to planar diffusion.

Of course the transitions between Categories 1, 2, 3 and 4 depend on the average distance between the particles, relative to their radius, as well as the dimensionless scan rate. The coverage of an electrode by spherical particles can be defined by

$$\theta = \frac{N\pi r^2}{A}, \tag{11.7}$$

where N is the number of particles of radius r on the supporting electrode of area A. Figure 11.7 shows how, for a fixed scan rate of $\sigma = 10^{-2}$, the voltammetry

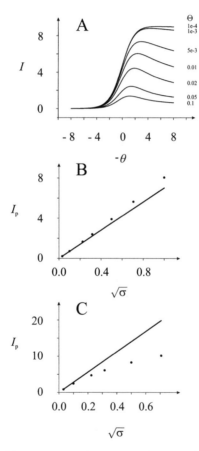

Fig. 11.7 (A) Simulated linear sweep voltammetry of a reversible electron transfer at a spherical particle modified electrode. Scan rate $\sigma = 0.01$, θ varies from 10^{-4} to 0.1. Peak current, I_p versus square root of the scan rate $\sigma^{1/2}$. Simulated data is shown as circles and the solid shows the Randles–Sevćik values for planar diffusion. (B) $\theta = 0.2$ and (C) $\theta = 0.05$. Reprinted with permission from Ref. [6]. Copyright (2007) American Chemical Society.

varies with θ. It can be seen that there is a transition from Category 2 to 3 to 4 as σ increases.

In contrast, for many applications, including those of electroanalysis, the coverage of the electrode with nanoparticles is likely to be relatively high so that responses close to those expected for the underlying supporting electrode, were it electrolytically active, would be expected. That said, if multi-layers, rather than a sub-monolayer of nanoparticles, are used to modify the electrode then porosity effects can alter the response as outlined in Section 6.5.

11.2 Nanoparticle Voltammetry: The Transport Changes as the Electrode Shrinks in Size

The equations and results presented in the previous section were derived assuming Fickian diffusion under fully supported conditions. If this is valid the conclusions will hold for particles of any radius r. Results were presented for nanoparticles of some tens of nanometres in size or larger and for these the assumptions are likely to be realistic. However, if one considers nanoparticles of a smaller size then the model will fail.

It fails for a variety of reasons. First, the application of the continuum approach implicit in Fick's laws of diffusion assumes that both the numbers of molecules are considered to be sufficiently large so as to remove statistical fluctuations, and also that the distances involved are large compared to molecular dimensions. As an electrode, and correspondingly its diffusion layer, is shrunk below 5–10 nm these assumptions can become compromised and the continuum description is replaced by one recognising individual molecular events. The seminar work of Bard and co-workers reported the measurement of the electrolytic current resulting from a single electroactive molecule trapped between two electrodes.[7,8] The experiment utilised a scanning electrochemical microscope (Section 6.7) with a small tip electrode of nanometre scale. The tip was insulated, as shown in Fig. 11.8, so that when brought up to the surface of a second substrate electrode, for example made of Indium Tin Oxide (ITO), the geometry of the tip and insulator provided a confinement volume of *ca* 10^{-18} cm^3 beneath the tip. If then solutions of concentrations of *ca.* 10^{-3} M are studied it is likely that on average just one molecule will be present in the confinement area.

The Einstein equation can be used to estimate that with a tip to substrate separation of *ca.* 10 nm, a molecule with a diffusion coefficient of 5×10^{-6} cm^{2}s^{-1} will take *ca.* 100 ns to diffuse between the two electrodes corresponding to 10^7 round trips per second. As the charge on a single electron is 1.6×10^{-19} C then if a

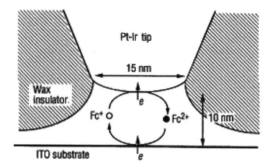

Fig. 11.8 Idealised schematic illustration of the top geometry and the tip-substrate config-uration used. Reprinted from Ref. [7] with kind permission of the AAAS.

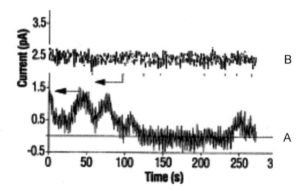

Fig. 11.9 Time evolution of the tip current observed at a tip potential of 0.55 V and a substrate potential of -0.3 V versus SCE. Curve A: tip-substrate separation of $\sim$10 nm in a solution containing 2 mM Cp_2FeTMA^+ and 2.0 M $NaNO_3$. Curve B: with the tip far from the substrate in the same solution as in curve A. Reprinted from Ref. [7] with kind permission of the AAAS.

redox event occurs at each collision with the tip, a current in the range of 10^{-12} A (picoAmps, pA) will flow.

Bard's experiments used a water-soluble ferrocene species, [(trimethylammo-nio)methyl] ferrocene, Cp_2FeTMA^+, (2mM) which underwent one electron oxi-dation at the tip. Figure 11.9 shows the current flowing for a tip potential of $+0.55$ V (vs. SCE) corresponding to the diffusion-controlled oxidation of the ferrocene species, and a substrate potential of -0.3 V (vs. SCE).

Imposed on the noise are peaks of 0.7 and 1.4 pA in addition to periods of zero average current. In the words of the authors: 'We believe that these represent current responses when one or two Cp_2FeTMA^+ molecules are trapped in the

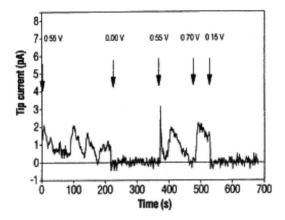

Fig. 11.10 Time evolution of tip current with a substrate potential of −0.3 V versus SCE at various tip potentials ET (indicated above the arrows). We set the initial tip current at −1.5 pA by adjusting *d*. The solution contained 2 mM Cp_2FeTMA^+ and 2.0 M $NaNO_3$. The data sampling rate was 1 s per point. Reprinted from Ref. [7] with kind permission of the AAAS.

10 nm gap between the tip and substrate and drift into or out of the tip region.'[7] In Fig. 11.9 curve B corresponds to the tip located far from the substrate and shows a consistent average current over a 300 s timescale. Figure 11.10 shows the effect of changing the tip potential between a value corresponding to the oxidation of the ferrocene (0.55 V, 0.70 V) and a value where the molecule is not oxidised (0.00 V, 0.15 V). The features attributed to the single molecule oxidation are only seen at the former potentials.

Finally, the experiments show the transition from single molecule to near continuum behaviour. Figure 11.11 shows current–voltage curves measured at different tip-substrate separations. Curve 1 corresponds effectively to bulk solution; curve 3 to the geometry used to record single molecule electrochemistry in Figs. 11.9 and 11.10; curve 2 was recorded with a *ca.* 15 nm greater separation between the tip and the substrate than curve 2. The transition from single to multiple molecule voltammetry is evident.

The transition of diffusional response from these of a few molecules to those predicted from the Fickian continuum model has also been addressed by Amatore *et al.*[9,10] They synthesised the electroactive fourth-generation PAMAM dendrimer capped by 64 ruthenium (II) bis-terpyridine redox moieties shown in Fig. 11.12.

They adsorbed *ca.* 10^6 such molecules onto an ultra-microelectrode and first studied the cyclic voltammetry at scan rates in excess of 1 MVs^{-1} (see Section 5.9). The resulting voltammograms are shown in Fig. 11.13.

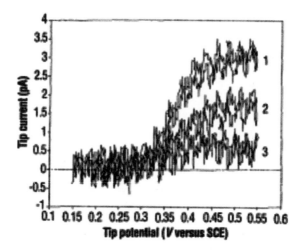

Fig. 11.11 Series of cyclic voltammograms taken at different tip-substrate separations in a solution containing 2 mM Cp_2FeTMA^+ and 2.0 M $NaNO_3$. The substrate potential was -0.3 V versus SCE, and the scan rate of the tip potential was 10 mV/s. Reprinted from Ref. [7] with kind permission of the AAAS.

The voltammetry switches between that expected for an adsorbed layer at low scan rates with the peak current, I_p, proportional to the voltage scan rate

$$I_p \propto v,$$

and with the forward and back peaks separated by only a small difference in potential, to a diffusional regime where

$$I_p \propto v^{1/2}$$

and a significant peak-to-peak voltage separation seen at faster scan rates. In Fig. 11.14 a plot of ($I_p \propto v^{1/2}$) against $v^{1/2}$ shows this transition.

The changeover reflects whether there is time for entire dendrimers to become fully oxidised, as happens at low scan rates, or whether the timescale of the voltammetry is so short that the charge is 'diffusing' (by hopping from redox centre to redox centre) over the surface of the dendrimers. This last situation occurs at the fastest scan rates and 'diffusional' voltammetry is seen. The 10^6 dendrimers on the electrode were (more than) enough to permit an accurate Fickian treatment of the charge diffusion. The authors then considered the chronoamperometric response that would result from different numbers of molecules adsorbed on the electrode surface. Theoretical calculations based on a random jump (stochastic) model of the electron transfer in comparison with the (statistical) Fickian limit

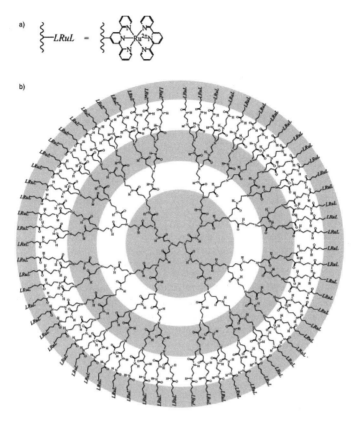

Fig. 11.12 (a) The ruthenium bisterpyridyl moiety [Ru(tpy)$_2$] (Ru). (b) The structure of the fourth generation PAMAM dendrimer with 64 pendant ruthenium bisterpyridyl moieties (Dend-Ru$_{64}$). The four inner concentric shaded areas represent each dendritic generation; the outermost represents the 64 Ru redox centers. Reprinted from Ref. [9] with permission from Wiley.

produced the results shown in Fig. 11.15 for different numbers of dendrimers on the electrode surface ranging from 1 to 7800 and correspondingly different electrode sizes from 5.5 nm to 510 nm. The difference between the stochastic and statistical results decreases rapidly as the number of dendrimers and electrode sizes increase.

We now return to a consideration of the reasons why Fickian diffusion does not accurately describe the rate of transport to small nanoparticles or nanodes.

A very significant change which occurs as an electrode is shrunk below the size of 5–10 nm is that the diffusion layer, typically of the order of the size of the electrode radius, becomes comparable in size with the diffuse layer surrounding the electrode even in the presence of quantities of (supporting) electrolyte which

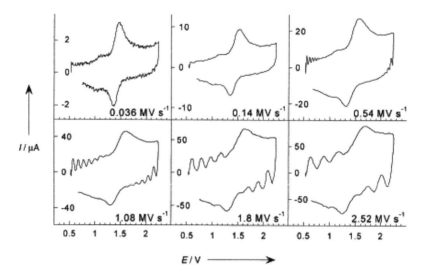

Fig. 11.13 A representative series of cyclic voltammograms of a saturated solution of Dend-Ru$_{64}$. Scan rates for each voltammogram are given at the bottom of each panel. Voltammetry was performed at 20°C in 0.6 M Et$_4$NBF$_4$/acetonitrile at a platinum disk electrode 5.0 ± 0.5 μm in radius. All potentials are versus a platinum pseudo-reference electrode. Reprinted from Ref. [9] with permission from Wiley.

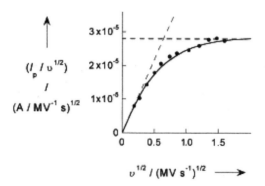

Fig. 11.14 Variation of the scan-rate normalized, anodic-peak current intensity $I_p/v^{1/2}$ as a function of scan rate $v^{1/2}$. Points are experimental data and the line is the predicted behaviour. Reprinted from Ref. [9] with permission from Wiley.

at a macroelectrode would ensure that the diffuse layer was very much smaller than the diffusion layer (see Chapter 10). As a result the physical picture associated with electrolysis is changed. Whereas at the macroelectrode, under well-supported conditions, the transport of the electroactive species to a distance near to the

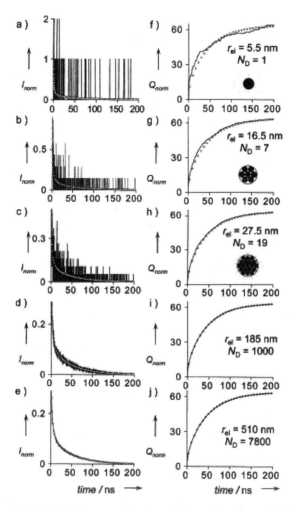

Fig. 11.15 (a–e) Vertical thin lines: predicted variations of the normalized stochastic current (I_{norm}) versus time as a function of the number of dendrimer molecules adsorbed onto the electrode surface without considering any instrumental distortions. In each panel, $I_{norm} = I/N_D$ represents the current, I, normalized by the number of dendrimers, N_D. Vertical units are chosen so that one electron transferred during the time window interval of the simulation gives a current $I_{norm} = 1$ for $N_D = 1$, so that I_{norm} reflects the sequential electron count for the average single molecule during one 0.1 ns time window. These variations are superimposed onto the statistical electrochemical normalized current predicted for an infinite number of adsorbed dendrimers (open circles; same curve in each panel a–e). (f–j) Same as (a–e) but the electrochemical Faradaic charge ($Q_{norm} = ne/N_D$) normalized per dendrimer is considered, in which ne is the actual number of electrons transferred from the beginning of the experiment. $N_D = 1$ (a,f), 7 (b,g), 19 (c,h), 1000 (d,i), 7800 (e,j). Reprinted from Ref. [10] with permission from Wiley.

electrode comparable with electron transfer via tunnelling occurs exclusively via diffusion, this is no longer the case when the diffusion layer is of the same scale as the diffuse layer (Debye length). At the well-supported macroelectrode the species undergoing oxidation or reduction is transported across the diffusion layer without experiencing an electric field or potential due to the electrode; the latter is screened from the diffusing molecule until it reaches close to the electrode. For the nano-electrode in contrast, the transport of the species occurs both via diffusion and, if the species is charged, by migration. The removal of products from the electrode is similarly influenced. Of course in any electrode reaction at least one of the reactants or products must be electrically charged so that the observed voltammetry will necessarily reflect transport by migration as well as diffusion.

A further effect is associated with the fact that molecules being transported to a small nano-electrode will experience the electrode potential and so, if they are charged, the concentration distributions of the species in solution, even without electrolysis, will be changed from that in bulk solution. In other words the screening of the electrode charge by the supporting electrolyte which ensures that at a macroelectrode a molecule in solution is 'unaware' of the electrode charge is in contrast lost at the nanoscale so that the attractive and repulsion of charged species can be significant. It is important to recognise that these 'population effects' can lead to major concentration differences from bulk solution for electrode potentials which deviate even modestly from the Potential of Zero Charge (PZC).

Finally we note that for very tiny electrodes or nanoparticles there can exist quantised charging and Coulomb staircase effects.[11–14] These effects can be illustrated by the following two seminal experiments.

Murray and co-workers studied electron transfer to 28 kDa core mass, hexanethiolate-protected gold cluster molecules.[11] A scanning tunnelling microscope (STM) tip was used to address individual clusters adsorbed on a gold-on-mica substrate under ultrahigh vacuum conditions at 83 K. Figure 11.16 shows a resulting current–voltage (I–V) curve for a single cluster; a 'Coulomb staircase' is observable with six charging steps regularly spread over the potential range studied.

Note that in the STM experiment it is the tunnelling current (I) between the tip and substrate which is recorded as a function of the voltage (V) applied between the two. In the words of Murray *et al.*: 'Each "step" in the staircase occurs at a particular bias voltage where it is energetically possible for an additional electron to reside on the cluster, and the current is very sensitive to the charge of this "middle electrode".'

Comparison was made with the cyclic (CV) and different pulse voltammetry (DPV, see Section 9.2) of the cluster dissolved (0.1 mM) in a solvent of toluene and acetonitrile with 0.05 M supporting electrolyte, as shown in Fig. 11.16, panel B.

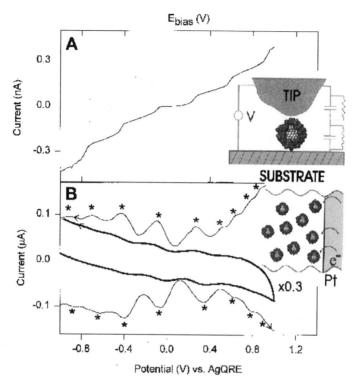

Fig. 11.16 Panel A: Au STM tip addressing a single cluster adsorbed on an Au-on-mica substrate (inset) and Coulomb staircase I–V curve at 83 K; potential is tip-substrate bias; equivalent circuit of the double tunnel junction gives capacitances $C_{upper} = 0.59$ aF and $C_{lower} = 0.48$ aF. Panel B: Voltammetry (CV, 100 mV/s; DPV, are current peaks, 20 mV/s, 25 mV pulse, top and bottom are negative and positive scans, respectively) of a 0.1 mM 28 kDa cluster solution in 2:1 toluene:acetonitrile/0.05 M Hx$_4$NClO$_4$ at a 7.9×10^{-3} cm^2 Pt electrode, 298 K, Ag wire pseudo reference electrode. Reprinted with permission from Ref. [11]. Copyright (1997) American Chemical Society.

The CV shows peaks which are better resolved in the DPV where up to nine peaks are seen, with the DPV voltammogram qualitatively reflecting the tunnelling data in Fig. 11.16, panel A. Note that in the latter experiments the signals result from the large number of clusters able to undergo electron transfer with the electrode: the voltage shows an 'ensemble Coulomb staircase'.

A second pioneering experiment in this area was carried out by Fan and Bard who achieved an 'electrochemical Coulomb staircase' using nanometre-sized electrodes.[14] The experimental arrangement is shown in Fig. 11.17 where R and O correspond to (Cp$_2$Fe)TMA$^+$ and (Cp$_2$Fe)TMA^{2+}, respectively.

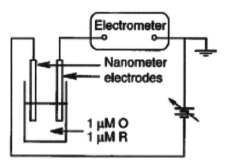

Fig. 11.17 Experimental setup for the electrochemical Coulomb staircase measurements. Reprinted from Ref. [14] with kind permission of the AAAS.

Experimental data is shown in Fig. 11.18 for a system with Ir-Pt ultra-microelectrodes of estimated radii 2.5 and 3.2 nm about 2.5 cm apart, with $1 \, \mu M$ concentration of the cations and also of NH_4^+ and SO_4^{2-} with $2 \, \mu M \, PF_6^-$. Note that the redox molecules serve as electron donors and acceptors and also along with the other ions to carry the charge between the two electrodes. Figure 11.18(A) shows the staircase shape which is emphasised if the data is plotted in differential form (dI/dV) as in Fig. 11.18(B). The peak spacing ΔV_{pp} was ~ 65 mV; a semi-classical model of the Coulomb staircase will give

$$\Delta V_{pp} = \frac{e}{C}$$

where e is the electron charge and C is the capacitance of the interface. The data in Fig. 11.18 suggests a relative value of C of around $10 \, \mu F \, cm^{-2}$.

11.3 Altered Chemistry at the Nanoscale

The previous section has illustrated the challenges of interpreting voltammetry at the nanoscale: coupled diffusion and migration, the breakdown of the Fickian laws of diffusion at the molecular scale, the possibility of quantum effects — all of which may need consideration when using nanodes or nanoparticle arrays. In addition, of course, the change to the nano-dimension may produce altered chemistry and the search for effective catalysts and electro-catalyst, for example for fuel cells, is a major driving force behind nano-electrochemistry. Such chemical effects at the nanoscale can derive from the altered electronic structures of nanomaterials compared to their bulk counterparts, and also from the changed surface of small particles compared to bulk materials. Identifying these chemical effects requires an appreciation of the physical effects described in the previous two sections.

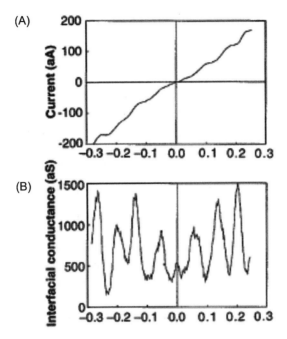

Fig. 11.18 (A) Experimental I–V characteristics of a two-interface system consisting of a pair of electrodes with radii of 2.5 and 3.2 nm immersed in a deaerated solution containing 1 μM each of $(Cp_2Fe)TMA^+$ and $(Cp_2Fe)TMA^{2+}$, NH_4^+ and SO_4^{2-} with 2 μM PF_6^-. (B) The corresponding differential conductance (dI/dV)–V plot. Reprinted from Ref. [14] with kind permission of the AAAS.

Perhaps the best example of change between the bulk and nanoscale is shown by the catalytic behaviour of gold.[15,16] Macroscale gold is relatively non-reactive with a limited chemistry. However, in the nanocrystalline form, as particles of a few hundred atoms, it becomes surprisingly active as a heterogeneous catalyst, for example with respect to selective oxidation reactions such as the epoxidation of alkenes, the oxidation of alcohols and the formation of hydrogen peroxide from molecular oxygen.[15,16] In the context of electrochemistry, gold atomic clusters[17] have been shown to bring about the four electron reduction of oxygen to water

$$O_2 + 4H^+ + 4e^- \longrightarrow 2H_2O,$$

whereas at bulk gold the predominant route is the two-electron pathway to hydrogen peroxide

$$O_2 + 2H^+ + 2e^- \longrightarrow H_2O_2.$$

Silver shows electrochemical differences between the nano- and macro-scales with altered rates of hydrogen evolution

$$H^+ + e^- \longrightarrow 1/2 H_2$$

being reported[17–20] along with the observations that the under potential deposition (upd) of metals such as Tl, Pb and Cd is seen on bulk polycrystalline silver and on large nanoparticles but is completely absent for nanoparticles of size less than 50 nm. Upd is the formation of monolayers or sub-monolayer of metal at potentials less negative than required to deposit the corresponding bulk metal:

$$Tl^+(aq) + e^- \rightleftarrows Tl(ads) \quad E_2$$

$$Tl^+(aq) + e^- \rightleftarrows Tl(bulk) \quad E_1$$

where $E_2 > E_1$. The difference likely reflects the changed surface topography where a small finite number of atoms are present in the nanoparticles, as well as differences in the electronic structure of the metal as the scale shrinks.

References

[1] D.W.M. Arrigan, *Analyst* **129** (2004) 1157.

[2] F.W. Campbell, R.G. Compton, *Anal. Bioanal. Chem.* **396** (2010) 241.

[3] I. Streeter, R.G. Compton, *J. Phys. Chem. C.* **111** (2007) 18049.

[4] F.W. Campbell, S.R. Belding, R. Baron, L. Xiao, R.G. Compton, *J. Phys. Chem. C.* **113** (2009) 9053.

[5] S.E. Ward Jones, F.W. Campbell, R. Baron, L. Xiao, R.G. Compton, *J. Phys. Chem. C.* **112** (2008) 17820.

[6] I. Streeter, R. Baron, R.G. Compton, *J. Phys. Chem. C.* **111** (2007) 17008.

[7] F-R.F. Fan, A.J. Bard, *Science* **267** (1995) 871.

[8] F-R.F. Fan, A.J. Bard, *J. Am. Chem. Soc.* **118** (1996) 9669.

[9] C. Amatore, Y. Bouret, E. Maisonhaute, J.I. Goldsmith, H. Abruña, *ChemPhysChem,* **2** (2001) 130.

[10] C. Amatore, F. Grün, E. Maisonhaute, *Angew. Chem. Int. Ed.* **42** (2003) 4944.

[11] R.S. Ingram, M.J. Hostetler, R.M. Murray, T.G. Schaaff, J.T. Khoury, R.L. Whetten, T.P. Bigioni, D.K. Guthrie, P.N. First, *J. Am. Chem. Soc.* **119** (1997) 9279.

[12] S. Chen, R.W. Murray, *J. Phys. Chem. B.* **103** (1999) 9996.

[13] J.R. Reimer, N.S. Hush, *J. Phys. Chem. B.* **105** (2001) 8979.

[14] F-R.F. Fan, A.J. Bard, *Science* **277** (1997) 1791.

[15] M.D. Hughes, Y.J. Xu, P. Jenkins, P. McMorn, P. Landon, D.I. Enache, A.F. Carley, G.A. Attard, G.J. Hutchings, F. King, E.H. Stitt, P. Johnston, K. Griffin, C.J. Kiely, *Nature* **437** (2005) 1132.

[16] G.J. Hutchings, *Chem. Commun.* **10** (2008) 1148.

[17] C. Jeyabharathi, S.S. Kumar, G.V.N. Kiruthika, K.L.N. Phani, *Angew. Chem. Int. Ed.* **49** (2010) 2925.

[18] F.W. Campbell, S.R. Belding, R. Baron, L. Xiao, R.G. Compton, *J. Phys. Chem. C.* **113** (2009) 14852.

[19] F.W. Campbell, Y-G. Zhou, R.G. Compton, *New J. Chem.* **34** (2010) 187.

[20] F.W. Campbell, R.G. Compton, *Int. J. Electroanal. Sci.* **5** (2010) 407.

Appendix

Simulation of Electrode Processes

The purpose of this appendix is to provide an insight into how simple numerical simulations of one dimensional diffusion problems can be carried out.

A.1 Fick's First and Second Laws

The voltammetry experiment conducted in quiescent solution considers mass transport of species by diffusion only. The flux of a species through a solution is mathematically described by Fick's First Law[1] which is described in Chapter 3. This is as follows:

Fick's First Law:

$$j = -D\frac{\partial c}{\partial x},$$ (A.1)

where j is the flux, D is the diffusion coefficient, c is the concentration of a species and x is the spatial coordinate. Equation (A.1) allows us to calculate the passage of flux in a steady-state system, where the concentration gradients are invariant in time. However, the nature of electrochemical systems is such that concentration gradients are usually constantly changing. Fick derived a second law (Chapter 3) to describe this change of concentration with time, t, in the form of a second order differential equation:

Fick's Second Law:

$$\frac{\partial c}{\partial t} = D\frac{\partial^2 c}{\partial x^2}.$$ (A.2)

As stated, these laws only describe a one-dimensional system. Generalised to three Cartesian directions, the 2nd law becomes:

$$\frac{\partial c(t, x, y, z)}{\partial t} = D\left(\frac{\partial^2 c}{\partial x^2} + \frac{\partial^2 c}{\partial y^2} + \frac{\partial^2 c}{\partial z^2}\right).$$ (A.3)

A.2 Boundary Conditions

The solution to a second order differential equation may only be solved with the introduction of boundary conditions. In physical terms, this represents the restrictions imposed on the electrolyte concentration (Dirichlet boundaries) or its derivative (Neumann boundaries e.g. flux) in time and space by the experiment.

A.3 Finite Difference Equations

Mass transport equations are partial differential equations. The concentration is a function of both distance, x and time, t. Since our model of cyclic voltammetry (Chapter 4) demands only one dimension in space, we can approximate this function as discrete points in time and in space. This process is known as 'discretisation'. Points in the x-direction are assigned values of $j = 0, 1, 2, 3, \ldots, NJ$ spaced Δx apart whilst those in the time are assigned values of $l = 0, 1, 2, 3, \ldots, Nl$, spaced Δt apart. Hence, any instantaneous point concentration is specified by values of l and j and we use the notation c_j^l to emphasise this. Finite difference equations are approximations of partial derivatives in terms of discretised values. Those applicable to our model are:

$$\frac{\partial c}{\partial x} = \frac{c_{j+1}^l - c_j^l}{\Delta x} \text{ (upwind differencing } - \text{ concentration gradient at } j + 1/2),$$

$$\frac{\partial c}{\partial x} = \frac{c_j^l - c_{j-1}^l}{\Delta x} \text{ (downwind differencing } - \text{ concentration gradient at } j - 1/2),$$

and

$$\frac{\partial c}{\partial x} = \frac{c_{j+1}^l - c_{j-1}^l}{2\Delta x} \text{ (central differencing } - \text{ concentration gradient at } j)$$

The upwind and downwind differencing equations may be combined to produce an approximation for the second derivative.

$$\frac{\partial^2 c}{\partial x^2} = \frac{\left(\frac{\partial c}{\partial x}\right)_{j+1/2} - \left(\frac{\partial c}{\partial x}\right)_{j-1/2}}{\Delta x} = \frac{c_{j+1}^l - 2c_j^l + c_{j-1}^l}{(\Delta x)^2}. \tag{A.4}$$

Also, for the derivative of concentration with time:

$$\frac{\partial c}{\partial t} = \frac{c_j^l - c_j^{l-1}}{\Delta t}.$$

A.4 Backward Implicit Method

The Backward Implicit (BI) method[2,3] is a powerful system for calculating a set of concentrations, in a one- or two-dimensional spatial system. Because the concentration profile is solved vector-by-vector, only three nodes can be spanned within one spatial coordinate, so limiting general application. However, it is well suited to producing quick solutions to 1D problems. The BI method involves the assimilation of a linear system of equations by discretisation of the mass transport equation, which can be re-arranged to form a matrix equation and subsequently solved.

To exemplify the use of the BI method, we will simulate the application of a potential step to a

$$A(aq) - e^- \rightleftarrows B(aq) \qquad (A.5)$$

redox couple, such that at a time, $t > 0$, the A species at the electrode surface is fully oxidised to B. The A species occupies the x-coordinate between $x = 0$ and $x = \delta$, where δ is a diffusion layer thickness large enough that semi-infinite diffusion operates, i.e.

$$[A]_{x=\delta} = [A]_{bulk}.$$

This problem was considered in Chapter 3. The mass transport for the A species is:

$$\frac{\partial [A]}{\partial t} = D \frac{\partial^2 [A]}{\partial x^2}$$

The BI methods follows the following four steps:

1. *Conversion of MT equations to finite difference form:*

$$\frac{[A]_j^l - [A]_j^{l-1}}{\Delta t} = D \frac{[A]_{j+1}^l - 2[A]_j^l + [A]_{j-1}^l}{(\Delta x)^2},$$

where $\Delta x = \dfrac{\delta}{NJ}$ and Δt is assigned arbitrarily.

2. *Rearrangement to give a set of linear equations:*

$$[A]_j^{l-1} = -\lambda [A]_{j-1}^l + (2\lambda + 1)[A]_j^l - \lambda [A]_{j+1}^l \quad (j = 1, 2, 3, 4, \ldots, NJ),$$

where $\lambda = \dfrac{D\Delta t}{(\Delta x)^2}$.

3. *Application of boundary conditions:*
   The boundary conditions applicable to this simulation are

(a) $t = 0;\quad 0 < x < \delta$
   $[A] = [A]_{bulk}$
   $[B] = 0$
(b) $t > 0;\quad x = \delta$
   $[A] = [A]_{bulk}$
   $[B] = 0$
(c) $t > 0;\quad x = 0$ (A species fully oxidised at surface)
   $[A] = 0$

The introduction of boundary conditions at the diffusion layer limit to the finite difference equations gives:

$$[A]_{NJ-1}^{l-1} = -\lambda[A]_{NJ-2}^{l} + (2\lambda + 1)[A]_{NJ-1}^{l} - \lambda[A]_{NJ}^{l}.$$

But since $[A]_{NJ}^{l} = [A]_{bulk}$,

$$[A]_{NJ-1}^{l-1} + \lambda[A]_{bulk} = -\lambda[A]_{NJ-2}^{l} + (2\lambda + 1)[A]_{NJ-1}^{l}.$$

At the electrode surface,

$$[A]_1^{l-1} = -\lambda[A]_0^{l} + (2\lambda + 1)[A]_1^{l} - \lambda[A]_2^{l}.$$

But since

$$[A]_0^{l-1} = 0,$$
$$0 = (2\lambda + 1)[A]_1^{l} - \lambda[A]_2^{l}$$

4. *Arrangement of linear equations into a matrix equation*

$$\begin{bmatrix} 0 \\ [A]_1^{l-1} \\ \vdots \\ [A]_{NJ-2}^{l-1} \\ [A]_{NJ-1}^{l-1} + \lambda[A]_{bulk} \end{bmatrix} = \begin{bmatrix} 2\lambda+1 & -\lambda & 0 & & \\ -\lambda & 2\lambda+1 & -\lambda & 0 & \\ & \ddots & \ddots & \ddots & \\ & 0 & -\lambda & 2\lambda+1 & -\lambda \\ & & 0 & -\lambda & 2\lambda+1 \end{bmatrix} \begin{bmatrix} [A]_0^{l} \\ [A]_1^{l} \\ \vdots \\ [A]_{NJ-2}^{l} \\ [A]_{NJ-1}^{l} \end{bmatrix}.$$

With the initial values of $[A]$ set to their initial bulk values, this matrix equation can be solved sequentially to give a concentration vector at every time node. After many iterations, the system will become steady-state, i.e. the concentration profile will be invariant with each new iteration. Equations of the above form can be solved using the Thomas Algorithm[4,5] devised by Laasonen[6] before the invention of computer. This allows the solution of a tridiagonal matrix equation:

$$\{d\} = [T]\{u\}$$

to be found implicitly, where $\{u\}$ is unknown, $\{d\}$ and $\{u\}$ being vectors of $J-1$ elements and $[T]$ being a tridiagonal matrix of the form:

$$[T] = \begin{bmatrix} b_1 & c_1 & 0 & & 0 \\ a_2 & b_2 & c_2 & & \\ \vdots & \vdots & \vdots & \vdots & \\ & & a_{j-2} & b_{j-2} & c_{j-2} \\ 0 & & & a_{j-1} & b_{j-1} \end{bmatrix}.$$

In solution to the mass transport equations, $\{d\}$ represents the set of known concentrations c_j^{l-1}, and $\{u\}$, the unknown concentrations c_j^l which are to be calculated in the next pass of the Thomas algorithm. Once calculated, the vector $\{u\}$ is set to $\{d\}$, and the new matrix $\{u\}$ calculated. This process is repeated up to the end time of the simulation. In this way, the concentration versus distance profile can be described as a function of time.

A.5 Conclusion

This appendix has introduced some elementary theory behind the mathematical modelling of electro-active species in solution, as well as its implementation to produce computational simulations.

References

[1] A. Fick, *Poggendorff's Annel. Physik.* **94** (1855) 59.
[2] J.L. Anderson, S. Moldoveanu, *J. Electroanal. Chem.* **179** (1984) 109.
[3] R.G. Compton, M.B.G. Pilkington, G.M. Stearn, *J. Chem. Soc. Faraday Trans* 1, **84** (1988) 2155.
[4] L.H. Thomas, *Elliptical Problems in Linear Difference Equations Over a Network*, Watson Sci. Comput. Lab. Rept. Columbia University, New York, 1949.
[5] G.H. Bruce, D.W. Peaceman, H.H. Rachford, J.D. Rice, *Trans. Am. Inst. Min. Engrs* **198** (1953) 79.
[6] P. Laasonen, *Acta Math.* **81** (1949) 30917.

Index